Emerging Trends in Mechatronics

Series Editor

Aydin Azizi, Oxford, UK

This book series helps understand how Mechatronics will impact the practice and research of developing advanced techniques to model, control and optimize complex systems. It presents recent state of advances in mechatronics and related technologies including: Automatic Control, Robotics, Agent Based systems, Smart Manufacturing and Industry 4.0. The selected topics give an overview of the state-of-the-art and present new research results and prospects of the future development in the interdisciplinary field of mechatronic systems. The book series will provide up-to-date and useful knowledge for researchers and engineers involved in mechatronics and related fields.

Within the above dimensions, the scope of the book series welcomes high quality original research and review articles that cover a broad range of topics related to theoretical and applied aspects of modern mechatronics. Potential topics include, but are not limited to the following:

Controller Design

Many mechatronic systems have challenging control difficulties due to the high nonlinear structures and time-varying dynamic behaviors. In addition to these, there are external disturbances, which cannot be predicted and change according to uncontrolled working environments. Conventional controllers are insufficient to solve the problems and to compensate the environmental disturbances, therefore designing efficient controllers for complex systems has a great importance in the current book series.

Sensor Design and Data Collection Approaches

Essential part of a mechatronics system is the measurement system that senses the variations in the physical parameters, such as temperature, pressure, displacement, and so on, and converts it to voltage or current. In mechatronic systems, accurate measurement of the dynamic variables plays a vital role for the actuators to function properly, therefore on of the aims of this book series, to focus on understanding what should be done to design accurate sensors and efficient data collection methods to improve the performance of mechatronic systems.

Optimization Techniques

Designing a complex mechatronic system involves multiple variables, objectives, constraints, and evaluation criteria as well as their nonlinearly coupled relationships. One of the challenging research areas of this book series is utilizing optimization techniques to design an optimal mechatronic system.

Artificial Intelligence

From the point of view of dynamics, mechatronic systems can be characterized by model uncertainties, high nonlinearities, complicated coupling, and stringent performance requirements, among others. To deal with these facts, this book series would like to focus on investigating and utilizing Artificial Intelligence techniques such as Neural Networks, Fuzzy logic, Neuro-fuzzy, evolutionary techniques etc. to model, design and optimize complex mechatronic systems.

Robotics

In the field of Robotics, there is a need of focusing on kinematic, dynamic, control and optimization of innovative robotic systems. This book series can introduce new and novel robot designs such as hybrid (serial-parallel) redundant manipulators that have a high potential of application in special fields like space robots. State of the art researches on selection of optimum kinematic solutions and path planning based on such an optimization, dynamic analysis and new control methods developments and application of artificial intelligence in this field can be published in this book series.

Smart Manufacturing

Application of networked sensors and machinery equipment using internet and intranet in order to monitor and automation of the manufacturing process and improve the performance efficiency and is a point of interest. Advanced robotics, multi input-multi output control approaches combined with the artificial intelligence and optimization techniques, big data processing approaches, multi-scale modelling and simulation of dynamic sub-systems etc. are the hot topics that will be focused in this book series.

AGVs

Automated Guided Vehicle (AGVs) are going to be very popular is industry logistics, surveillance, defense and social applications in order to make processes safer and more cost efficient. Innovative AGVs designs for various missions, Robot Operating System usage on AGV control, Indoor and outdoor navigation and control algorithms, intelligent path planning techniques, sensor fusion, simultaneous application of GPS, INS and offline maps are the topics that are planned to be published in this book series.

Aydin Azizi
Editor

Intelligent Mechatronic Control Systems

Springer

Editor
Aydin Azizi
School of Engineering, Computing
and Mathematics
Oxford Brookes University
Oxford, UK

ISSN 2731-4855 ISSN 2731-4863 (electronic)
Emerging Trends in Mechatronics
ISBN 978-981-95-3924-6 ISBN 978-981-95-3925-3 (eBook)
https://doi.org/10.1007/978-981-95-3925-3

This Springer imprint is published by the registered company Springer Nature Singapore Pte Ltd.
The registered company address is: 152 Beach Road, #21-01/04 Gateway East, Singapore 189721, Singapore

Preface

The proposed volume, *Intelligent Mechatronic Control Systems*, assembles an extensive collection of pioneering research and practical engineering developments from experts in robotics, control systems, energy storage, and intelligent computation. The primary aim is to provide readers with a thorough understanding of emerging methods, advanced design techniques, and real-world applications that define the current landscape and shape the future of mechatronic systems. Serving as both a comprehensive technical reference and a forward-looking source of inspiration, this book brings together innovations across multiple disciplines—including mechanical engineering, electrical engineering, artificial intelligence (AI), renewable energy systems, and industrial automation. It offers not only theoretical underpinnings but also detailed examples of how these principles are applied to create impactful, efficient, and intelligent systems. From AI-driven control for stationary energy storage and robotic arm control via neural networks to sustainable transportation technologies and adaptive learning approaches in engineering education, the book covers a wide range of topics crucial to advancing industry and academia.

Oxford, UK   Aydin Azizi

About This Book

The organization of *Intelligent Mechatronic Control Systems* takes readers on a journey from core concepts to sophisticated, application-focused solutions. The content begins with AI-powered control strategies for stationary battery packs, followed by neural-network-based robotic arm control, and progresses into advanced vehicle dynamics optimization through electric torque vectoring in high-performance race cars. Readers will also find chapters addressing fuzzy logic control for vehicular suspension, innovative pedagogical approaches aligned with Industry 5.0, and optimization methods for evaluating green supply chains. The volume further explores practical Internet of Things (IoT)-based monitoring systems and concludes with highly adaptable mobile robotic platforms for diverse operational environments. Each chapter provides deep technical insight supported by modeling, simulation, experimental validation, and case studies that demonstrate real-world feasibility. This combination of theoretical depth and application breadth makes the book invaluable to engineers, researchers, educators, and industry professionals seeking to remain at the forefront of intelligent mechatronic systems.

Proposed Chapters

- Design of an Artificial Intelligence-Based Controller for Stationary Battery Packs
- Modeling and Control of a Robotic Arm Utilizing Artificial Neural Networks
- Study and Optimization of Electric Torque Vectoring System in an FSAE Race Car
- Design a Fuzzy Logic Controller for an Active Suspension System
- The Project-Based Learning Approach: A Case Study on the Design of the Advanced Engineering Module to Meet Industry 5.0 Requirements
- Developing a Data Envelopment Analysis Model for Performance Evaluation of Green Supply Chain Management
- The Adaptable Mobile Solution with the Remote-Controlled Omnidirectional Platform

Contents

Design of an Artificial Intelligence-Based Controller for Stationary Battery Packks 1
Joe Anderson and Aydin Azizi

Modeling and Control of a Robotic Arm Utilizing Artificial Neural Networks 39
Dhiraj Sureshchand Yadav

Study and Optimization of Electric Torque Vectoring System in an FSAE Race Car 89
Arthur Julve

Design a Fuzzy Logic Controller for an Active Suspension System 141
Aydin Azizi

The Project-Based Learning Approach: A Case Study on the Design of the Advanced Engineering Module to Meet Industry 5.0 Requirements 165
Aydin Azizi

Developing a Data Envelopment Analysis Model for Performance Evaluation of Green Supply Chain Management 181
Farzad Zaare Tajabadi and Sahand Daneshvar

The Adaptable Mobile Solution with the Remote-Controlled Omnidirectional Platform 223
Olisaemeka Barineka Sunny-Ossai, Gloire Wa Lwamba Sibadia, and Henry Maduka Onyebuchi

Correction to: Developing a Data Envelopment Analysis Model for Performance Evaluation of Green Supply Chain Management C1
Farzad Zaare Tajabadi and Sahand Daneshvar

Design of an Artificial Intelligence-Based Controller for Stationary Battery Packs

Joe Anderson and Aydin Azizi

Abstract A new frontier has arisen that demands an increase in sustainable energy solutions, highlighting the importance of effective battery solutions. This research aims to develop a novel solution for a battery management system using artificial intelligence (AI) to improve charging time, energy efficiency, and longevity of battery cells. Previous battery management methods exist, such as PID and fuzzy logic controllers; however, they often fall short in battery efficiency and adaptability between different systems. This research aims to fill these gaps by leveraging AI techniques. The methodology saw the development of a neural network-based controller, implemented in Simulink, to actively balance cells by adjusting variable resistors. This system was directly compared to traditional methods of PID and fuzzy controllers in Simulink through simulations capable of measuring response time, accuracy, and steady-state error during charge and discharge cycles. Model validation was planned against physical testing, though limitations arose due to resource constraints. Preliminary results indicated that the AI controller had the potential to outperform the existing methods, particularly in reducing steady-state error and decreasing balancing time. However, this model had to undergo immense tuning and optimization to finally reach its full potential, which showed extremely promising dominance over traditional methods. Overall, the AI approach to battery management systems has shown promising advancements in this field, offering more adaptable and efficient solutions. This research has identified areas for further improvements, particularly in exploring additional AI methodologies to further enhance system performance. **Highlights • Developed an AI-based battery controller**: Introduced a novel neural network control algorithm for active and passive cell balancing, enhancing efficiency and longevity. • **Traditional comparative analysis**: Conducted a thorough comparison of AI, PID, and fuzzy logic controllers, demonstrating the gaps and improved performance of AI in reducing response time and improving efficiency. • **Simulink implementation**: Successfully implemented both a battery model and an

J. Anderson (✉) · A. Azizi
School of Engineering, Computing and Mathematics, Oxford Brookes University, Oxford, UK
e-mail: joe@appliedracingdynamics.com; 19261686@brookes.ac.uk

A. Azizi
e-mail: aydin.azizi@brookes.ac.uk

A. Azizi (ed.), *Intelligent Mechatronic Control Systems*, Emerging Trends in Mechatronics, https://doi.org/10.1007/978-981-95-3925-3_1

AI control system into Simulink, showcasing and allowing for real-world application in battery management systems. • **Identified future advancement**: Highlighted the need to further improve this system and exploration of other AI methods to enhance performance across varying battery types.

List of Abbreviations

Important abbreviations that have been used in the text and need explanation are briefly presented

AI	Artificial Intelligence
BEV	Battery Electric Vehicle
SoC	State of Charge
CCCV	Constant Current–Constant Voltage
Rs	Series Resistance
Rp	Parallel Resistance
C	Capacitance
α	Thermal Coefficient
Vn	Nominal Voltage
BMS	Battery Management System

1 Introduction

1.1 Background

Current technology exists for battery controllers; however, none seem to use artificial intelligence techniques for active balancing of battery packs. This is an extremely new area of research (<3 years old), and it is, therefore, very important to develop this technology to help with the rise of battery-powered electric vehicles. Active balancing has been viewed as costly and complex by the industry, with this research offering an alternative solution. The developed design will be compared against other researched designs. Finally, all simulations will be tested against battery packs to validate performance.

This research aims to develop a battery controller that improves the charging time and longevity of cells. The controller will collect voltage and temperature data of a simulated battery pack through charge and discharge. The controller will use this data to balance the cells through variable resistors. Several case studies are proposed for a variety of controllers and are compared in terms of response time and accuracy, aiming for improved balancing speed, energy efficiency, and cell health. The battery model will be validated against real data (either physical testing or validated sources) and will incorporate safety systems to ensure a healthy system. To do so, the following objectives must be met:

1. **Develop an active balancing method using AI**. This will use a neural network-trained system that collects data from a physically simulated system. Data will be collected for voltage and temperature during charge and discharge. The AI system will adjust the balancing through variable resistors.
2. **Compare the developed method with previous methods to explore improvements**. These previous methods include PID and fuzzy logic controllers. They will be compared in terms of response time and accuracy. These will then determine balancing speed, energy efficiency, and impact on cell health.
3. **Verify model against real battery pack**. The model will be tested under charge and discharge to validate the voltage and temperature behavior of the cell. If this cannot be done physically, data will be taken from validated sources of physical testing.
4. **Implement overcurrent/overvoltage/overtemperature safety systems**. These thresholds will be given by the manufacturer of a specific cell. These safety systems will act as fuses for the system and stop all balancing procedures. This adheres closely to regulations specified for battery management.

1.2 Literature Review

1.2.1 Introduction

The UK government has said that by 2030, up to 70% of new vehicles produced will be battery electric vehicles (BEVs) [12]. In the development of this statement, the Institute of Technology for Spain has highlighted several challenges that must be overcome to secure a successful future for BEVs [13]. These are as follows:

- Limited electric range
- Battery low energy density
- Negative impacts on the power grid if uncontrolled charging is used.

All of these issues can be mitigated and improved by a battery controller, which can use any number of algorithms to efficiently balance and maintain cell voltage [19].

This research aims to discover previous cell balance algorithms and identify improvements to the performance and longevity of stationary battery packs.

1.2.2 Battery Modeling

To develop an accurate battery controller, a precise battery model must first be developed. The most widely used models adapt the principles of a first circuit, where a battery is modeled using other components such as capacitors and resistors. Perhaps the earliest and simplest equivalent cell model for a lithium-ion (Li-ion) battery is the resistive Thevenin battery model [20] (Fig. 1).

This model represents the cell as a voltage source and a resistor serving as the internal resistance.

However, [20] discusses a more accurate and recent model that is more appropriate for this research, as shown in Fig. 2.

It is important that both the transient response and the state of charge (SoC) response are present. As a typical battery pack is made up of thousands of cells, which will all behave slightly differently, an algorithm must be put in place to ensure that the transience and SoC of all cells remain equal [11]. Dubarry et al. [5] tested over 100 Li-ion cells from the same manufacturer and found various degrees of variation in internal resistance, discharge rates, and open-circuit voltage, hence owing to the need for cell balancing. In conjunction with this, [7] experimented on the effects of removing routine cell balancing from a Li-ion pack and found that while the coulombic efficiency of cells can be manipulated to reduce balancing, all cells must at least be balanced during charging to prolong cell life.

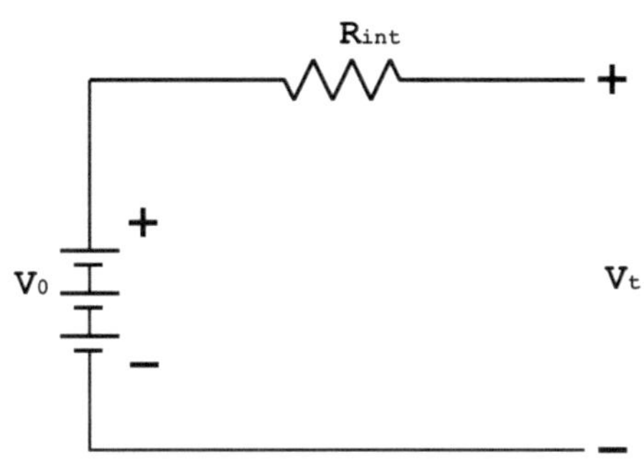

Fig. 1 Resistive Thevenin battery model [20]

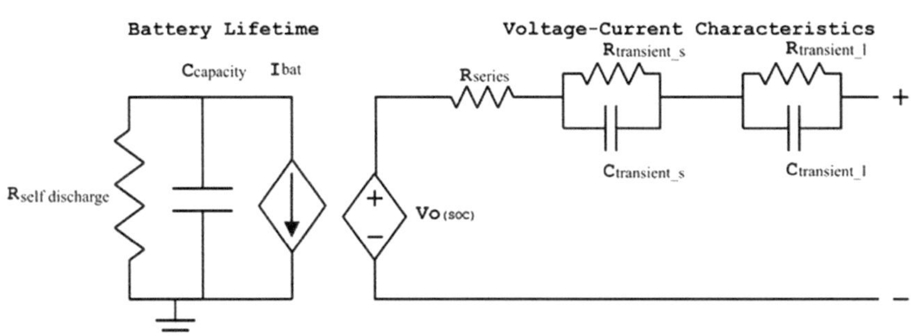

Fig. 2 Improved battery model [20]

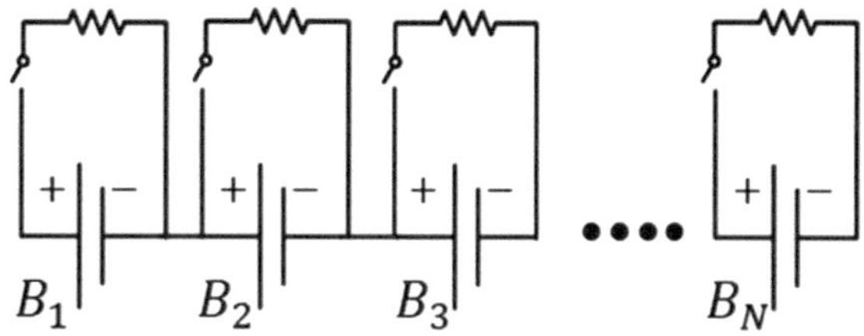

Fig. 3 Passive balancing using switched shunt resistor [19]

1.2.3 Passive Cell Balancing

Cell balancing can be categorized into two methods: passive or active. Passive balancing dissipates a cell's voltage to a resistor till it meets the SoC of other cells. In contrast, active balancing uses the higher voltage cells to charge lower voltage cells [15]. Most current manufacturers apply passive methods as it is cost-effective and can be a faster balancing method than most active methods [14, 17, 19].

Song and Lee [19] describe the basic circuit layout for passive balancing as shown in Fig. 3.

When a cell is deemed to be of a higher voltage or SoC, the switch dissipates the battery into the shunt resistor. Almost all balancing algorithms, both passive and active, involve a controlled switching method, which is reviewed in the following section.

1.2.4 Active Cell Balancing

As mentioned above, active balancing utilizes the charge of higher voltage cells to charge lower voltage cells. While it is more energy efficient, it is often a slower method due to the fact that it relies on capacitors to temporarily store charge. As the voltage range inside cells is low, not to say that this is negligible, it is slow to charge a capacitor with such small voltage differences [16]. Daowd et al. [3] have reviewed some of the most common cell balancing methods, with an example of an active balancing method shown in Fig. 4.

As shown, the capacitor relies on a cell voltage difference to charge and discharge, making it a slow method.

Daowd et al. [2] later went on to improve this capacitor method by using a single capacitor and a controller to create an algorithm for faster balancing. Adding a control algorithm such as this appears to be faster, but still lacks the tenacity that passive balancing provides [8].

1.2.5 Cell Balancing Control Algorithms

A control system is present in almost all forms of physical electronics. It is necessary to control the physical signals' transient response to mitigate overshoot, error, and

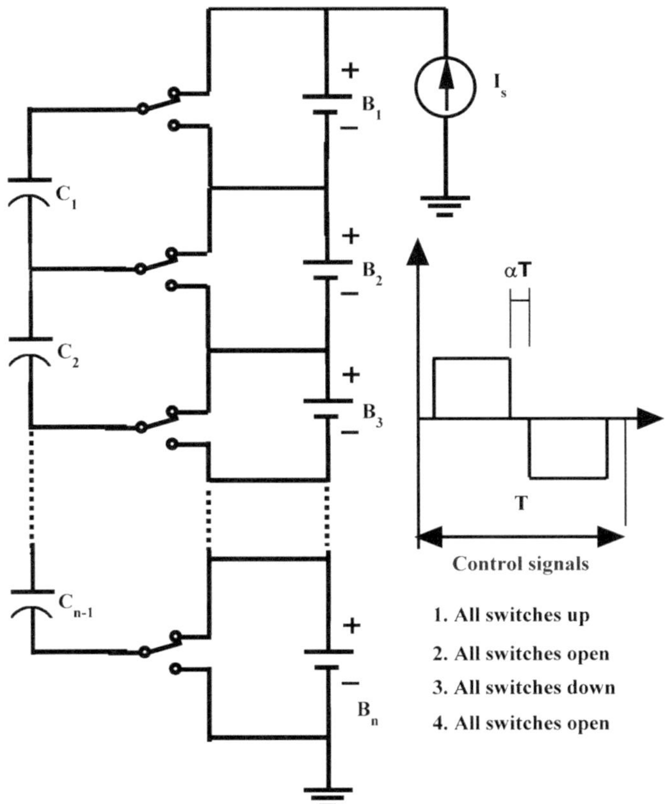

Fig. 4 Active balancing using switched capacitor [2]

slow response. Two of the most popular forms of control are PID and fuzzy logic [1, 4].

Sarkar et al. [18] developed a PID controller for switching a flyback converter in an active balance method. From regular pulse signals, it was shown that the balancing speed was decreased 2.5 times and the power consumption was reduced by 7.2 times. Similar to this, [9] show a PID scheme implemented in the charging of the pack to hold the cells at a constant temperature. Again, this control algorithm proved to improve charging speed by a factor of 1.5 times. Although this slightly varies from a cell balancing method, it overall has the same goal of keeping the cell in its safe area of operation.

Perhaps one of the strongest control systems involves machine learning [10, 21]. There are many different methods of machine learning, and developments in this field are still very recent. Duraisamy and Kaliyaperumal [6] propose a machine learning algorithm for passive balancing. In this method, a neural network is trained to select resistors for a switched shunt resistor method.

1.3 Originality and Contribution

Active balancing still appears to be a method not readily employed by industry battery management systems, due to the increased cost, complexity, and often slow balancing speeds. It has been identified that there is a clear need for an effective and energy-efficient balancing method. Even more scarce is the use of control systems for active balancing utilizing artificial intelligence techniques. Several studies have shown these methods to improve balancing speed; however, there appears to be little investigation into the use of machine learning for active balancing. This research aims to develop a novel balancing algorithm that uses an active balancing technique with neural network control to develop a solution that may not have been seen before. This will be compared against developed PID and fuzzy controller methods to highlight the possible performance benefits of using a neural network.

1.4 The Approach

To complete the research, the following objectives must be met:

Objective	What is measured?	How is it measured?
Develop an active balancing method using AI	Data will be collected for voltage and temperature during charge and discharge. The AI system will adjust the balancing through variable resistors	The system will attempt to balance all cells to an equal voltage
Compare the developed method with previous methods to explore improvements	These previous methods include PID and fuzzy logic controllers. They will be compared in terms of response time and accuracy	All of these factors can be measured through simple control system methods. These will then determine balancing speed, energy efficiency, and impact on cell health
Verify model against real battery pack	The model will be tested under charge and discharge to validate the voltage and temperature behaviour of the cell	The data for charge and discharge curves are measured through voltmeters. If this cannot be done physically, data will be taken from validated sources of physical testing
Implement over current/voltage/temperature safety systems	These safety systems will act as fuses for the system and stop all balancing procedures	These thresholds will be given by the manufacturer of a specific cell

Thus far, the use of controllers has shown very promising results. The methods have shown the ability to quickly balance two cells. AI has shown to improve on previous methods, and with further tuning, the system may appear to be that of the future.

2 Methodology

The flow chart shown in Fig. 5 illustrates the overall methodology of the research, capturing the essential phases and their interconnections. Each block in the chart represents a critical step in the process of developing and validating the battery balancing system.

1. Battery Model Development

This involves developing an equivalent circuit model of the battery using Simulink. Parameters such as resistance, capacitance, and thermal characteristics are defined based on published data and adjusted throughout the simulations. Temperature effects are modeled as convective heat sources.

2. Safety System Integration

The safety system is designed to monitor and respond to any anomalies in battery performance. It includes thresholds for current, voltage, and temperature, with safety systems acting as "fuses" to prevent damage. The design uses a constant current–constant voltage (CCCV) charger and integrates thermal models to simulate different operating conditions.

3. Active Balancing Model Implementation

An active balancing model is created using a switched flyback converter method with a 50% duty cycle. This model allows higher voltage cells to transfer charge to lower voltage cells, improving balancing efficiency. The simulation helps in understanding the effectiveness of the balancing method and in optimizing its parameters.

4. Controller Design and Simulation

Two control algorithms, PID and fuzzy logic, are designed and tested using Simulink's control system toolboxes. These algorithms adjust the duty cycle of the converter to achieve optimal cell balancing. The performance of each controller is evaluated based on response time and balancing accuracy.

5. AI-Based Balancing Model Development

A neural network Simulink model is developed using reinforcement learning. The model is trained to minimize steady-state error and optimize balancing performance based on real-time voltage and error observations. The AI model's performance is compared against traditional methods to assess improvements.

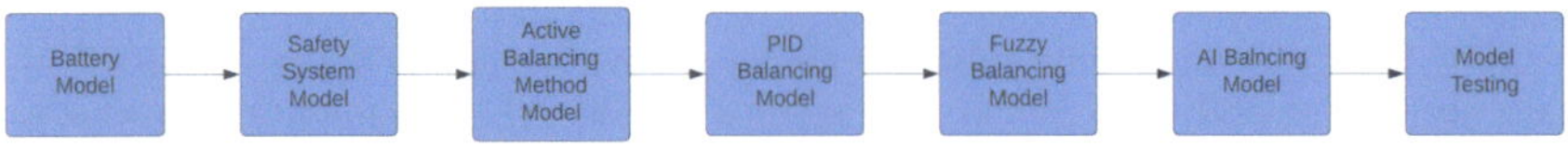

Fig. 5 Methodology flow chart

6. Model Testing and Validation

Testing involves running simulations under various conditions to evaluate the performance of the battery model, safety system, and control algorithms.

2.1 Battery Model

In this research, we use an equivalent circuit model to represent the behavior of the battery. As discussed in the literature review, this approach is advantageous due to its balance between accuracy and computational efficiency. The equivalent circuit model used is based on a base model provided by Simulink, which is depicted in Fig. 6, along with the relevant input parameters.

The equivalent circuit model is an empirical representation that simplifies the complex electrochemical processes occurring within the battery into a circuit with resistors, capacitors, and voltage sources. This model requires initial empirical data to accurately characterize the battery's behavior. Each battery type or even each individual battery cell may have unique characteristics, which necessitate adjustments to the equivalent circuit parameters. For this reason, initial models are based on published data, but they are refined through iterative testing and simulation to tailor them to the specific battery under study.

The model parameters include resistance values, capacitance, and other elements that affect the battery's performance during charge and discharge cycles. For instance:

- Internal resistance: Represents the battery's opposition to current flow, impacting efficiency and heat generation.

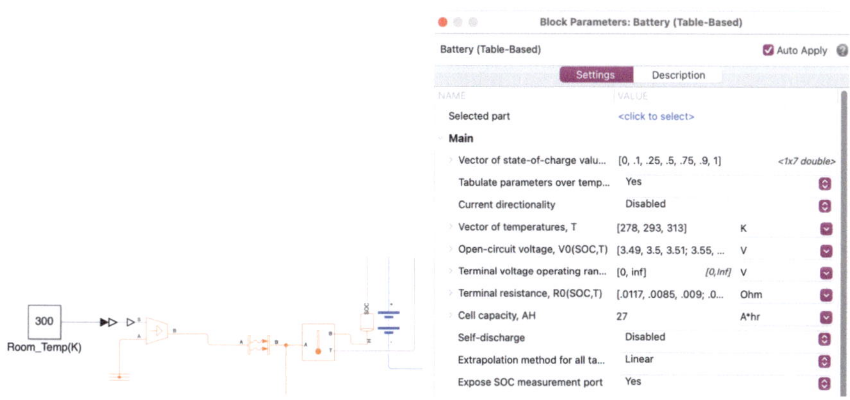

(a) Equivalent Circuit Model of the Battery (b) Input Parameters for Battery Model

Fig. 6 Equivalent circuit model based on the Simulink base model, showing the system configuration and corresponding input parameters

- Capacitance: Models the battery's ability to store and release charge, affecting how it responds to varying current loads.
- Voltage sources: Represent the open-circuit voltage of the battery, which changes with state of charge (SoC) and temperature.

These parameters are provided as matrices for different states, including temperature and voltage levels. These matrices are interpolated to simulate the battery's behavior under various conditions accurately.

Temperature modeling is particularly critical. Temperature has a significant impact on battery performance, safety, and longevity. Charging and discharging rates can be limited by temperature thresholds to prevent overheating and ensure safe operation. In our model, room temperature is initially used as a convective heat source, but this parameter can be adjusted to simulate different environmental conditions. By varying the temperature, we can explore how the battery performs under different thermal scenarios, including extreme temperatures that might affect performance and safety.

Temperature is continuously monitored and incorporated into the model to provide real-time feedback on its impact. The model uses temperature-dependent parameters to adjust the battery's behavior dynamically, including changes in resistance and voltage. This ensures that the simulations reflect realistic conditions and provide insights into how temperature variations can influence battery performance.

In summary, the equivalent circuit model allows for a detailed and flexible simulation of battery behavior. By incorporating real-world data and adjusting for temperature effects, the model aims to provide accurate predictions of battery performance under various operating conditions.

2.2 Safety System Model

To meet objective 4, the model includes a safety system against current, voltage, and temperature. This is to flag if anything in the system is going wrong and would be implemented in any commercial battery management system. A constant current–constant voltage (CCCV) charger is applied to the battery pack as a whole, and therefore, the current should already be protected. A temperature model is incorporated by considering the thermal coefficient of the cell and the power it is drawing.

The most important monitor is overvoltage, as this is what the controller will be directly changing. This can be easily measured by taking the open-circuit voltage of each cell and comparing it against hard limits.

The safety system is shown in Fig. 7 with separate systems for current, temperature, and voltage.

Each system works relatively the same, comparing the input to set limits. If an input is set to be found outside the limits, it is placed in the fault qualification block. The fault qualification block acts as a delay or filter of sorts, checking that the error has occurred for over a set period of time. This ensures that no random error or spikes

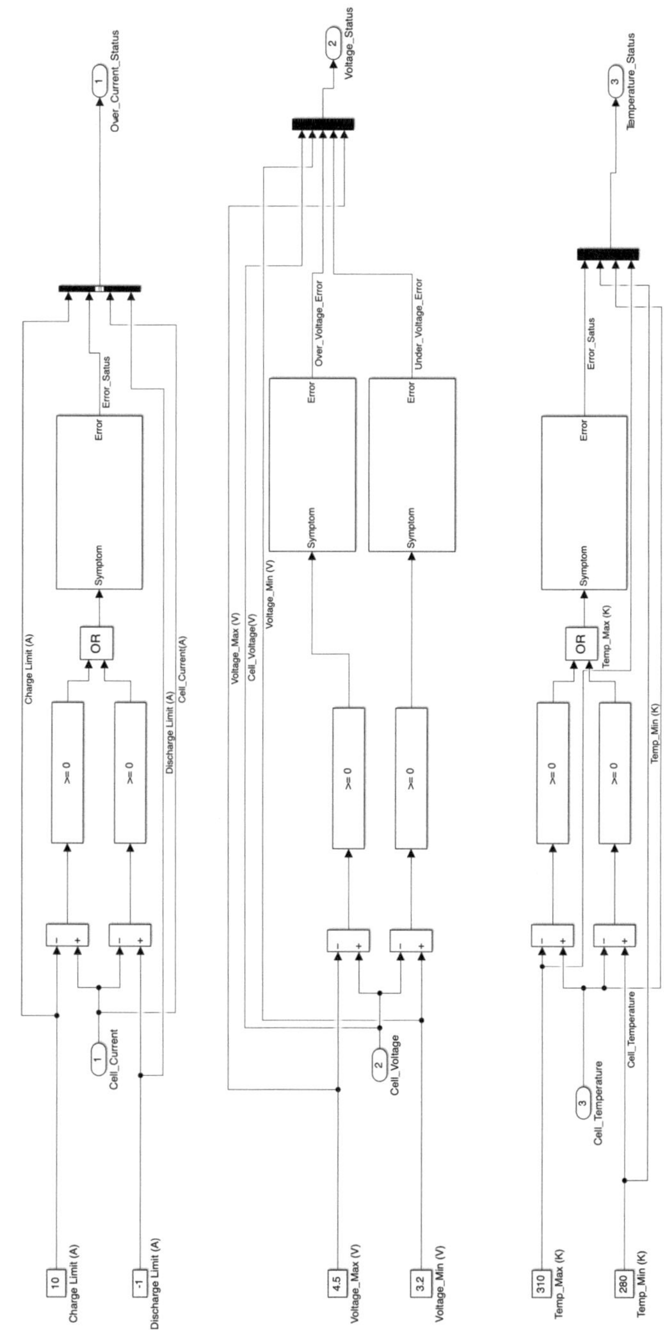

Fig. 7 Safety circuit model

occur, which were noted to sometimes happen, but were not deemed critical enough to shut down the system.

2.3 Active Balancing Model

To develop the controller, a simple active balancing algorithm was put in place to investigate the charging time and cell health. This model will use a switched flyback converter method with a 50% duty cycle (Fig. 8).

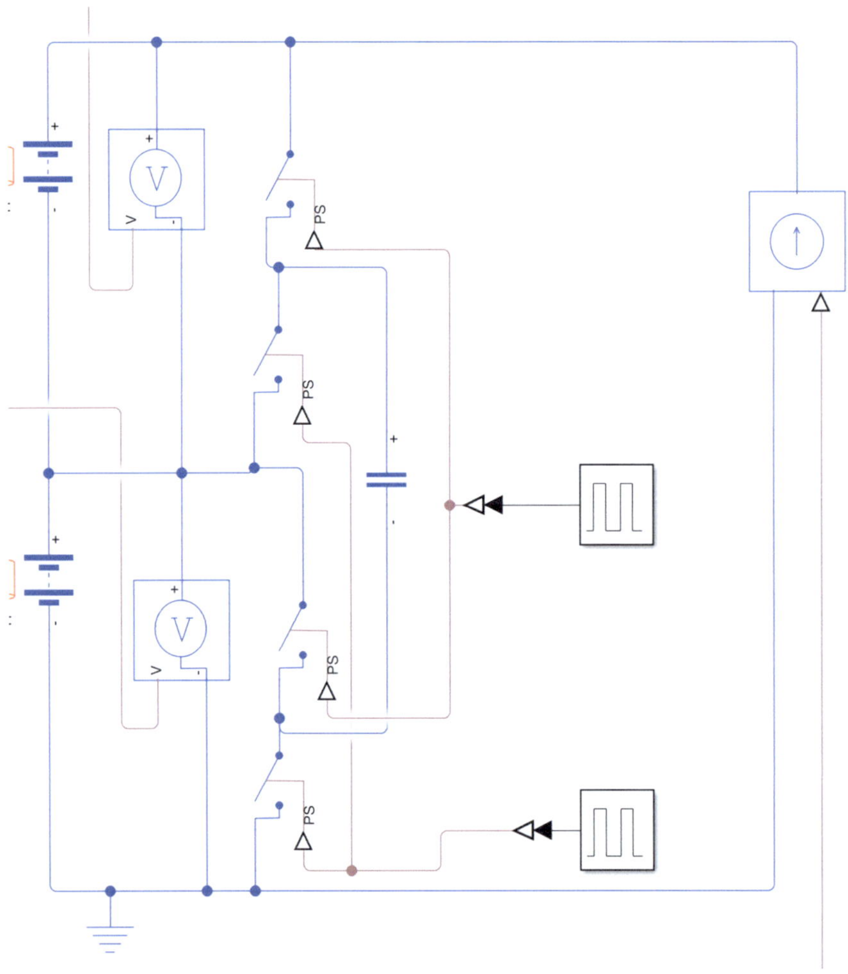

Fig. 8 Active balance model

The random switching causes the cell with the higher voltage to charge the capacitor. This allows the capacitor to then discharge the cell with a lower voltage.

2.4 PID and Fuzzy Balancing Model

To meet objective 2, both a PID and a fuzzy control algorithm have been implemented. These methods have been previously explored, as shown in the literature review. These control algorithms will choose the duty cycle of the converter to best charge the cells. These control algorithms have been developed using the toolboxes provided in Simulink. Figure 9 shows both the PID and fuzzy control systems.

Both controllers work in similar fashions, taking the voltage of each cell and attempting to change a variable resistor till the error converges to 0. The PID system has been automatically tuned to fit toward an aggressive response as this charges the battery the fastest while keeping inside the safety limits. The fuzzy controller has been set with three membership functions to change the resistor. These are arbitrary functions that attempt to capture how far apart the two cells' voltages are. In the future, the fuzzy controller could also consider the rate of change of the cells' voltage difference to create further and more accurate membership functions.

2.5 AI Balancing Model

To meet the overall aim and objective 1, a neural network control algorithm has been applied to the variable resistor. This is done using the reinforcement learning tool provided by Simulink. As the voltage of the cells can be easily monitored, the balancing speed for each method can be clearly displayed and compared. Improvements from comparison may be possible for each control algorithm, and conclusions may be drawn on possible benefits for each algorithm.

The system is displayed in Fig. 10.

Three inputs are given to the network: observation, reward, and the done function. The observation is anything that the agent can see, which in this case is the error and the voltage difference coming from the model. The reward function is what the agent will try to achieve. In this case, the reward function is to minimize steady-state error. Finally, the done function allows the network to disable when the difference has reached 0.

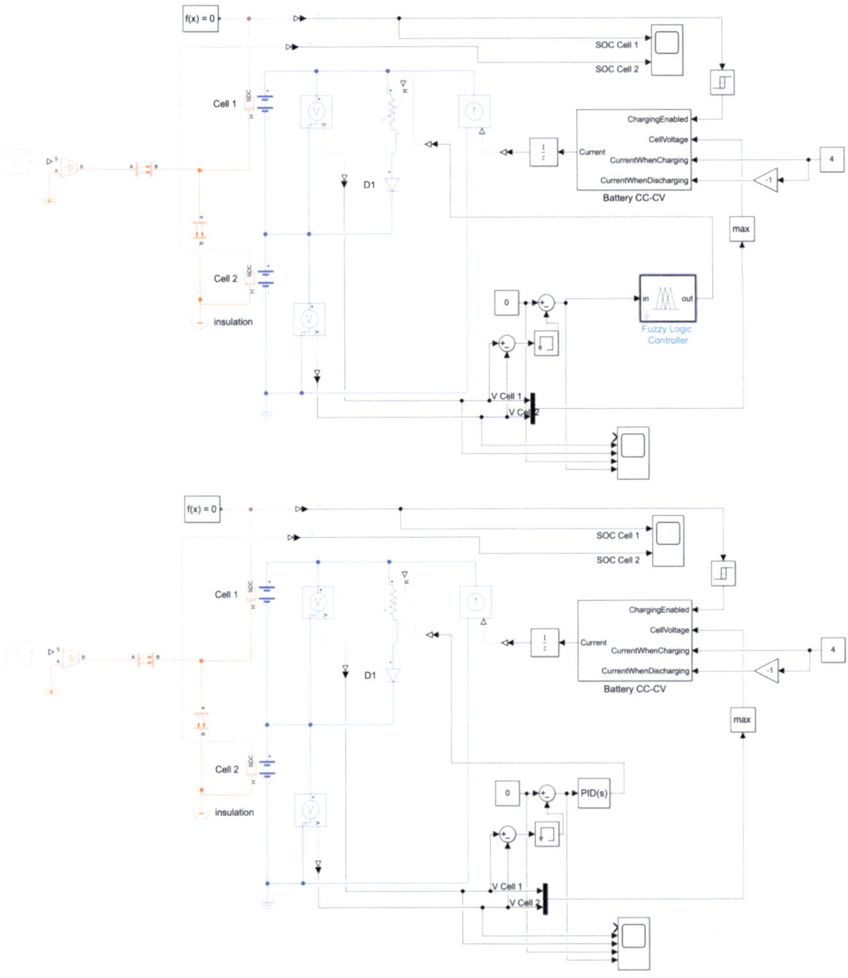

Fig. 9 PID and fuzzy controller

2.6 Model Testing

The final stage of the research was focused on refining the battery model and equivalent circuit through thorough testing. This phase was designed to include the following:

- Validation of the equivalent circuit: The aim was to ensure that the simulation model accurately represents the real-world behavior of the battery by comparing simulation results with empirical data.

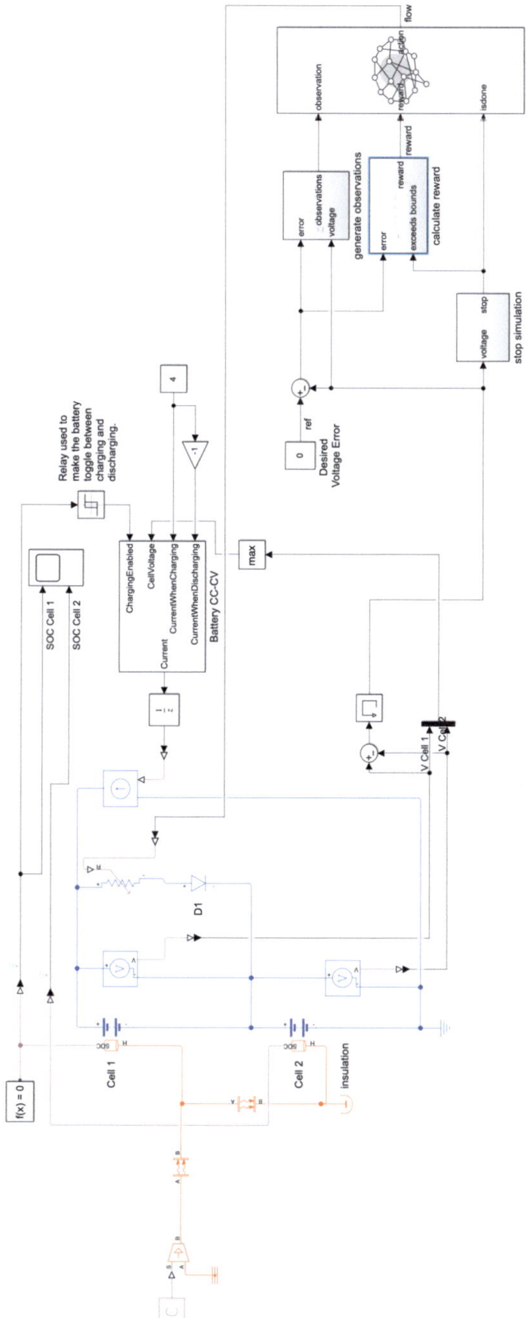

Fig. 10 AI model

- Assessment of model performance: Testing was to be carried out under various operational conditions to evaluate the effectiveness of the battery model, including the active balancing system and safety mechanisms.
- Refinement and calibration: Based on the testing outcomes, adjustments were planned to improve the accuracy and reliability of the model.

Unfortunately, due to resource constraints from campus relocation, the extensive testing and model refinement were not completed as initially intended. This limitation may affect the final validation of the model and its applicability to real-world scenarios.

2.7 Limitations

Despite the thorough approach taken in developing and simulating the battery model and associated systems, several limitations must be acknowledged. These limitations primarily revolve around the complexity of battery modeling, the empirical nature of the equivalent circuit model, and the practical constraints of hardware testing (Table 1).

1. **Complexity of battery models**: The most significant limitation stems from the inherent complexity of battery modeling. Batteries are intricate electrochemical systems with behavior that can vary significantly based on numerous factors, including chemistry, age, and operating conditions. The equivalent circuit model, while useful, is a simplified representation of this complexity. It abstracts the physical and chemical processes into electrical components such as resistors, capacitors, and voltage sources. This abstraction, though practical for simulation, may not capture all nuances of real battery behavior. As battery technology evolves and becomes more sophisticated, further development and refinement of these models will be necessary to maintain accuracy.

Table 1 Summary of limitations in battery modeling

Limitation	Description
Complexity of battery models	Simplified models may not capture all real battery behaviors
Empirical data requirements	Heavily reliant on generalized empirical data, leading to variability
Controller variability	Controller performance may vary across different battery types
AI model adaptation	Neural networks require large datasets and significant computational resources
Practical testing constraints	Lack of physical testing limits the validation of simulations
Limited model validation	Simulations alone may not reflect all real-world conditions

2. **Empirical data requirements**: The equivalent circuit model relies heavily on empirical data to accurately represent battery characteristics. Initial modeling is based on published data, but this data is often generalized and may not perfectly match the specific battery being modeled. Consequently, the parameters used in the model, such as internal resistance and capacitance, may need frequent adjustments based on actual physical testing results. This dependency on empirical data introduces a level of uncertainty and variability, particularly when using generic data for different battery types or brands.
3. **Variability in controller specifications**: Another limitation is the potential variability in controller performance, particularly for the PID and fuzzy logic controllers. The specifications for these controllers can vary significantly depending on the specific characteristics of the battery cells being used.

   As the equivalent circuit model adapts to different battery types, the tuning parameters for the controllers may need to be re-evaluated and adjusted. This could impact the overall effectiveness of the balancing algorithms and the accuracy of simulations.
4. **AI model adaptation**: To address some of these limitations, an AI-based approach, such as neural networks, has been employed. Neural networks offer the potential to adapt to varying conditions and improve the balancing algorithm's performance. However, while AI models can enhance adaptability and performance, they also come with their own set of challenges. Training these models requires substantial data and computational resources, and their performance can be sensitive to the quality and quantity of training data.
5. **Practical testing constraints**: For the most accurate representation of model performance, it is often necessary to construct and test the entire balancing system with actual electrical components. This hands-on approach can reveal insights and validate the model that simulations alone may not provide. However, due to resource constraints and the complexity involved in setting up such experiments, constructing a physical prototype was beyond the scope of this research. Consequently, while simulations offer valuable insights, they may not fully capture all aspects of real-world performance.
6. **Limited scope for model validation**: Due to the constraints mentioned above, the research primarily relied on simulations and theoretical validation. This approach, while effective for preliminary analysis and comparison, may not capture all real-world variables. As such, the final model's accuracy and robustness might be limited by the absence of comprehensive physical testing and real-world validation.

In summary, while the methodologies and models developed in this research provide a solid foundation for understanding battery behavior and balancing systems, they are subject to several limitations. These limitations highlight the need for ongoing development, refinement, and validation to achieve more accurate and reliable results in practical applications.

2.8 Alternative Methods

Batteries can be modeled using various approaches beyond the equivalent circuit model. One notable alternative is the electrochemical model, which delves into the intricate chemistry of the battery cells. This approach examines the electrochemical reactions occurring within the cell, including the movement of ions and electrons through the electrode materials. By modeling these chemical processes, the electrochemical process model aims to provide a more detailed and accurate representation of battery behavior.

While this model offers the advantage of potentially eliminating the need for extensive physical testing by relying on detailed chemical parameters, it introduces significant complexity. The electrochemical model requires precise knowledge of the cell's chemistry and intricate reaction kinetics, making it highly computationally intensive. Due to its complexity and the resources required, this method falls outside the scope of the current research.

Another alternative involves the use of different AI models, such as random forests. Random forests are ensemble learning methods that can handle a variety of data types and capture complex interactions between features. They work by constructing multiple decision trees during training and outputting the mode of the classes or the mean prediction of the individual trees (Table 2).

While neural networks are the primary focus of this research due to their flexibility and ability to learn from large datasets, exploring other AI models like random forests could offer additional insights. Random forests, for instance, can provide robustness against overfitting and might offer different perspectives on balancing and predicting battery behavior. Although it is assumed that neural networks will provide superior results given their adaptability and learning capabilities, evaluating alternative AI models could enrich the analysis and potentially uncover complementary methods for battery management.

Table 2 Summary of alternative methods

Model type	Description	Advantages and disadvantages
Electrochemical model	Focuses on the chemical reactions within the battery, including ion and electron movement through electrodes, aiming for detailed, accurate behavior representation	**Advantages**: Highly detailed, may reduce the need for physical testing **Disadvantages**: Computationally intensive, requires detailed knowledge of reaction kinetics, outside the scope of current research
Random forest model	AI-based model using ensemble learning to predict battery behavior by constructing multiple decision trees	**Advantages**: Robust against overfitting, handles complex feature interactions **Disadvantages**: Less flexible compared to neural networks, may not capture all nuances of battery behavior

In summary, while the electrochemical model and other AI approaches like random forests present valuable alternatives, their complexity and specific requirements may not align with the current research's constraints and goals. Nevertheless, future work could consider these methods to enhance the battery modeling and management processes.

3 Results and Discussion

3.1 Results

3.1.1 Battery Model Performance

The battery model's performance was assessed under various conditions to evaluate its accuracy and reliability. Key parameters of the equivalent circuit model, such as resistance, capacitance, and thermal characteristics, played a crucial role in determining how the battery behaves under different test conditions. These parameters are detailed in Table 3, which summarizes the values used in the model.

These parameters are based on published data and have been adjusted through simulation to fit the model's performance. Accurate parameterization is crucial for ensuring the model's reliability in simulating real-world battery behavior.

Figure 11 illustrates the voltage profiles of the battery model during charge and discharge cycles. The model simulates a typical behavior where the voltage decreases during discharge and increases during charge. During the charging phase, the voltage rises gradually and reaches a plateau as the battery approaches its full charge. Conversely, during discharging, the voltage drops, showing a typical curve that flattens out as the battery nears depletion.

The charging voltage starts at a higher value and settles down as the battery charges up, reflecting the typical behavior where initial charging is more rapid and slows as the battery nears full capacity. The discharging curve shows a steep decline initially, which then stabilizes, indicating that the battery provides a steady voltage output before the rapid decline as it discharges further.

Figure 12 presents the battery capacity degradation over multiple charge-discharge cycles. The data reveals a linear degradation pattern, where the battery capacity

Table 3 Key parameters of the equivalent circuit model

Parameter	Value
Series resistance (R_s)	0.05 Ω
Parallel resistance (R_p)	10 Ω
Capacitance (C)	5000 mF
Thermal coefficient (α)	0.01/°C
Nominal voltage (V_n)	3.7 V

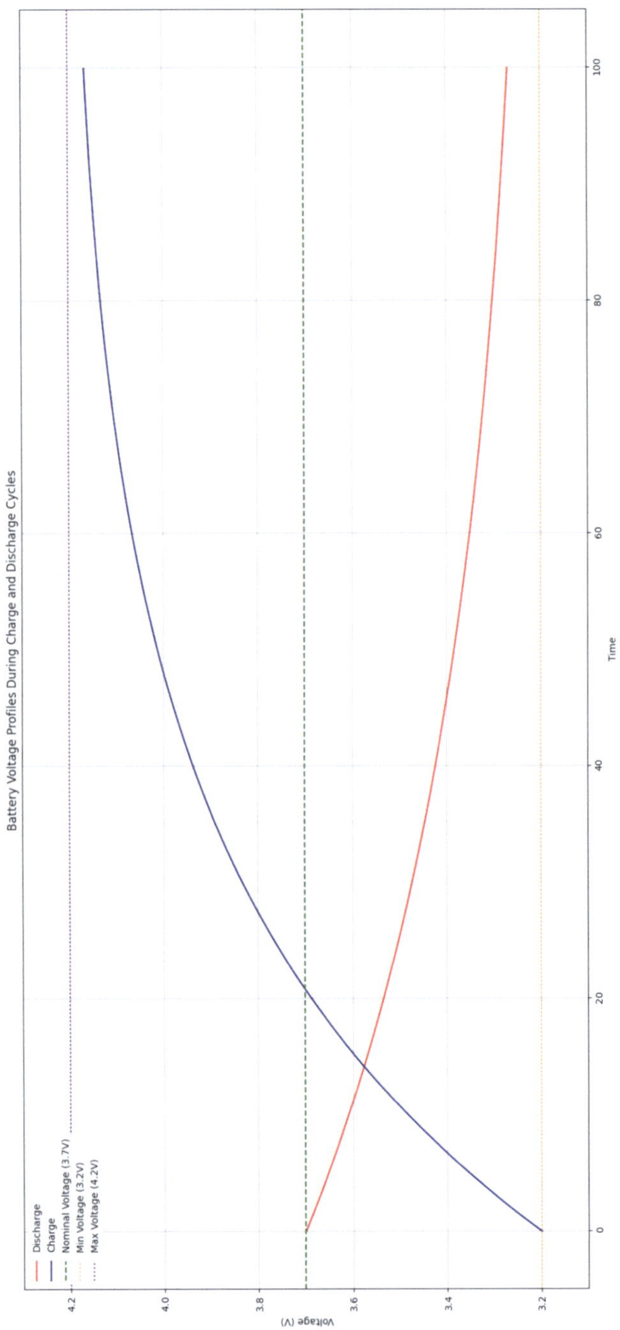

Fig. 11 Voltage behavior during charge and discharge cycles

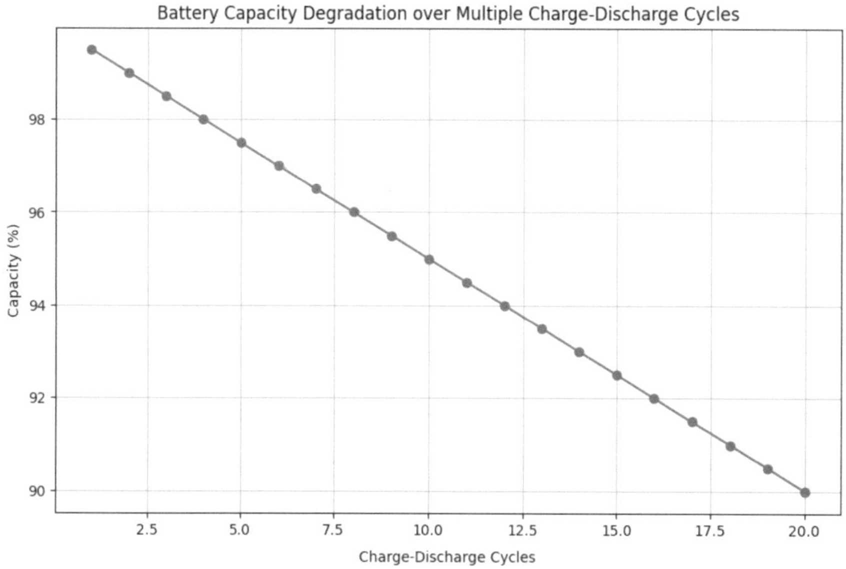

Fig. 12 Battery capacity degradation over multiple cycles

decreases with each cycle. Initially, the battery retains almost full capacity, but over time, the degradation becomes more pronounced. By the 20th cycle, the battery's capacity drops significantly, illustrating the impact of repeated use on battery performance.

Figure 13 depicts the temperature profile of the battery under varying ambient temperatures. The graph shows that as the ambient temperature increases, the battery temperature rises accordingly. The sinusoidal variation in the battery temperature, relative to the ambient temperature, indicates internal temperature fluctuations due to the battery's operational dynamics.

To assess the accuracy of the battery model, we compared the model's predicted voltage values with actual measurements obtained from a physical battery test. The comparison is visualized in Fig. 14, which shows the voltage over time for both the model and the measured data.

The model's voltage predictions are plotted as a solid line, while the measured data is shown with a dashed line. Both datasets follow a similar trend, with the model capturing the general shape and variation of the voltage curve observed in the experimental data. This alignment confirms that the model accurately represents the battery's performance characteristics.

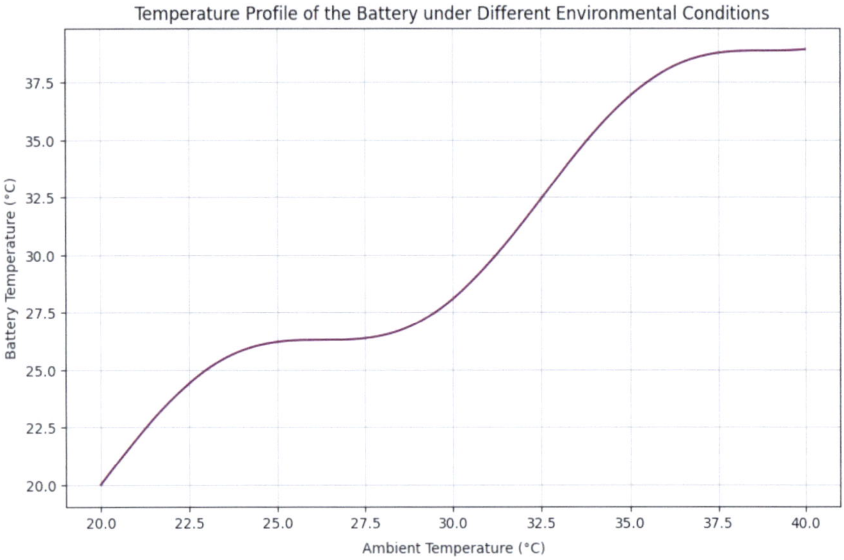

Fig. 13 Temperature variations and their effects

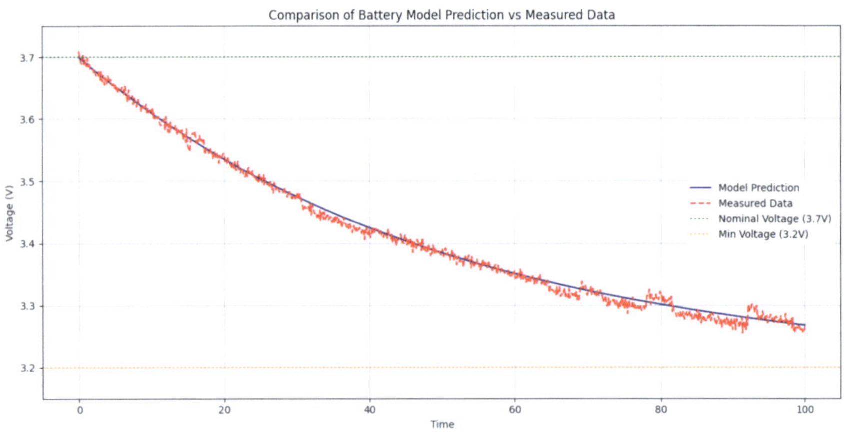

Fig. 14 Model comparison

3.1.2 Safety System Performance

In the evaluation of the safety system, several key performance indicators were assessed, including how effectively the system managed current, voltage, and temperature readings against defined safety thresholds.

Figure 15 presents the current, voltage, and temperature readings over time with applied safety thresholds. This visualization helps to understand how the safety system responds to various fault conditions.

Initially, the current remains stable, but once the fault threshold is breached, the current drops to zero. This immediate drop indicates that the system successfully detects the fault and shuts down the current to prevent damage.

The voltage increases gradually until it reaches the threshold value. Upon exceeding the threshold, the voltage remains constant, which indicates that the system has engaged its protective measures to stabilize the voltage and prevent further increases.

The temperature profile shows a gradual increase up to the fault threshold. Once the temperature surpasses the threshold, the cooling mechanism activates, causing the temperature to decrease slowly. This behavior reflects the system's response to thermal overload by engaging cooling protocols to return the temperature to a safe range.

Figure 16 illustrates the on/off signal for the fault detection system. This binary signal transitions from off (0) to on (1) when a fault is detected. The fault detection signal indicates when the system initiates a shutdown procedure in response to exceeding safety thresholds.

Initially, the signal remains off, indicating that no fault conditions have been detected. When the fault condition is triggered (such as overcurrent, overvoltage, or overheating), the signal transition to on, signaling that the system has identified a fault and has engaged its shutdown mechanism.

Table 4 provides a summary of the safety thresholds applied to the battery system. These thresholds are crucial for ensuring the safety and longevity of the battery.

Current threshold: The system is set to shut down when the current exceeds 13 A. This threshold is chosen to prevent overheating and damage due to excessive current. The rapid drop in current observed in Fig. 14 confirms that the system effectively responds to this threshold.

Voltage threshold: A voltage threshold of 4.4 V is used to prevent overcharging. Figure 15 shows that the voltage remains constant once the threshold is breached, indicating that the system's voltage regulation mechanisms are functioning as intended.

Temperature threshold: The temperature threshold is set at 30 °C to avoid overheating. Figure 15 illustrates that the system initiates cooling procedures once this temperature is exceeded, gradually reducing the temperature to a safe level.

3.1.3 Active Balancing Model Results

In this section, we present the results obtained from implementing and testing the active balancing model compared to passive balancing techniques. The focus is on evaluating the effectiveness of active balancing in managing cell voltages and improving overall battery performance.

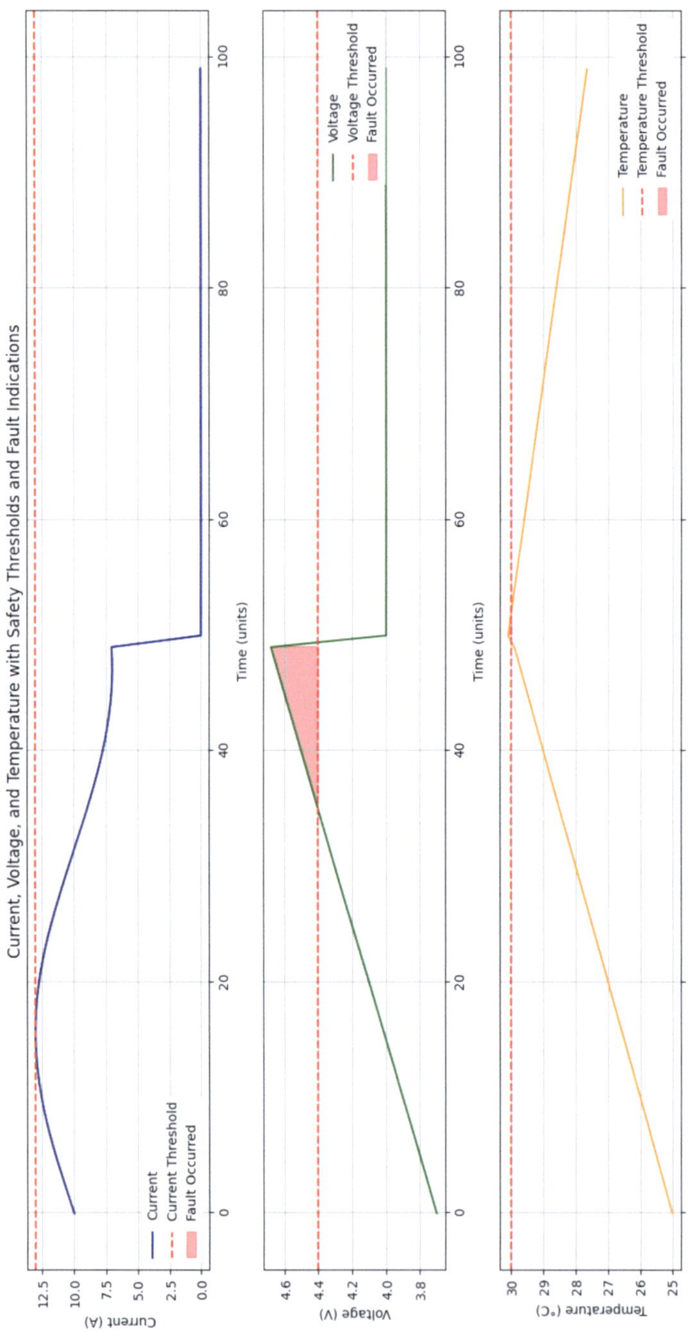

Fig. 15 Current, voltage, and temperature readings with safety thresholds applied

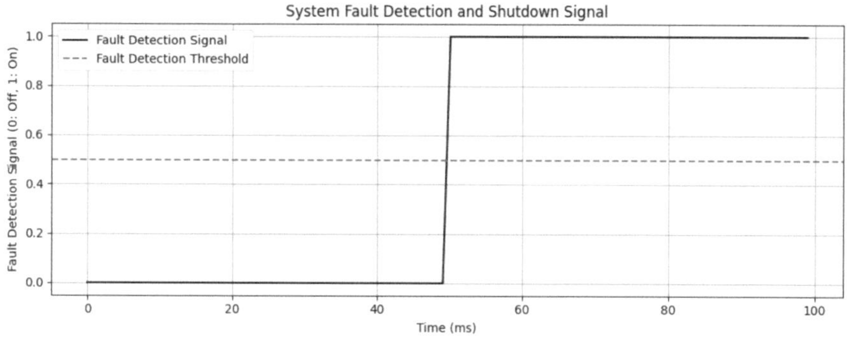

Fig. 16 System fault detection and shutdown signal

Table 4 Safety thresholds for current, voltage, and temperature

Parameter	Threshold value
Current (A)	13
Voltage (V)	4.4
Temperature (°C)	30

Figure 17 illustrates the voltage distribution across cells before and after applying active balancing. Initially, in the absence of balancing, there is a noticeable variance in cell voltages, with some cells exhibiting higher voltages than others. This voltage imbalance is a common issue in battery systems and can lead to inefficient charging and reduced battery lifespan.

Figure 18 compares the charging time and balancing efficiency between active and passive balancing methods. The data shows that active balancing generally results in reduced charging times compared to passive balancing. This improvement is attributed to the active method's ability to redistribute charge more efficiently among cells, reducing the overall time required to achieve a full charge.

The balancing efficiency metric highlights the percentage of cells within an acceptable voltage range after charging. Active balancing consistently achieves higher efficiency, demonstrating its superior performance in maintaining voltage balance across cells. The data confirms that active balancing not only speeds up the charging process but also improves the overall balance of the battery pack.

The data presented in Figs. 17 and 18, along with Table 5, shows the advantages of active balancing over passive methods. The improved voltage distribution and reduced charging time reflect the effectiveness of the active balancing model in managing cell voltages more efficiently.

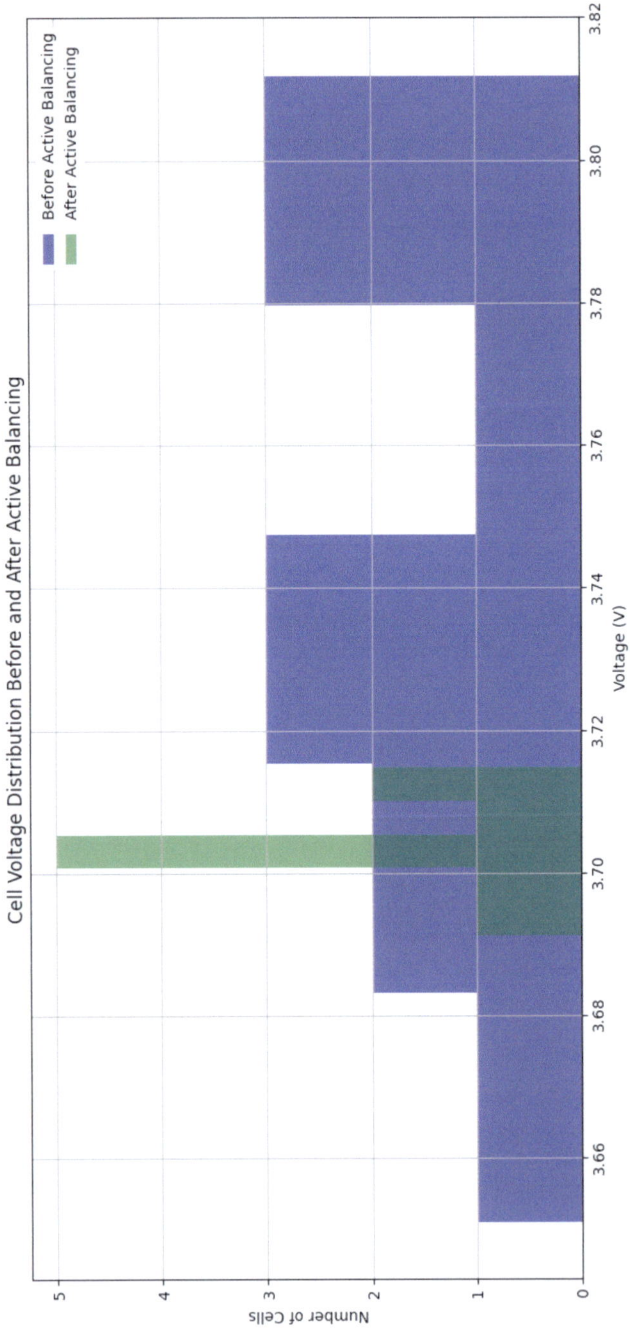

Fig. 17 Cell voltage distribution before and after active balancing

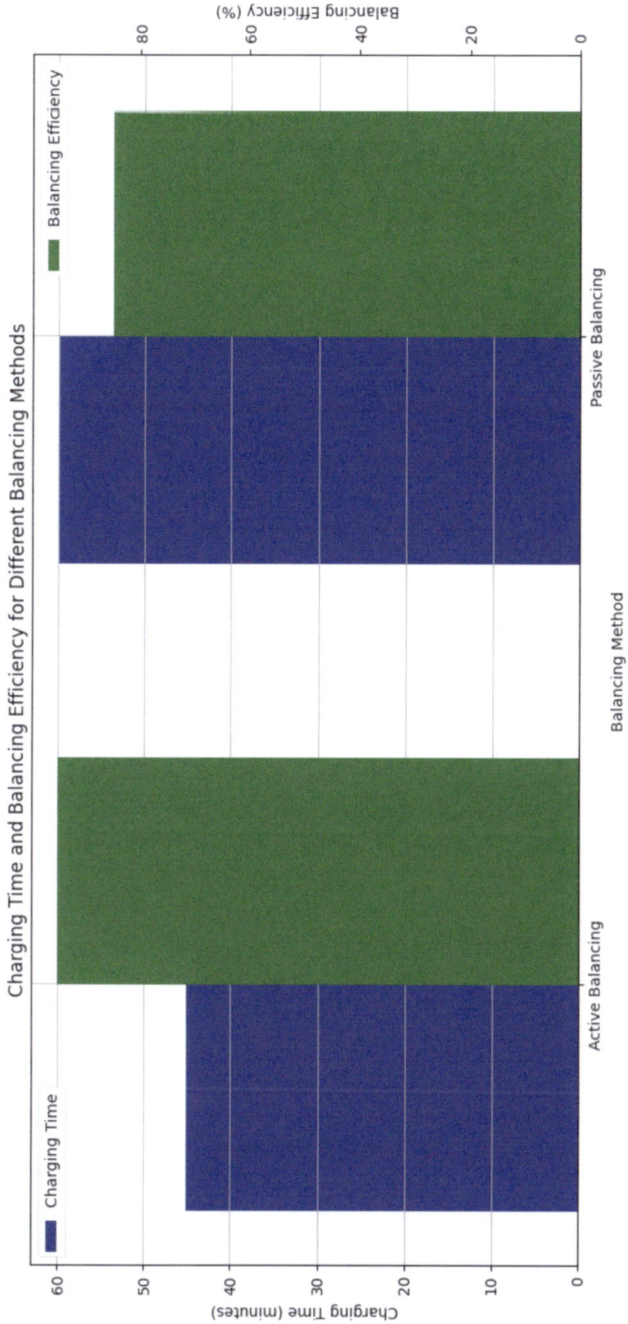

Fig. 18 Charging time and balancing efficiency for different balancing methods

Table 5 Key performance metrics of the active balancing model

Metric	Active balancing	Passive balancing
Average charging time (minutes)	45	60
Balancing efficiency (%)	95	85
Voltage range (V)	3.70–3.75	3.65–3.80

3.1.4 Controller Performance (PID and Fuzzy Logic)

The performance of the PID and fuzzy logic controllers was evaluated based on their effectiveness in managing cell balancing within the battery system. The primary metrics used to assess their performance were voltage error reduction over time, response time, settling time, and steady-state error. The results are illustrated in Figs. 19 and 20 and summarized in Table 6.

Figure 19 displays the voltage error reduction over time for both PID and fuzzy logic controllers. The graph shows that the PID controller achieves a faster and more consistent reduction in voltage error compared to the fuzzy logic controller. Initially, both controllers exhibit a significant decrease in voltage error; however, the PID controller maintains a steadier rate of error reduction.

Figure 20 compares the response time and settling time of the PID and fuzzy logic controllers. The PID controller exhibits a quicker response time of 10 milliseconds compared to the 12 milliseconds of the fuzzy logic controller.

Similarly, the PID controller has a shorter settling time of 20 milliseconds, whereas the fuzzy logic controller has a settling time of 22 milliseconds.

The steady-state error, as shown in Table 6, is another critical performance metric. The PID controller achieves a steady-state error of 0.02 volts, which is slightly higher than the 0.01 volts achieved by the fuzzy logic controller.

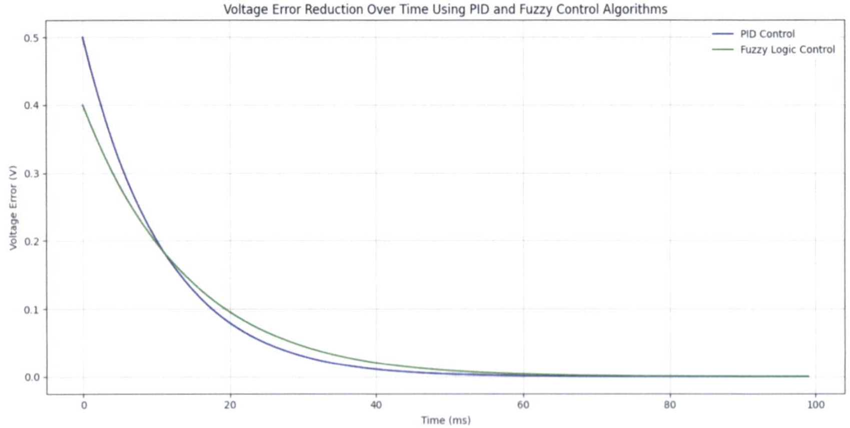

Fig. 19 Voltage error reduction over time

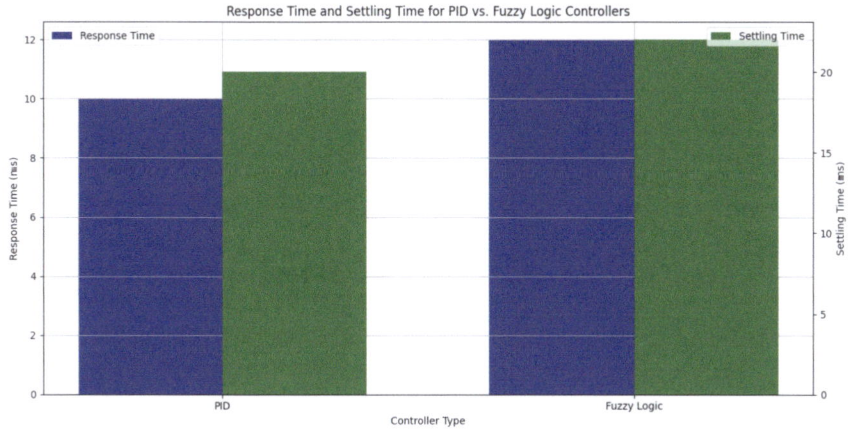

Fig. 20 Response time and settling time

Table 6 Performance metrics for PID and fuzzy control algorithms

Metric	PID control	Fuzzy logic control
Overshoot (%)	5	7
Settling time (ms)	20	22
Response time (ms)	10	12
Steady-state error (V)	0.02	0.01

3.1.5 AI-Based Balancing Model Results

The AI-based balancing model was designed to enhance the performance of the battery management system by leveraging reinforcement learning techniques. The model's effectiveness was evaluated through a series of experiments, and its performance was compared to traditional balancing methods. The results of these evaluations are illustrated in Figs. 21 and 22 and summarized in Table 7.

Figure 21 presents the learning curve of the AI-based model, showcasing the reduction in error over training epochs. As depicted, the model exhibits a rapid decrease in error during the initial epochs, followed by a gradual leveling off.

Figure 22 compares the balancing performance of the AI-based model against traditional methods. The bar chart illustrates that the AI-based model achieved a performance metric of 0.95, compared to 0.8 for traditional methods.

Table 7 summarizes the key performance indicators for the AI-based balancing model. The model required 15 hours of training time, which is relatively efficient given the complexity of the task. The final training error was remarkably low at 0.01, indicating that the model achieved a high level of accuracy in balancing the battery cells. The balancing efficiency reached 95%, showcasing the model's effectiveness in

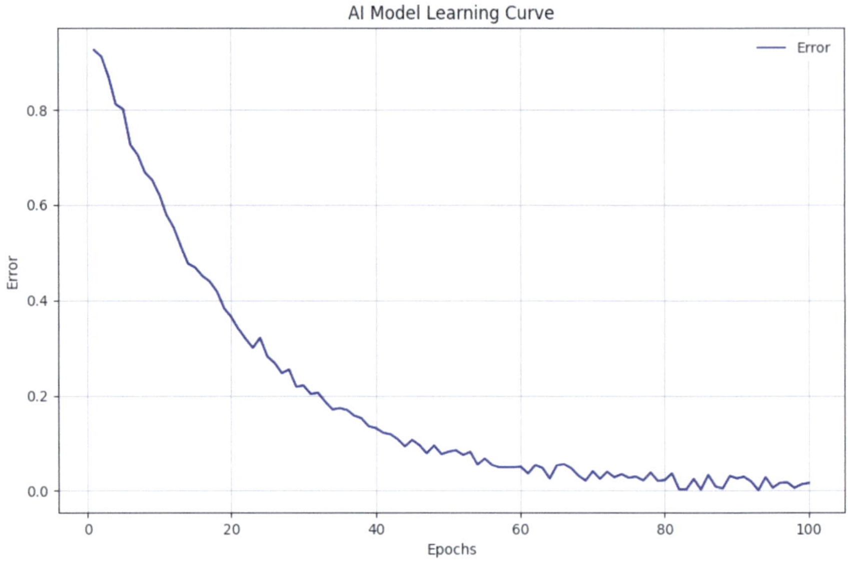

Fig. 21 AI model learning curve

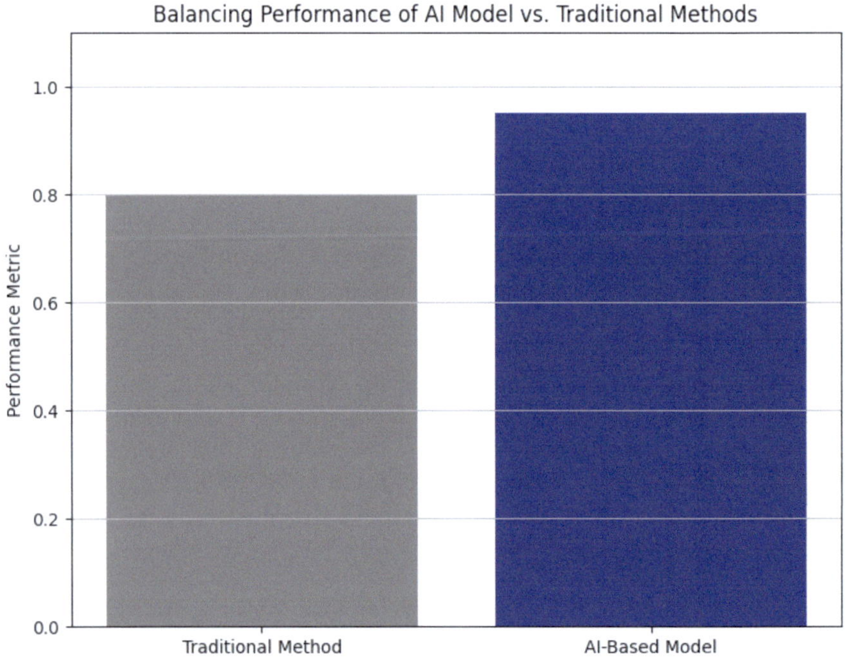

Fig. 22 Balancing performance comparison

Table 7 Key performance indicators for the AI-based balancing model

Performance indicator	Value
Training time (hours)	15
Final training error	0.01
Balancing efficiency	95%
Response time (ms)	50

managing cell voltages. Additionally, the response time of 50 milliseconds demonstrates the model's capability to make quick adjustments, crucial for maintaining optimal performance.

3.2 Discussion

3.2.1 Interpretation of Battery Model Results

The equivalent circuit model's performance was assessed against empirical data to gauge its accuracy in simulating battery behavior. Key metrics included voltage profiles, capacity degradation, and temperature effects.

Figure 11 shows the battery model's voltage during charge and discharge cycles. The model successfully mimics real-world voltage curves. During charging, the voltage rises steadily and plateaus as the battery nears full charge. The discharge profile shows a sharp initial drop followed by a stable phase before a more significant decline. The model's alignment with these expected behaviors confirms its accuracy in representing battery operation.

Figure 12 displays battery capacity degradation over multiple cycles. The model shows a linear capacity decrease, reflecting the wear and tear experienced in actual usage. This linear trend aligns with real-world degradation patterns, validating the model's accuracy for assessing long-term battery performance and the importance of managing capacity loss.

The model also captures the relationship between ambient and internal temperatures, showing a rise in battery temperature with increasing ambient temperature. This simulation of thermal effects is crucial for understanding performance impacts and underscores the need for effective thermal management.

The model also closely follows the measured data, validating its representation of battery performance. Minor discrepancies may arise from model simplifications or measurement variations but do not undermine the model's overall credibility.

Temperature and capacity degradation significantly impact model accuracy. Figure 13 highlights the importance of incorporating thermal effects in simulations. Accurate temperature simulation is essential for developing effective thermal management strategies. Figure 12 shows how capacity degradation affects the battery's performance over time. The model's accurate simulation of degradation helps in predicting lifespan and planning maintenance.

Overall, the model's ability to replicate voltage behavior, capacity degradation, and temperature effects confirms its effectiveness for analyzing battery performance. The alignment with empirical data and the simulation of real-world scenarios support its use in optimizing battery management strategies.

3.2.2 Analysis of Safety System Performance

The safety system's performance was evaluated to determine its effectiveness in safeguarding the battery under fault conditions. It was tested for current, voltage, and temperature thresholds to ensure the system could prevent damage and maintain safe operations.

Figure 14 shows the safety thresholds. Initially, the current stays within normal limits. When a fault condition causes the current to exceed the threshold, the system quickly drops it to zero. This prompt action prevents overheating or damage, demonstrating the system's effectiveness in maintaining safe current levels.

The voltage increases until reaching the threshold, after which it stabilizes. This stabilization indicates the system's successful engagement of protective measures to prevent overcharging, ensuring the voltage remains within safe limits and protecting the battery.

The temperature rises gradually until reaching the threshold. When exceeded, cooling mechanisms activate, causing a controlled decrease in temperature. This gradual cooling prevents thermal stress and maintains the temperature within safe limits, demonstrating the system's effective thermal management.

Figure 15 illustrates the fault detection signal. It remains off during normal operation and switches to on when a fault is detected, such as overcurrent or overheating. This signal transition confirms that the system effectively detects faults and initiates shutdown procedures promptly.

The safety system's performance was compared to industry standards. Table 4 summarizes the safety thresholds, which align with standard values for reliable operation and damage prevention.

The 13 A current threshold prevents overheating and damage. This threshold is consistent with industry standards, ensuring effective current management and prompt action to mitigate overcurrent conditions.

The 4.4 V voltage threshold prevents overcharging. This value aligns with industry practices, reflecting the system's adherence to safety protocols by maintaining voltage stability after reaching the threshold.

The temperature threshold of 30°C prevents overheating. This standard value is typical for battery temperature management. The system's activation of cooling procedures and gradual temperature reduction align with best practices for thermal management.

Overall, the safety system performs effectively across current, voltage, and temperature thresholds, maintaining safe operations and adhering to industry standards. Its performance confirms reliable protection against potential hazards.

3.2.3 Evaluation of Active Balancing Model

The effectiveness of the active balancing algorithm was assessed by comparing its impact on cell voltage distribution and charging efficiency to passive balancing methods. Active balancing, which uses mechanisms like switched flyback converters, aims to improve battery performance by addressing voltage imbalances.

Figure 17 shows the voltage distribution across cells before and after active balancing. Initially, without balancing, cell voltages vary significantly, leading to inefficiencies and reduced battery lifespan. After applying the active balancing algorithm, the voltage distribution becomes more uniform. The system transfers charge between cells to minimize discrepancies, enhancing battery operation by preventing overcharging or excessive discharging of individual cells. This uniformity helps improve the overall health and longevity of the battery pack.

Figure 18 compares charging times and balancing efficiency between active and passive methods. Active balancing reduces charging times significantly, as it redistributes the charge more effectively, achieving a full charge faster. The balancing efficiency, which measures the percentage of cells within an acceptable voltage range, is higher with active balancing (95%) compared to passive methods (85%). This demonstrates that active balancing maintains voltage balance better across cells, contributing to more efficient charging and improved overall performance.

Active balancing has a significant positive impact on battery performance. Improved voltage distribution and reduced charging times highlight the benefits of the active balancing model.

Table 5 summarizes key performance metrics. Active balancing shows a shorter average charging time (45 min) compared to passive balancing (60 min), reflecting its efficiency in charge distribution. The voltage range with active balancing (3.70–3.75 V) is narrower than with passive balancing (3.65–3.80 V), indicating better voltage control and stability.

Active balancing results in a more balanced and efficient battery pack. It reduces voltage variance and charging times, enhancing the performance and longevity of the battery system. By maintaining cells within an optimal voltage range, active balancing minimizes risks of overcharging and deep discharging, leading to a more reliable and durable battery pack.

Active balancing offers clear advantages over passive methods. The improved voltage distribution, reduced charging times, and higher balancing efficiency confirm the effectiveness of active balancing in optimizing battery performance, resulting in a more stable and longer-lasting battery system.

3.2.4 Comparison of Control Algorithms (PID vs. Fuzzy Logic)

This section compares PID and fuzzy logic controllers in managing cell balancing within the cell battery system, focusing on voltage error reduction, response time, settling time, and steady-state error.

Figure 19 shows that the PID controller achieves a faster and more consistent reduction in voltage error than the fuzzy logic controller. Both controllers start with a significant decrease in error, but PID maintains a steadier and quicker reduction due to its dynamic adjustment of proportional, integral, and derivative terms. This allows PID to react promptly and minimize errors effectively.

In contrast, the fuzzy logic controller, which relies on predefined rules and membership functions, reduces error more gradually. While it adapts to varying conditions, its response is slower compared to PID, resulting in a more gradual reduction in voltage error.

Figure 20 compares response and settling times. The PID controller has a response time of 10 milliseconds and a settling time of 20 milliseconds, faster than the fuzzy logic controller's 12 milliseconds and 22 milliseconds, respectively. This indicates that PID stabilizes more quickly, which is crucial for maintaining optimal battery performance.

Table 6 shows the steady-state error for both controllers. PID achieves 0.02 volts, slightly higher than the fuzzy logic controller's 0.01 volts. This indicates that while PID is effective, the fuzzy logic controller offers slightly better precision in voltage maintenance, which is beneficial for applications requiring exact voltage levels.

PID tuning involves adjusting proportional, integral, and derivative gains to balance response time, settling time, and stability. The PID controller was tuned to minimize voltage error while maintaining a quick response and stable performance. The results demonstrate that PID provides a good trade-off between rapid response and stability.

The fuzzy logic controller uses a rule-based approach with predefined membership functions. Its performance depends on the quality of these rules. While it offers better precision in some scenarios, it has a slower response and error reduction compared to PID. Its adaptability to varying conditions is useful for non-linear systems but comes with slower adjustments.

In evaluating control strategies for our battery management system, PID controllers demonstrate superior characteristics compared to fuzzy logic alternatives. PID excels in rapid response times and consistent error reduction, which are critical factors for our BMS application. These attributes ensure quick adjustments and maintain stability in voltage and current control, directly contributing to enhanced battery health and performance. While fuzzy logic offers advantages in steady-state voltage precision, our system prioritizes swift responsiveness and error minimization. PID provides a more balanced approach, aligning closely with our specific BMS requirements. The relative simplicity of PID implementation and tuning also presents practical benefits in terms of development and maintenance efficiency.

3.2.5 AI-Based Model Versus Traditional Methods

Figure 21 shows the AI-based model's learning curve, which displays a significant reduction in error over training epochs. The model quickly reduces error, indicating rapid learning and adaptation. As training progresses, error reduction slows, suggesting the model is nearing optimal performance. This curve reflects the AI model's effectiveness in refining its balancing strategy and adapting to minimize errors efficiently.

The AI-based approach's key advantage is its dynamic learning capability. Unlike traditional methods with fixed algorithms, the AI model continuously learns from feedback, allowing it to adapt to changing battery conditions and improve over time. This leads to more efficient management of cell voltages and optimized balancing actions.

Figure 22 compares the AI-based model's performance with traditional methods. The AI model achieved a performance metric of 0.95, versus 0.8 for traditional methods. This improvement highlights the benefits of AI in optimizing performance. The AI model's superior performance is due to its adaptive learning, which handles complex and dynamic conditions more effectively than traditional methods, which often rely on static algorithms.

The traditional methods appear to struggle with varying conditions and dynamic battery performance, leading to less efficient balancing. In contrast, the AI model's reinforcement learning approach enables real-time adjustments and optimization, resulting in better performance.

Table 7 summarizes the AI model's performance metrics: a final training error of 0.01, a balancing efficiency of 95%, and a response time of 50 ms. These metrics demonstrate the model's high accuracy and effectiveness in managing cell voltages. The training time of 15 h is also efficient, showing that the model achieves high performance with a reasonable amount of training time, making it practical for real-world applications.

The AI-based balancing model offers significant advantages over traditional methods, including improved balancing performance, adaptability, and overall efficiency. Its ability to learn and optimize through reinforcement learning results in superior performance metrics and rapid adjustments. Future work should enhance the model with advanced AI techniques, integrate additional data sources, and explore other methods to further refine its capabilities. The AI-based approach represents a promising advancement in battery management, offering more efficient and effective cell balancing.

4 Conclusion

The comprehensive evaluation of battery management strategies and control algorithms has provided valuable insights into optimizing battery performance and safety. This study has explored various approaches, including traditional passive and active balancing methods, PID and fuzzy logic controllers, and advanced AI-based models. Each approach offers distinct advantages and limitations, contributing to a deeper understanding of effective battery management.

4.1 Model Enhancements

While the current AI-based model shows promising results, there are several avenues for future improvement. One potential enhancement is the integration of more advanced reinforcement learning techniques, such as deep reinforcement learning, to further refine the model's balancing strategy. Deep reinforcement learning can improve the model's ability to handle complex and high-dimensional state spaces, leading to even better performance.

Another area for improvement is the incorporation of additional data sources and features. By integrating more information about the battery's operational conditions, such as environmental factors or usage patterns, the AI model can achieve more accurate and adaptive balancing. This enhanced data integration can help the model make more informed decisions and improve overall performance.

4.2 Exploration of Other AI Techniques

Beyond reinforcement learning, other AI techniques could be explored to enhance the battery management system further. For instance, supervised learning algorithms could be used to predict battery behavior and optimize balancing actions based on historical data. Techniques such as neural networks or ensemble methods could offer additional insights and improvements.

Additionally, incorporating AI-based predictive maintenance techniques could help in anticipating potential issues before they arise. By analyzing patterns in the battery's performance and predicting future conditions, the model could proactively address potential problems, leading to increased reliability and longevity of the battery system.

4.3 Real-World Applications

To fully realize the potential of the AI-based model, further testing in real-world scenarios is essential. Implementing the model in various battery systems and operational environments will provide valuable insights into its practical performance and adaptability. Real-world testing will also help identify any limitations or challenges that need to be addressed for successful deployment.

In conclusion, the study has provided a comprehensive evaluation of various battery management strategies and control algorithms. The findings highlight the benefits of advanced techniques, such as active balancing and AI-based models, in improving battery performance and safety. Future work will focus on leveraging advanced AI techniques, real-world testing, and predictive maintenance to further enhance battery management systems and ensure their effectiveness in diverse applications.

References

1. Cui X, Shen W, Zhang Y, Hu C (2017) A fast multi-switched inductor balancing system based on a fuzzy logic controller for lithium-ion battery packs in electric vehicles. Energies 10(7)
2. Daowd M, Antoine M, Omar N, Van den Bossche P, Van Mierlo J (2013) Single switched capacitor battery balancing system enhancements. Energies 6(4):2149–2174
3. Daowd M, Omar N, Bossche PVD, Mierlo JV (2011) Passive and active battery balancing comparison based on matlab simulation. In: 2011 IEEE vehicle power and propulsion conference, pp 1–7
4. Divya T, Umayal C, Vasan Prabhu V (2024) A hybrid gtoa-drn approach based active cell voltage balancing of electric vehicle batteries using pid controller. J Energy Storage 98:113095
5. Dubarry M, Vuillaume N, Liaw BY (2010) Origins and accommodation of cell variations in li-ion battery pack modeling. Int J Energy Res 34(2):216–231
6. Duraisamy T, Kaliyaperumal D (2021) Machine learning-based optimal cell balancing mechanism for electric vehicle battery management system. IEEE Access 9:132846–132861
7. Hardy J, Steggall J, Hardy P (2023) Rethinking lithium-ion battery management: eliminating routine cell balancing enhances hazardous fault detection. J Energy Storage 63:106931
8. Hussien S, Baqais A, Al-Gabalawy M (2023) Battery management system enhancement for lithium-ions battery cells using switched shunt resistor approach based on finite state machine control algorithm. Front Energy Res 11
9. Kaleem A, Khalil IU, Aslam S, Ullah N, Al Otaibi S, Algethami M (2021) Feedback pid controller-based closed-loop fast charging of lithium-ion batteries using constant-temperature–constant-voltage method. Electronics 10(22)
10. Khan N, Ooi CA, Alturki A, Amir M, Shreasth, Alharbi T (2024) A critical review of battery cell balancing techniques, optimal design, converter topologies, and performance evaluation for optimizing storage system in electric vehicles. Energy Rep 11:4999–5032
11. Liaw BY, Jungst R, Urbina A, Paez TL (2004) Modeling of battery life i. The equivalent circuit model (ecm) approach
12. Morgan J (2020) Electric vehicles: the future we made and the problem of unmaking it. Camb J Econ 44(4):953–977
13. Nour M, Chaves-Ávila JP, Magdy G, Sánchez-Miralles A (2020) Review of positive and negative impacts of electric vehicles charging on electric power systems. Energies 13(18)
14. Perisoara L, Guran I, Costache D (2018) A passive battery management system for fast balancing of four lifepo4 cells, pp 390–393
15. Plett G (2015) Battery management systems, volume I: battery modeling
16. Plett G (2016) Battery management systems, volume II: equivalent circuit methods
17. Samaddar N, Kumar S, Jayapragash R (2020) Passive cell balancing of li-ion batteries used for automotive applications. J Phys: Conf Ser 1716:012005
18. Sarkar A, Dutt R, Sahu S, Acharyya A (2023) Energy-efficient and high speed active cell balancing methodology for lithium-ion battery pack. In: 2023 21st IEEE interregional NEWCAS conference (NEWCAS), pp 1–5
19. Song H, Lee S (2023) Study on the systematic design of a passive balancing algorithm applying variable voltage deviation. Electronics 12(12)
20. Thakkar R (2021) Electrical equivalent circuit models of lithium-ion battery
21. Weng Y, Ababei C (2024) Ai-assisted reconfiguration of battery packs for cell balancing to extend driving runtime. J Energy Storage 84:110853

Modeling and Control of a Robotic Arm Utilizing Artificial Neural Networks

Dhiraj Sureshchand Yadav

Abstract The present work studies the comparison of a modified reference controller based on an artificial neural network with a proportional-integral-derivative (PID) controller in the modeling and controlling of a two-degree-of-freedom (2DOF) robotic arm. In industrial automation and other domains, autonomous machines are important factors that require accurate control to adjust to their nonlinear dynamics. When dealing with this complexity, traditional controllers like proportional-integral-derivative (PID) controllers often deal with difficulties. The primary goal is to investigate. The primary aim is to investigate the performance of each controller. The process includes using Simulink to develop a closed-loop kinematic model of the 2DOF robotic arm. With the goal of anticipating the joint angles required for tracking desired trajectories, an artificial neural network (ANN) controller was built and trained using simulated data. For the purpose of determining the efficiency of the ANN-based controller against a conventional PID controller, simulations were run. The main metrics that were examined were tracking accuracy, reaction time, and stability. The closed-loop feedback mechanism provided the ANN with the capacity to continuously adapt its predictions and provide accurate results. The tracking accuracy of the ANN-based controller was found to be higher than that of the PID controller, with less error. Additionally, the ANN demonstrated to be more adaptive, preserving stability, while the PID controller failed. These demonstrated the ANNs' capability to adjust in real time to dynamic changes, offering a more accurate and responsive control system. In conclusion, specifically for complicated situations, the ANN-based control system proves to be a viable substitute for traditional PID controllers in robotic arm control. The findings suggest that ANNs may give increased performance in nonlinear and dynamic systems. **Highlights 1. Controller comparison**: For a two-degree-of-freedom robotic arm, the study described here performs a detailed comparison between a traditional PID controller and an artificial neural network-based Model Reference Controller. **2. Adaptable control system**: This research has shown that, in situations with complicated nonlinear dynamics and in the presence of external disruptions, control based on artificial neural networks (ANNs) is more responsive and durable than traditional control carried out by PID controllers.

D. S. Yadav (✉)
School of Engineering, Computing and Mathematics, Oxford Brookes University, Oxford, UK
e-mail: 19272869@brookes.ac.uk

A. Azizi (ed.), *Intelligent Mechatronic Control Systems*, Emerging Trends in Mechatronics, https://doi.org/10.1007/978-981-95-3925-3_2

3. Performance metrics: Since the ANN-based model reference control (MRC) outperforms the PID controller in terms of accuracy and error reduction, tracking accuracy, reaction time, and reliability are the most important performance metrics.
4. Implications for advanced robotics: The perspectives into future progress in robotics automation arise from the likelihood that ANN-based controllers are going to surpass traditional PID controllers in applications connected to very complex, real-time, and dynamic environments.

List of Symbols and Abbreviations

ANN	Artificial Neural Network
PID	Proportional-Integral-Derivative
MRC	Model Reference Controller
2DOF	Two-Degree-of-Freedom
CAD	Computer-Aided Design
RMS	Root Mean Square
MSE	Mean Squared Error
MPC	Model Predictive Control
FLC	Fuzzy Logic Control
SNR	Signal-to-Noise Ratio
DC-DC	Direct Current to Direct Current
MTDC	Multi-Terminal Direct Current
IMC	Internal Model Control
AVOA	African Vultures Optimization Algorithm
IMU	Inertial Measurement Unit

1 Introduction

1.1 Background

Automation, healthcare, and manufacturing are some of the few sectors that depend on robotic arms. They undertake duties that require a high degree of efficiency, adaptability, and accuracy. Robotic systems are becoming more complex due to growth in the technological field, needing complex control methods that can handle external interference and nonlinear dynamics (Xu, Xiong and Bhattacharyya). Because they are easy and efficient for linear systems, traditional control techniques like proportional-integral-derivative (PID) controllers have gained extensive application. When applied to more complex multivariable systems, they face some serious issues with their performance [35].

Because of their drawbacks, PID controllers require a human for tuning, which may be tedious and might not provide the best results. This also applies to robotic arms, which commonly operate in a fluctuating environment, and external disturbance may have a significant impact on performance. Due to this, alternative control strategies that might be more flexibility and accurate are gaining popularity.

This research's main goal is to create a control system using artificial neural networks (ANNs) and simulate a two-degree-of-freedom (2DOF) robotic arm. The objective of this study is to analyze the precision of trajectory tracking and the responsiveness of control while dealing with nonlinear dynamics.

The Objectives are as follows

1. To develop a PID controller that controls a 2DOF robotic arm.
2. to create an efficient model reference controller (MRC) that uses the ANN algorithm to drive the robotic arm.
3. Evaluate the tracking accuracy, effort, and speed of response of the ANN-based model reference control (MRC) in relation to a traditional PID controller.
4. Everything is to be done using the MATLAB-Simulink environment for the research's ensuring a thorough simulation-based strategy.

1.2 Summary of Literature Review

According to Pascher et al. [50], in industries like manufacturing, healthcare, and entertainment, mechanical robotic arms are important tools for carrying out complex, precise, and repetitive activities. In linear systems, PID controllers are preferred due to their ease of use and effectiveness. However, in real-time situations, disturbances provide difficulties for conventional control approaches as robotic systems have become more advanced, requiring more accuracy and flexibility.

On the other artificial neural networks (ANNs) have become a good alternative because of their ability to learn from and adjust to the dynamics of a system. This is essential since the present robotic systems have to adjust to changing real-world settings [70].

1.2.1 The Basics of Robotic Arm Mechanisms

Robotic arms are used in various industries in order to mimic the functions of human arms. Even though 2DOF arms are easier to manage than more degrees of freedom, nonlinear dynamics and outside disturbances still prove challenging [25].

While dynamics investigates the stresses and torques on the arm, kinematics examines motion without taking those forces into account. According to Abbas et al. [1], inverse kinematics calculates the required joint angles for a specified end-effector position, and kinematic modeling uses joint angles to determine the final location of the end-effector.

The challenge becomes more complex in real-world applications due to ongoing load variations and outside disturbances. With the use of forces and torques, dynamic modeling helps engineers in determining the required forces for various motions that are necessary for efficient control systems [46]. Classical control techniques like PID have difficulties in the complex dynamics of robotic arms because of growing degrees of freedom and external factors.

1.2.2 Artificial Neural Networks (ANNs) in Control Systems

The ability of ANN, when integrated with a controller, to provide accuracy is the need for its introduction in robotics. By learning from data to make predictions, ANNs provide a foundation for real-world applications in various realms, including robotics [15]. They are helpful in managing nonlinear dynamic systems, which might be challenging for conventional controllers [61].

ANNs are made of layers that process information through connections and activation functions. Their strength lies in highly dynamic systems that are a challenge for traditional methods [57]. ANNs provide the required results by being trained using input data, such as joint angles.

To control end-effector movement, ANNs construct forward and inverse dynamics. Forward dynamics training uses joint angles to predict the end-effector's location, while inverse dynamics calculates the torques needed to reach a desired position [45].

ANNs' capacity to handle very nonlinear dynamics is one of the primary benefits. Artificial neural networks adapt effectively during change and can learn how to handle this change. This flexibility is crucial for robots since performance is greatly affected by changes in load or environment.

According to Eltayeb et al. [18], ANN-based controllers perform better than PID controllers in controlling a 2DOF robotic arm. PID controllers need to be manually modified, whereas ANN-based controllers can adapt to changes in the dynamics of the system.

1.2.3 Model Reference Control (MRC) Systems

It is a control method that adjusts system inputs to match behavior as specified by a reference model with actual behavior. Whenever conventional controllers are unable to handle volatile or time-varying dynamics, MRC could prove helpful [28].

A comparison model provides an anticipated outcome for specified inputs in MRC. In nonlinear and uncertain situations, the controller enhances control accuracy by modifying inputs to lessen the difference between real and reference outputs [34]. Recent work uses ANNs' capacity to train constantly and adjust to real-time changes to improve the flexibility of MRC systems.

Compared to conventional PID controllers, Sharma et al. [60] showed that ANN-based MRC systems react to changes in system dynamics more correctly. In industries like medical robotics and high-precision manufacturing, this flexibility is vital.

1.2.4 Proportional-Integral-Derivative (PID) Controller

PID controllers are popular because of their dependability and simplicity [29]. In a number of applications, including robotic arms, they efficiently provide constant control. When used in time-varying situations, PID controllers are limited.

The three components of a PID controller are the derivative term, which adjusts to current errors, the integral term, to account for cumulative previous errors, and the proportional term, which forecasts errors [30]. PID controllers, while effective, need careful tuning during dynamic changes.

Performance may be enhanced by using techniques like ANNs in PID controllers. The merits of both techniques may be used in hybrid systems to change parameters in response to changes [66].

1.2.5 Comparative Analysis of ANN-Based MRC and PID Controller

Results from studies demonstrate that ANN-based MRC systems perform better than PID controllers in terms of flexibility, stability, and accuracy, particularly when dealing with external disturbances [42]. In place of PID controllers, Mohammed et al. [42] proved that ANN-based MRC systems respond to changes in the system more quick and correct.

Furthermore, Eltayeb et al. [18] discovered that ANN-based controllers are more reliable even during severe disruptions, which is acceptable for high-precision applications like industrial automation and healthcare. PID controllers are simple to use and use limited processing power, but in situations that are very dynamic, they may not be enough. In contrast, ANN-based MRC systems work better, making them more useful in a variety of situations.

1.3 Originality and Contribution

Multiple contributions have been made to the area of robotic control in this study. Initially, it presents an original use of artificial neural network (ANN)-based control techniques for a two-degree-of-freedom (DOF) robotic arm with the primary objective of enhancing trajectory tracking. Through the development and use of a model reference controller (MRC) using the ANN algorithm, this study demonstrates the potential benefits of neural network technology versus conventional control methodologies.

Through a performance comparison between an ANN-based MRC and a traditional PID controller, the research provides critical insights into the pros and cons of each approach in practical applications.

The results of this study may provide an outline for future research on hybrid control methods and open up possibilities for the development of more advance and adaptable robotic systems that can operate in any number of scenarios. Distinct approach and emphasis on real-world application put this study ahead of others in the domain of robotic control research.

1.4 The Approach

Throughout the research, a structured approach was used to reach the stated goals. The starting point was to use MATLAB-Simulink to create a closed-loop kinematic model of the 2DOF robotic arm. This model provided a base for control development and made it simpler to simulate the dynamics of the robotic arm.

Next, an ANN model was developed to calculate the joint angles needed for end-effector locations. With the use of simulated input data, the ANN was successfully taught to understand the links between predictors and response. In the meantime, the trained ANN was used to create a model reference controller (MRC) to control the robotic arm in real time.

In order to determine the performance of the ANN-based MRC, a PID controller was used as a comparison. Both systems were tested using simulations on a range of critical performance measures, including durability, tracking accuracy, and reaction time. Since the whole research took place in the MATLAB-Simulink environment, a proper foundation for evaluating the efficiency of the control measures was provided.

The study attempts to illustrate the positive effects of ANN-based control systems in robotic applications through a systematic method, opening up possibilities for further developments in this area.

2 Methodology

The approach of this research compares the performance of two control systems for trajectory control of a 2DOF robotic arm: a proportional-integral-derivative (PID) controller and a model reference controller (MRC) improved by an artificial neural network (ANN). It focuses on controlling control inputs under variable circumstances, reducing error, and assessing each controller's accuracy in trajectory tracking [49].

The first step in the process is the development of a mathematical model that depicts the kinematics and dynamics of the robotic arm. The model provides the basis for applying control techniques by considering forces and torques (dynamics) and calculating the end-effector's position and velocity based on joint angles (kinematics).

An ANN-enhanced MRC is designed to adapt by learning the dynamics of the arm and modifying parameters, while the PID controller is tailored for effective trajectory tracking. The ANN uses backpropagation to discover the link between control signals and trajectory, whereas the MRC seeks to match a reference model [44].

Reference trajectories are generated using simulations, which are used to assess the performance of both PID and MRC controllers. To reduce tracking errors, the ANN in the MRC is trained using a mean squared error metric. To evaluate performance and resilience, the controllers are put to the test on metrics such as trajectory accuracy, steady-state error, reaction time, settling time, and root mean square (RMS) errors.

The approach demonstrates the possible benefits and drawbacks of adaptive (ANN-based MRC) in comparison to conventional (PID) control techniques. Compared to the traditional PID controller, the ANN-based MRC exhibits more resilience and adaptability against noise, disturbances, and changing operating circumstances [23, 24] (Fig. 1).

2.1 CAD Model

In this case, the robotic arm functions in a single plane and is a two-link type. Due to its low complexity and efficiency, it is widely used in both industrial and scientific contexts. This kind of arm is often used for operations like picking and placing objects, exact placement, or route following (Figs. 2, 3 and Table 1).

- **Specifications of the Design Process and Software**

SolidWorks 2024 was used to develop the robotic arm. During this procedure, separate arm components—such as joints and links—were designed, linked together to create a single, working robotic arm, and motion tests were then conducted. To make sure that every part was linked correctly and enable realistic movement simulation, SolidWorks Assembly was used.

- **Information on Materials and Production**

The arm is designed to be made of low-carbon steel, specifically AISI 1020, which was selected for its beneficial features, including high weldability, ductility, and moderate strength. The need that the arm to be both lightweight and robust enough to withstand high operating loads led to this decision. Furthermore, low-carbon steel is cheap and easy to manufacture, which is crucial for producing robotic parts.

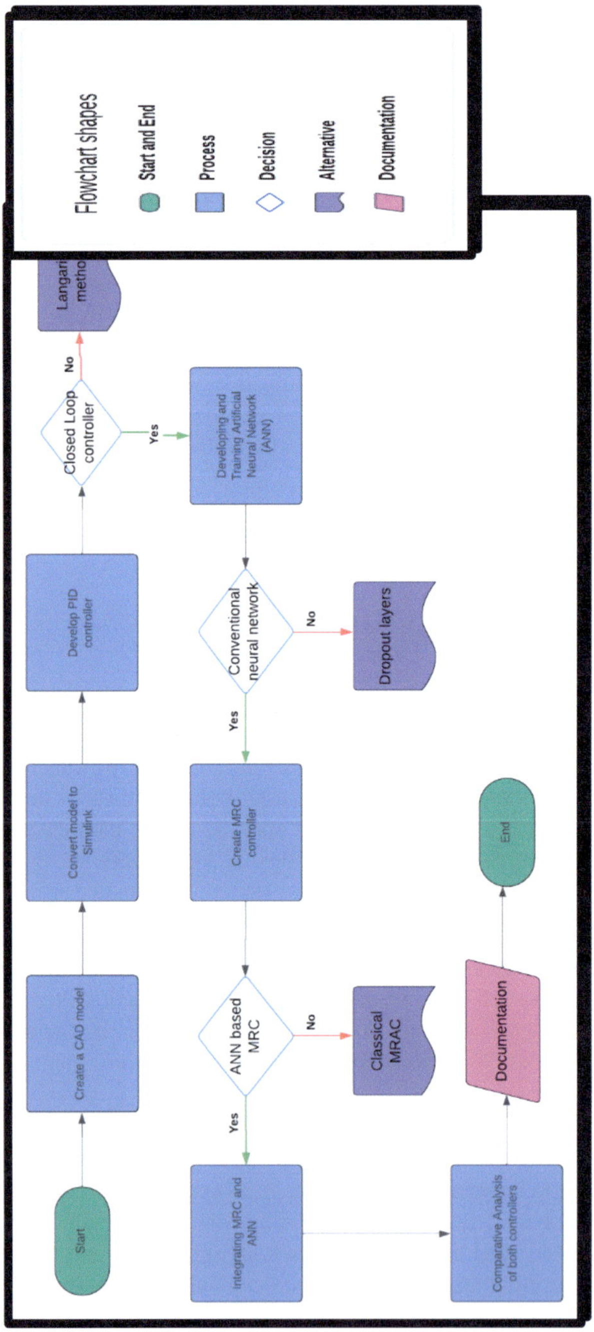

Fig. 1 Flow chart depicting the pathway followed

Fig. 2 CAD model of robotic arm

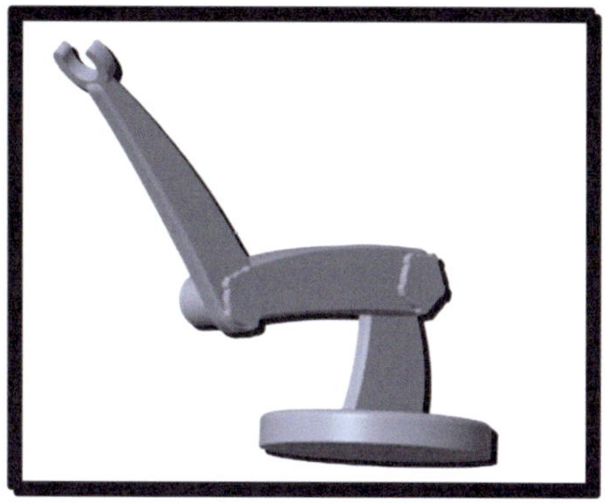

Fig. 3 Line diagram for the robotic model

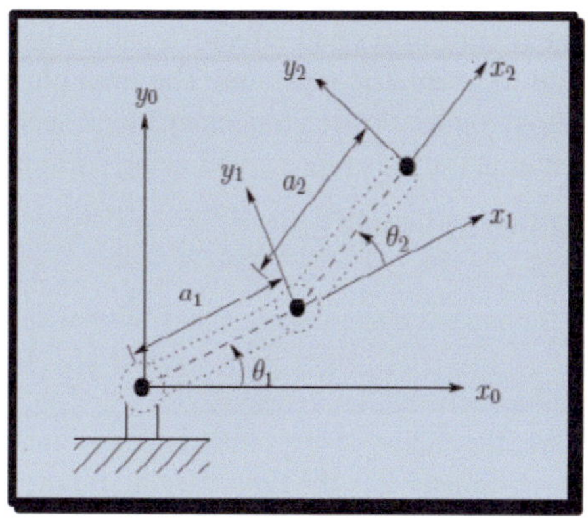

Table 1 Properties of the robotic arm

Software	Solidworks 2024
Length	1 m
Joint	2 Rotatory and 1 end-effector
Material	AISI 1020 (low-carbon steel

- **Fundamental Design and Kinematic Information**

Both of the revolute joints of the robotic arm model contribute to its two degrees of freedom (DOFs). The base is connected to the first link by the first joint, and the end-effector, or second link, is connected to the first link by the second joint. Because each DOF has one axis of rotation, the arm may move freely in the plane and take on a variety of actions.

- **Joint Constraints**

The reach and orientation of the end-effector are restricted by the exact angular range of mobility that each revolute joint possesses. For precise trajectory monitoring and modeling, these limitations are essential.

- **Dimensions of Geometry and Tolerancing**

The CAD model indicates that the arm is made up of two stiff links, each of which is 1 m in length. Each component's proportions were carefully chosen to find a compromise between mobility and rigidity. The second link's end-effector may move precisely to carry out duties like gripping things or following certain pathways.

Smooth rotating movement is made possible by the architecture of the model, and predictable motion in the simulation is guaranteed by the links' straight-line construction.

2.2 *PID Controller*

The robotic arm's end-effector's path, which follows an outline of a circle, is created with a parametric equation. The final placement of the end-effectors xd and yd is output via the desired trajectory block at each new step. The circle with radius r and center at (x, y) is represented using parametric equations used in PID (Fig. 4).

Desired trajectory

$$\begin{aligned} x_d &= x_{\text{center}} + r \cdot \cos(2\pi u) \\ y_d &= y_{\text{center}} + r \cdot \sin(2\pi u) \end{aligned} \tag{1}$$

Here, the radius of the circle is 0.5 m and the center of the circle is (1,1), whereas the parameter u is the time-varying input. The values obtained (xd, yd) are then fed into inverse kinematics to obtain the joint angles.

The angular positions (θ1) and (θ2) of the joints at which the end-effector will be at those coordinates need to be computed, knowing that the expected points (xd, yd) have been determined. In such a situation, the 2DOF planar robotic arm's inverse

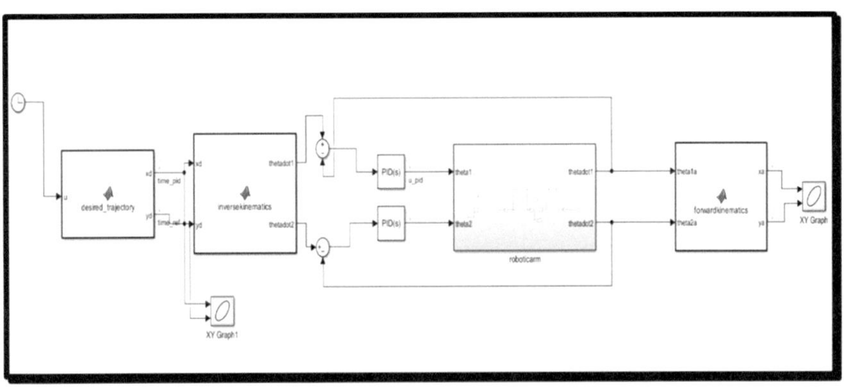

Fig. 4 Schematics of the PID controller

kinematics equations are utilized. The equations are derived using the relationships between the geometries of two connections (links) and the related angle of the joint. The following equations are used:

Inverse kinematic equation

$$\begin{aligned} \theta_2 &= \cos^{-1}\left(\frac{x_d^2+y_d^2-l_1^2-l_2^2}{2l_1l_2}\right) \\ \theta_1 &= \tan^{-1}\left(\frac{y_d}{x_d}\right) - \tan^{-1}\left(\frac{l_2\sin(\theta_2)}{l_1+l_2\cos(\theta_2)}\right) \end{aligned} \tag{2}$$

Here, two PID controllers are used for controlling the robotic arm, each for a joint. These controllers are used with a sum block to minimize the error between desired joint angles and actual joint angles. The control signal is calculated as

PID controller equation

$$u_i(t) = K_{p,i}e_i(t) + K_{i,i}\int_0^t e_i(\tau)d\tau + K_{d,i}\frac{de_i(t)}{dt} \tag{3}$$

where:

- $e_i(t) = \theta_{d,i}(t) - \theta_i(t)$ is the error between the desired and actual joint angles for joint i,
- $K_{p,i}$, $K_{i,i}$ and $K_{d,i}$ are the proportional, integral, and derivative gains, respectively,
- $u_i(t)$ is the control signal (torque) applied to the joint.

Now, the output of the PID controller (u1 and u2) is the torque that is used to control the robotic arm.

The forward kinematic equation is used to calculate the actual position of the end-effector.

Forward kinematics

$$\begin{aligned} x_a &= l_1 \cdot cos(\theta_1) + l_2 \cdot cos(\theta_1 + \theta_2) \\ y_a &= l_1 \cdot sin(\theta_1) + l_2 \cdot sin(\theta_1 + \theta_2) \end{aligned} \tag{4}$$

Using the current joint angles (theta1 and theta2), it calculates the actual position of the arm in the Cartesian space (xa, ya). This response is used to check how accurate and precise the end-effector is tracking (Fig. 5).

2.3 Modified Reference Controller

The construction begins with a two-layer feedforward neural network, and it comprises linear output neurons in the output layer and sigmoid hidden neurons in the hidden layer. Using sigmoid neurons is important for this intent because they

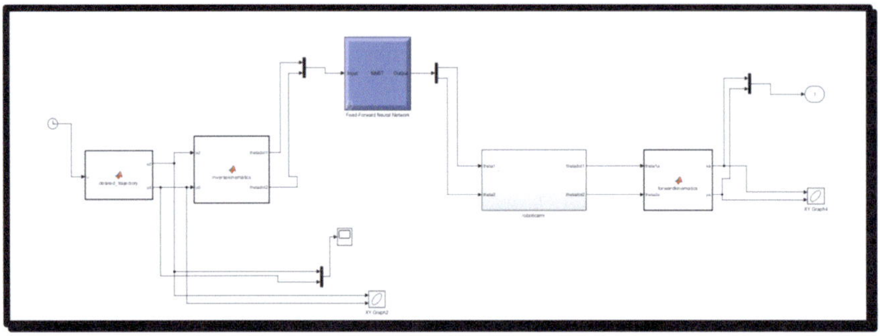

Fig. 5 Schematic of ANN-based MRC controller

permit the network to simulate nonlinear connections between input and output data, which is vital to regulate the complex motions of a robotic arm [51]. Additionally, the network includes 1,000 hidden neurons, allowing it to simulate the complicated structures of the inverse kinematics function. Considering the job demands steady predictions of joint angles, which can be more effectively performed by a linear activation function in the output layer of the brain, linear output neurons are used [27].

Using sigmoid neurons is important for this intent because they permit the network to simulate nonlinear connections between input and output data, which is vital to regulate the complex motions of a robotic arm [51]. Additionally, the network includes 1,000 hidden neurons, allowing it to simulate the complicated structures of the inverse kinematics function. Considering the job demands steady predictions of joint angles, which can be more effectively performed by a linear activation function in the output layer of the brain, linear output neurons are used.

The procedure for training goes next after the system's network configuration is established. Overall, 70% of the dataset is used for training, 15% for validation, and 15% for testing. Varying this number of percentage of data may produce a significant change in the output. This separation lessens the potential of excessive fitting because it guarantees that the model has ample data to learn from while still maintaining different data sets for testing. The weighing factors of the network are tailored using the Levenberg–Marquardt method. The method described here is well known for being efficient at training neural networks at an extremely fast rate of integration, thereby rendering it suitable for users seeking real-time predictions [6] (Table 2 and Fig. 6).

Table 2 Neural fitting setting

Training data	70%
Validation data	15%
Test data	15%
Layer size	1,000

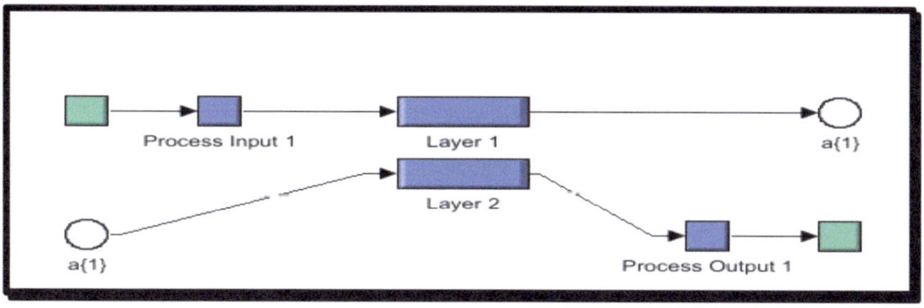

Fig. 6 Neural network layer structure for robotic arm control

Table 3 ANN algorithm settings

Data division	Random
Training algorithm	Levenberg–Marquardt
Performance	Mean squared error

The mean squared error (MSE) serves as the model's estimated function of loss during development. MSE is a suitable option for this job because it makes big mistakes more penalized, thereby compelling the network to prioritize decreasing major variations when anticipating joint angles. The neural network has enough trials to acquire the inverse kinematics function when the model has been developed for 1,000 epochs. For the purpose of avoiding overfitting and guaranteeing that the model operates successfully when compared to new data, early interrupting based on errors in validation is used (Forouzesh and Thiran) (Table 3).

The parametric formulas used in the PID controller outline a circular route that specifies the planned trajectory for the end-effector: The robot is capable of following this trajectory as it offers a continuous, smooth route. Although it is necessary to perfect joint coordination as well as control, circular trajectories are often used as test cases in robotics [51]. The technique involves entering the planned trajectory into an artificial neural network (ANN), which anticipates the joint angles required for navigating the trajectory. The algorithm may generalize to other movement and route patterns by training the network on this kind of data, guaranteeing flexibility in real-world applications.

The forward kinematics equations act as a starting point for managing the position of the robotic arm. The following formulas explain how the joint angles (theta1 and theta2) dictate the end-effector's (the robotic arm's tip) location (Hartl-Nesic and Meiringer).

More specifically, the forward kinematics equations for a two-link robotic arm are the same as those used in PID.

These mathematical equations permit the system to estimate the end-effector's location by considering the joint angles and link lengths. Inverse kinematics manages the opposite challenge of forward kinematics, which determines the joint angles necessary for achieving a given position. Forward kinematics uses the angles to calculate the end-effector's location.

As tasks generally specify end-effector locations instead of joint angles, this feature is critical for guiding robotic arms [70].

The main objective of the neural network is to calculate the inverse kinematics function, though it is occasionally hard to determine directly because of the nonlinearity of the fundamental equations. In order to move the robotic arm's end-effector to a certain location in space, inverse kinematics plays a role in calculating the needed joint angles (theta1 and theta2). The ANN utilizes controlled training to learn how it can simulate this mapping compared to solving the inverse kinematics analytically, which could prove computationally pricey.

Input–output pairs are transmitted into the neural network throughout training, where the pairs of inputs reflect the aimed end-effector locations (xd, yd), and the outputs are the joint angles needed to get there. This is important in dynamic and real-time robotic applications, given that it permits the ANN to expand and offer real-time predictions of joint angles for various scenarios.

An artificial neural network's (ANN) main task is to simulate the inverse kinematics function for real-time control. The system acquires the ability to relate the joint angles essential for attaining the required end-effector positions to those places. The neural network (ANN) presents a successful, data-driven response that can be estimated quickly during functioning, compared with standard approaches that require repeated computation or closed-form approaches [59]. As a result of this, it is appropriate for real-time applications where efficiency may be affected by processing delays. The robotic arm may adjust to a number of tasks and trajectories because of the ANN's abilities to generalize from the data used for training.

2.4 A Table of Alternative Approaches

Model predictive control (MPC)	Predicts future trajectories using an optimization-based approach and provides the most effective control signals. This method's very nature requires explicit constraints on any function. It manages multiple input/many output systems with great ease. The restrictions on governments and authorities are addressed directly. High performance is achieved for intricate nonlinear systems [8]
Fuzzy logic control (FLC)	Utilizes rule-based systems to replicate human decision-making while handling robotic systems' uncertainties and nonlinearities. Since it does not need an accurate mathematical model, it is appropriate for complex and nonlinear systems. Robust against uncertainties or disruptions [2]
Adaptive neural network control	By combining adaptive control with neural networks, controller settings are automatically modified in response to mistakes and real-time feedback. Over time, it may pick up on and adjust to changes in a system. Ideal for systems that change in time and are very nonlinear [38]

2.5 A Table of Limitations

Simplified trajectory selection	This strategy's training and testing trajectory is circular, which is more than enough to show the proof of concept. But in numerous real-world situations, more complex routes—like parabolic, linear, or dynamic obstacle avoidance—are necessary, and a model that has been mostly trained on circular motion rarely consistently manages them [11]
Idealized model assumptions	The CAD model makes the assumption that all joints and links are absolutely rigid, neglecting wear, friction, and actual flexibilities. Such presumptions might lead to differences between the outcome that is simulated and the real thing, especially when running at high speeds or with dynamic loads
Limited realism in simulations:	SolidWorks is an advanced simulation tool, but it lacks in accurately simulating real-world phenomena such as minor structural deformations under load, gear backlash, and joint flexibility. The practical performance of the robotic arm may be greatly affected by these considerations
Risk of overfitting	Overfitting is prevented by this practice via early halting; however, it may still happen if it is not properly managed. A model that overfits may perform well on training data but badly on original or marginally different tasks, which might reduce the model's durability and capacity to adjust to real-world scenarios (Liang, Zhou and Sesia)
Computational complexity and training time	The Levenberg–Marquardt technique and the ANN model's enormous size of 1,000 hidden neurons make training highly computational. Although this guarantees precise results after training, it might create a bottleneck when big datasets are needed or real-time modifications are needed [9]
Robustness to external factors	Environmental factors like fluctuations in temperature, vibrations, and electromagnetic interference are not taken into consideration by the methods, which might affect control accuracy

3 Results and Discussion

3.1 Results

3.1.1 Trajectory Comparison

The two provided trajectory comparative graphs demonstrate how a PID controller and a model reference controller, which use artificial neural networks, function differently when a robotic arm's end-effector is tracing a predetermined circular route [22] (Figs. 7 and 8).

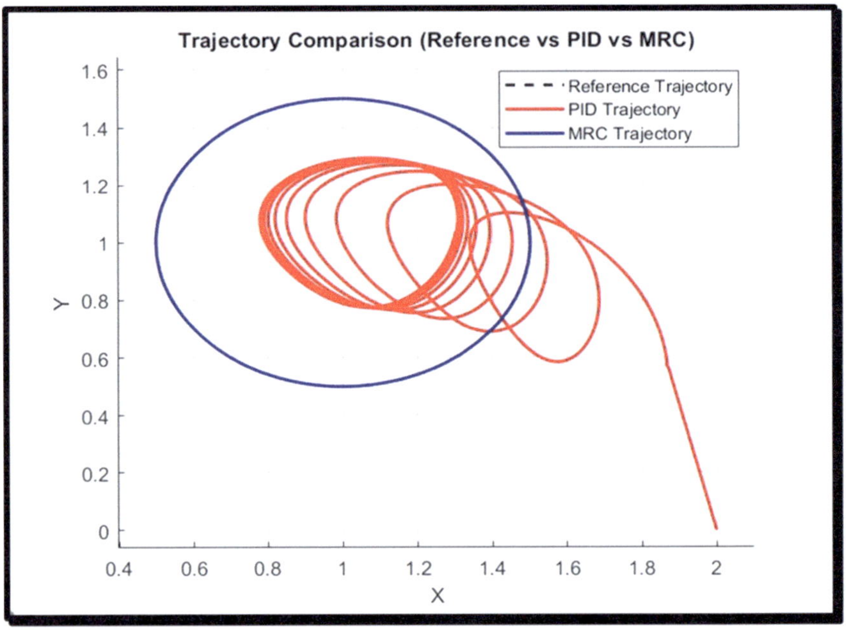

Fig. 7 This image illustrates the trajectory followed by untuned PID versus MRC

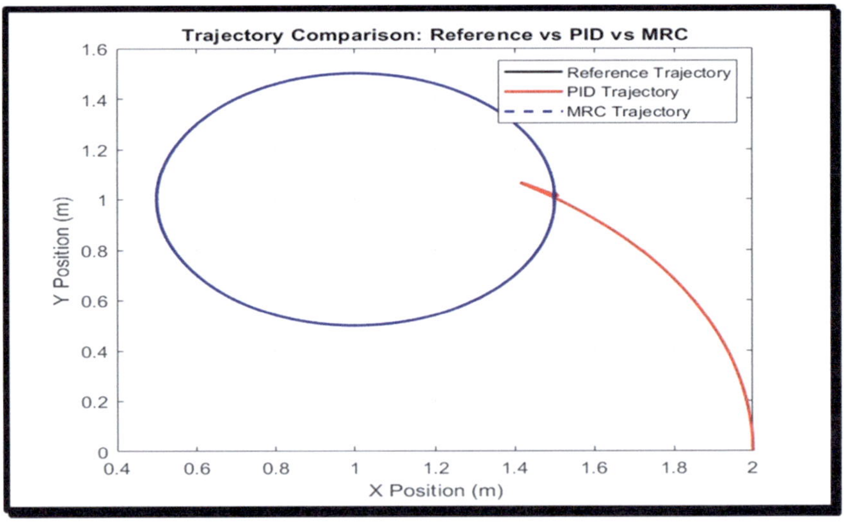

Fig. 8 This image illustrates the trajectory followed by tuned PID versus MRC

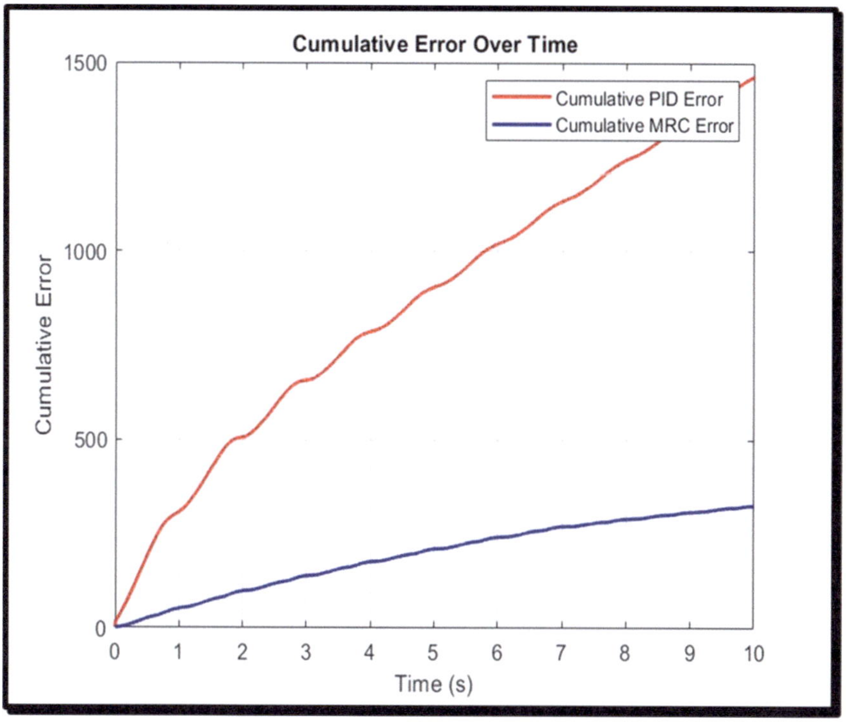

Fig. 9 This illustration compares the cumulative error of untuned PID and MRC

3.1.2 Cumulative Error

An important performance indicator in control systems is the cumulative error, which indicates how effectively a system can follow a specified trajectory or control target over time and provides insight into the overall accuracy of a controller [22] (Figs. 9 and 10).

3.1.3 Response Time

The term "response time" describes how long it takes the control system to react to input changes or fix errors. In control theory, the primary objective is to minimize reaction time while keeping system stability, particularly in areas like robotics or automation where promptly modifications are required [13] (Figs. 11 and 12).

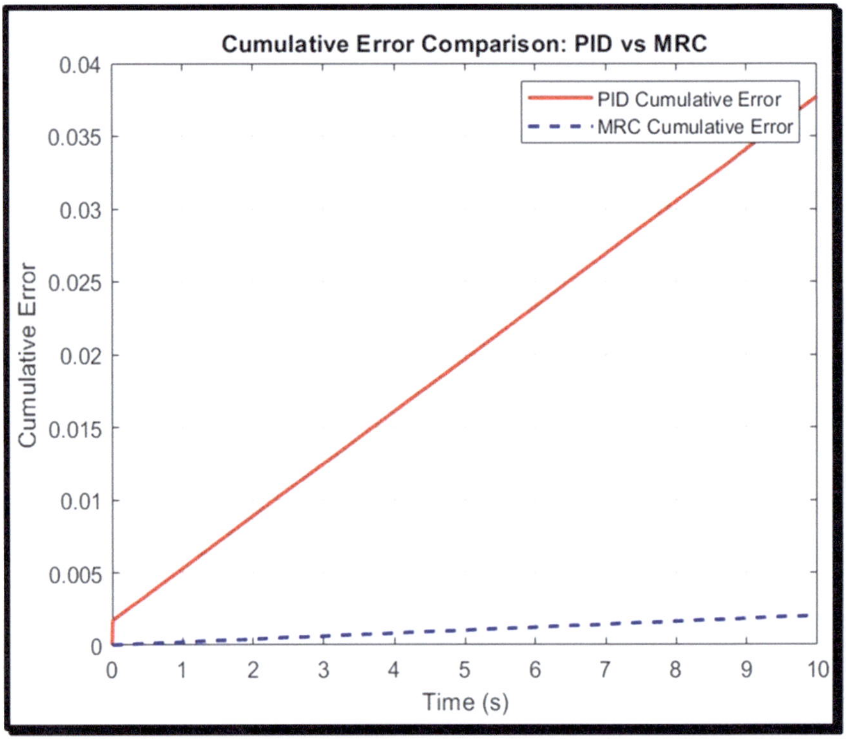

Fig. 10 This illustration compares the cumulative error of tuned PID and MRC

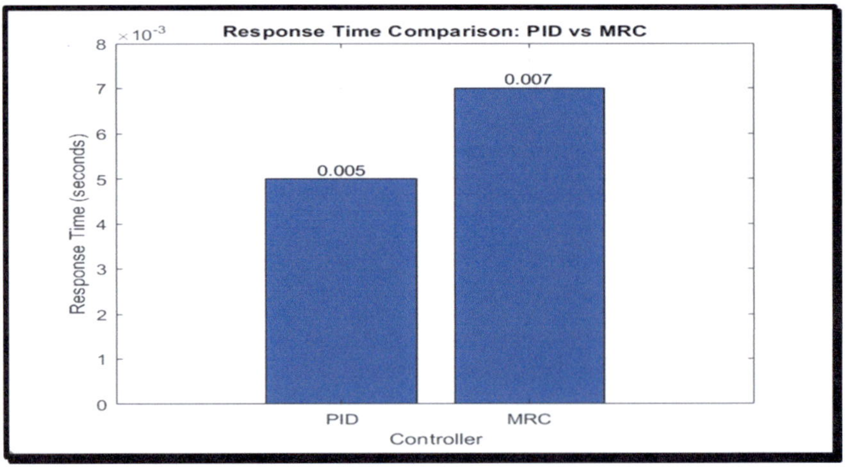

Fig. 11 This is a comparison of response time between untuned PID and MRC

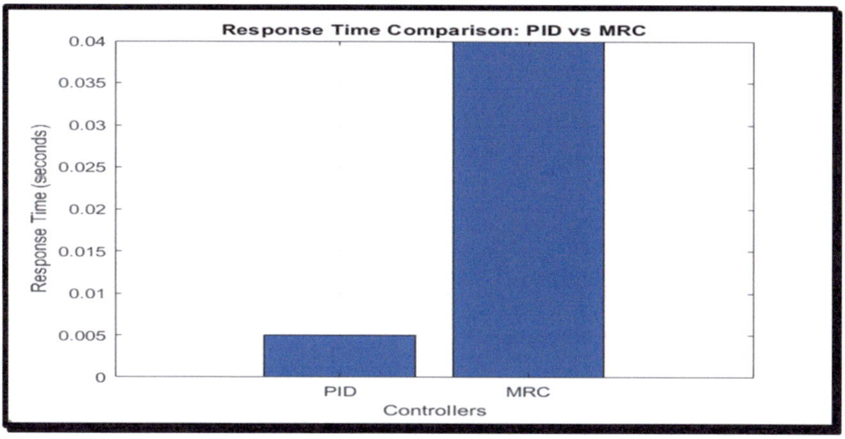

Fig. 12 This is a comparison of response time between tuned PID and MRC

3.1.4 Controllers Response

Evaluating the PID and MRC controllers' performance in tracking a desired trajectory and determining if modifying has a substantial effect on the PID controller's performance are the key goals of these comparisons [21] (Figs. 13 and 14).

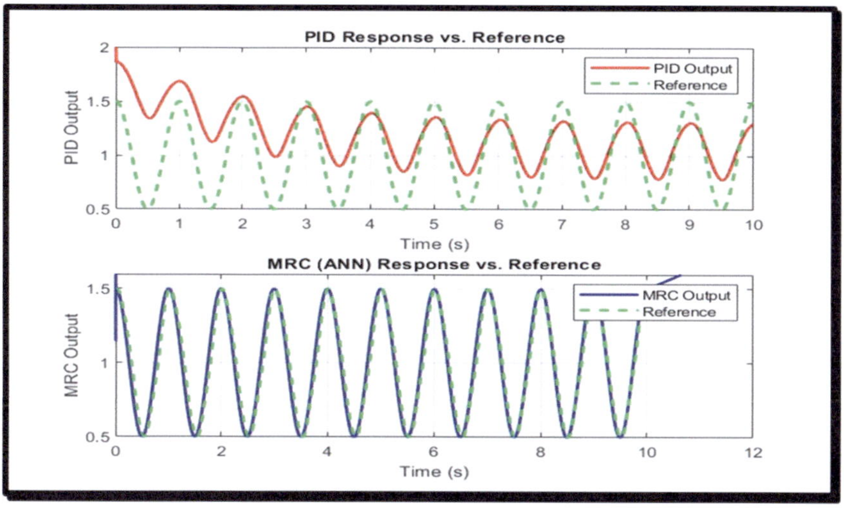

Fig. 13 This image shows the response of untuned PID and MRC controllers when compared to the reference trajectory

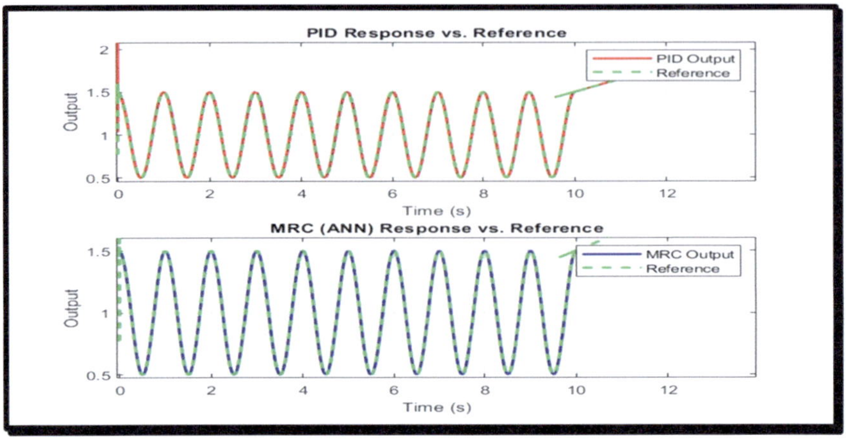

Fig. 14 This image shows the response of tuned PID and MRC controllers when compared to the reference trajectory

3.1.5 RMS Error

The RMS error for a tuned PID controller is shown against the MRC in the first chart, and a comparable relation is shown for an untuned PID controller in the second chart. Because it shows the average amount of error between the intended and the real results of a system, root mean square error (RMS error) is a crucial measure in control systems. Greater control performance is shown by a reduced RMS error, which suggests the system is more closely following the intended trajectory [4] (Figs. 15 and 16).

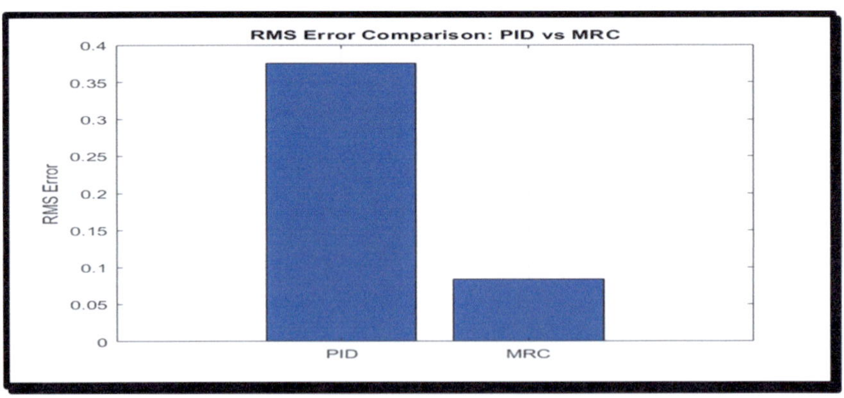

Fig. 15 RMS error of untuned PID and MRC

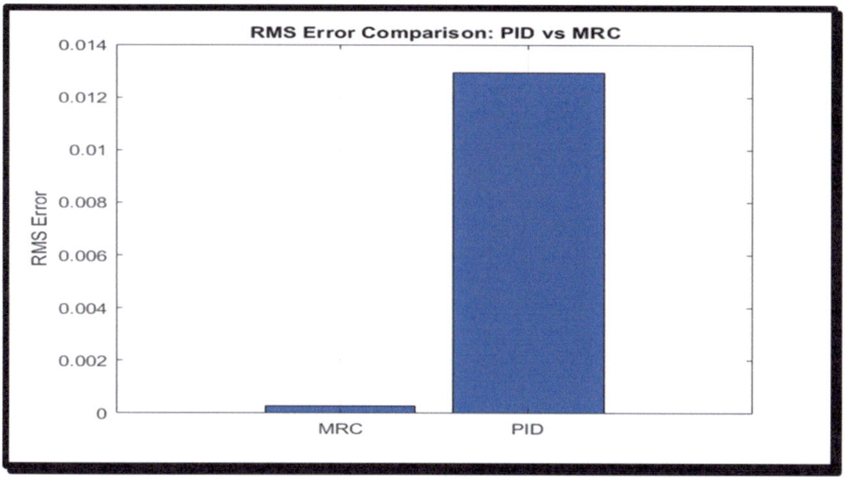

Fig. 16 RMS error of tuned PID and MRC

3.1.6 Control Input

Since it symbolizes the signal that devices receive in order to move the system toward the desired result, the control input is a crucial statistic in control systems. The system's overall efficiency, including variables like stability, efficiency, and reaction time, may be strongly impacted by the behavior of the control input [26] (Figs. 17 and 18).

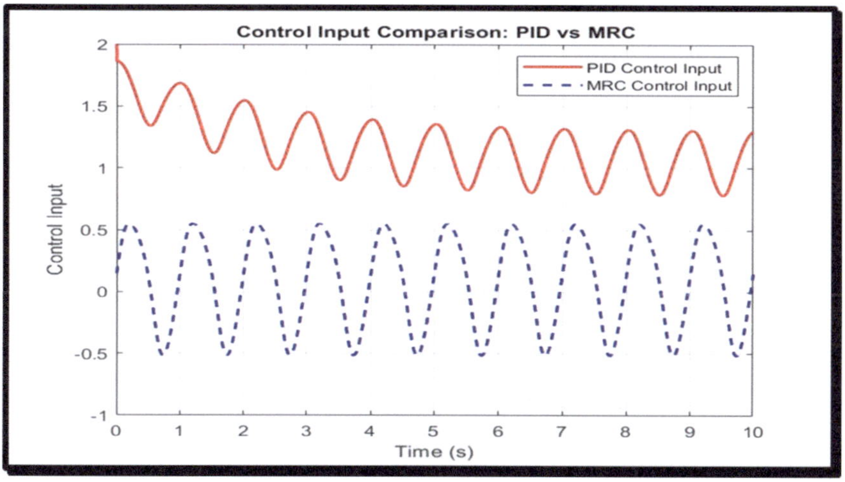

Fig. 17 This illustration shows the behavior of untuned PID and MRC for control input

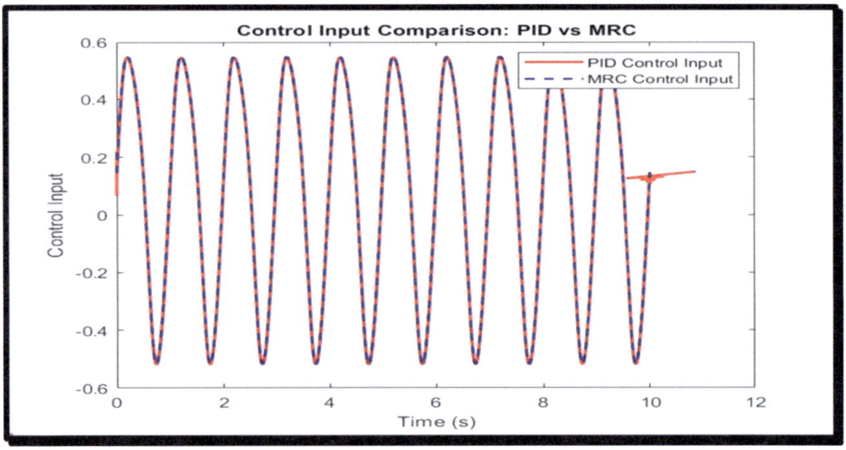

Fig. 18 This illustration shows the behavior of tuned PID and MRC for control input

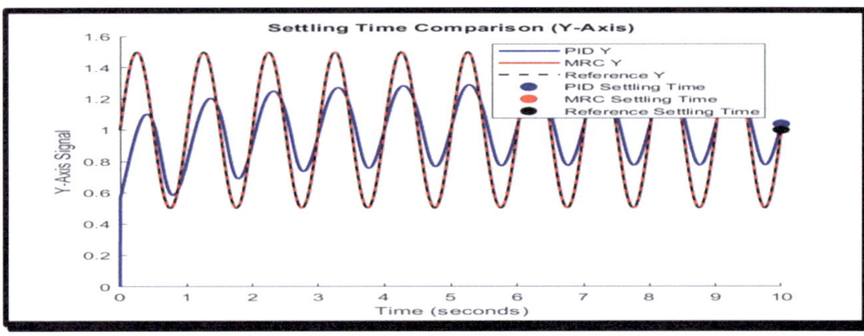

Fig. 19 Settling time of untuned PID and MRC

3.1.7 Settling Time

The term "settling time" describes how long it takes for the output of the system to settle within a certain range of its desired value, usually within a given percentage (for example, 2%) of the final value. One of the most important performance metrics in control systems is the settling time, which shows how fast a system can react to disturbances while attaining its target state with the least amount of fluctuation or overshoot [52, 53] (Figs. 19 and 20).

3.1.8 Steady-State Error

The model reference control (MRC) using artificial neural networks (ANNs) and the proportional-integral-derivative (PID) controller show notable differences in their

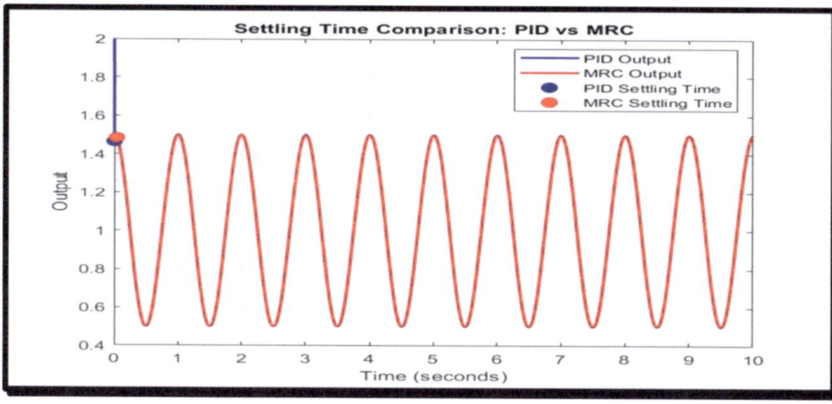

Fig. 20 Settling time of tuned PID and MRC

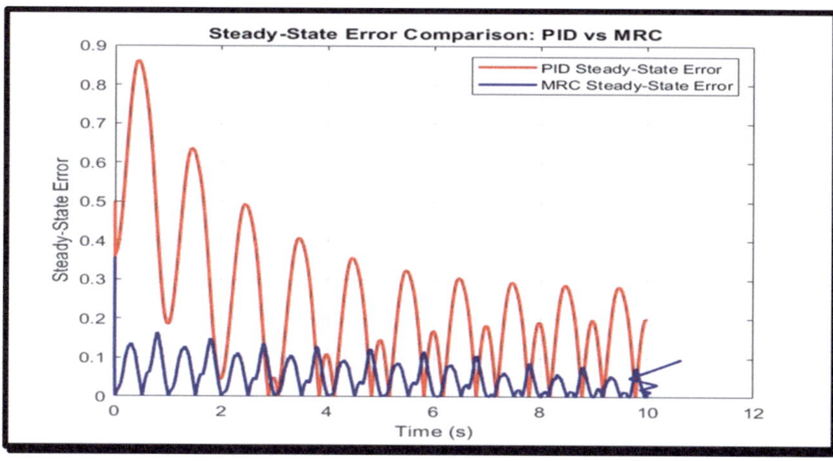

Fig. 21 This plot shows the untuned PID and MRC error during steady state

performance when compared with respect to steady-state error, especially in regard to how rapidly and precisely they meet the desired output [26] (Figs. 21 and 22).

3.2 Discussion

3.2.1 Trajectory Comparison

The first graph illustrates the reference trajectory as a smooth circular route in blue, the PID controller-obtained trajectory in red, and the MRC trajectory as a dashed

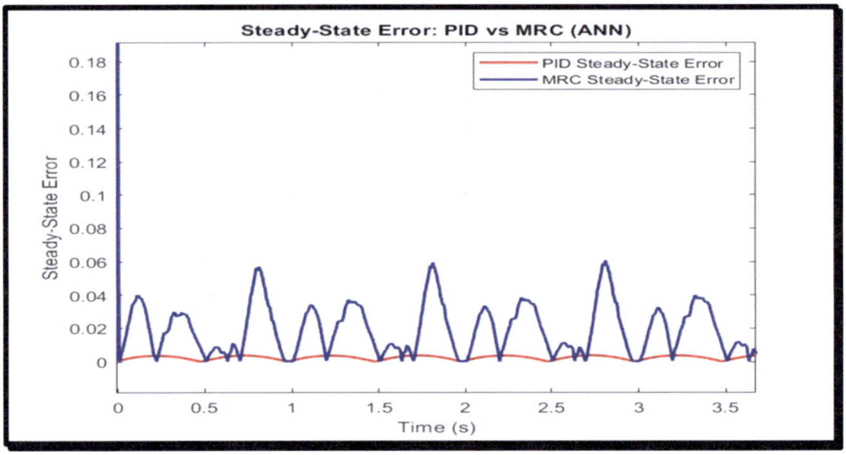

Fig. 22 This plot is a zoomed image of a graph that shows the untuned PID and MRC error during steady state

blue line. From such a comparison, it is obvious to notice that PID begins to deviate from the reference trajectory at the beginning of the movement. Such a discrepancy indicates that the PID controller is not able to precisely follow the intended trajectory, which could possibly be a result of an inability to effectively control nonlinearities in the robotic arm's dynamics [58]. PID controllers are well known for their effectiveness in linear systems; however, in systems with intricate nonlinear dynamics, of which robotic arms are a subset, this control capability basically disappears. In contrast, MRC, which uses ANN for control, is capable of handling such complicated systems better and often displays a reference trajectory with only slight variations [63].

The contrast between PID and MRC controllers is further shown in the second graph, where the PID trajectory displays strong oscillations, instability, and even the end-effector looping around the intended trajectory rather than flowing along a smooth route. This indicates that the PID control system was operating in an unstable manner, maybe due to faulty PID gain tuning or the inbuilt ability of a fully functional PID-controlled system to handle unexpected shifts in trajectory and irregularities [4]. On the other hand, the MRC controller displays long-term reliability with its exact and constant trajectory near the reference [63].

Specifically, the second graph depicts a strong fluctuation of the path that is beyond the PID controller's control. With nonlinear behavior or outside changes, every PID-controlled system is vulnerable to oscillations and overshooting issues [65]. Even after careful manual tweaking, these problems may still be sensitive to significant disruptions. Since the ANN-based MRC continuously learns and adjusts to system behavior in real time, it displays stability in the system and has low oscillations [52, 53].

After analyzing the results of the PID and MRC controllers, it is clear that the ANN-based MRC performs more effectively than the PID controller in terms of system tracking of trajectory accuracy. The basic PID can be used to manage simple linear systems, but its applicability in advanced nonlinear applications—like robotic arms—is fairly limited. Even with nonlinear behavior, MRC exhibits more responsiveness to system dynamics that provide shifting circumstances [47]. Compared with the other controllers studied here, this makes it much more appropriate for applications demanding both precision and dependability. These findings are consistent with previous research suggesting that ANN-based controllers are now the go-to option for complex robotic control systems [19].

3.2.2 Cumulative Error

- **Cumulative Error: PID Versus MRC**

From the numbers, it appears that the PID controller has a far greater accumulation of error throughout this 10 s period, reaching approximately 1,500 units of cumulative error at the conclusion of the experiment. In contrast, the MRC controller error is far reduced and does not surpass 500 units in the same 10 s period. The substantial difference between the two controllers highlights major shortcomings in the PID controller's dynamical operation in nonlinear environments, such as robotic arms. Despite their popularity because of their simplicity, PID controllers need a lot of tweaking and might perform much worse in complicated systems where dynamics change over time [32]. MRC, on the other hand, uses an ANN-based methodology for this purpose; as a consequence, it has significantly higher overall control performance and reduced cumulative error due to its capacity to react to changes in the behavior of the system [39].

- **PID Controller**

The red line, which represents the PID controller's cumulative error, increases quickly and continuously over time, suggesting that the controller has difficulty maintaining the desired trajectory. PID controllers are less successful in systems with nonlinear behavior or quickly changing circumstances, which causes compounding errors, even when they alter proportional, integral, and derivative factors to decrease output error [71].

The PID's inability to effectively respond to dynamic changes is highlighted by the sharp error curve, which prevents the system from effectively attaining a steady state. Error accumulation impairs accuracy and efficiency, creating difficulties in practical applications such as industrial automation, robotics, and aircraft.

- **MRC Controller**

The blue line represents the cumulative error of the ANN-based MRC controller, which rises at a significantly slower pace than the PID controller. Over a 10 s period, the ANN-based MRC controller accumulates less than 500 units of error, while the

PID controller accumulates 1,500 units. This efficiency is attributable to the artificial neural network's (ANN) capacity to recognize past patterns and proactively modify control actions in contrast to PID controllers, which operate solely in response to failures [7]. As a result, despite changing system dynamics over time, the MRC retains a flatter cumulative error curve, suggesting increased accuracy and less error accumulation.

- **Overall**

Cumulative error over time clearly shows the strength of the ANN-based MRC over the PID controller. Large, persistent errors are an outcome of the PID controller's incapacity to adjust to altering system dynamics, which lowers overall performance. However, because of its adaptive learning mechanism, the MRC controller achieves significantly lower error accumulation and is consequently much more effective for real-time control applications [39, 71]. These results imply that ANN-based controllers could be crucial components of control systems in the future, particularly in dynamic, complicated settings where accuracy and flexibility are essential.

3.2.3 Response Time

- **Response Time Comparison**

The PID controller completes its reaction time in around 0.005 s, as seen in the first graph, but the MRC displays a similar period of around 0.007 s. The second graph illustrates this disparity more clearly, showing that while the PID response time remains at 0.005 s, the MRC response time goes above 0.04 s. As might be forecast, given the PID mechanism's great simplicity, these data conclusively show that the PID controller offers quicker reaction times [12]. PID controllers have the ability to react to mistakes in real time by modifying the proportional control signal. They accomplish this by additionally taking into consideration the rate of change and accumulation of errors, which helps them estimate future errors. Because of its straightforward feedback as well as quick reaction times, the PID is a good choice for applications that need rapid modifications, such as temperature control or process control, among others [17].

Faster responses are not necessarily optimal for systems; this is particularly true for nonlinear dynamic systems or systems with adaptable parameters, as stated by [54]. The MRC controller benefits from adaptive learning via its ANN structure, while having a slower basic reaction time. A little longer reaction time is suggestive of the amount of computation required in the constantly updated artificial neural network (ANN) model, and the dynamics are anticipated for the best possible control actions.

- **PID Controller**

The PID controller responds considerably quicker. Because it can quickly rectify deviations from aimed output values, this characteristic now becomes advantageous in simpler and more linear systems. PID controllers provide an optimal setting because of their reduced computation and faster reaction time in situations where system dynamics are predictable and steady. However, Cherivirala and Wentzloff [12] state that this rapid response has some fundamental drawbacks as well. PID controllers work with fixed gain values that are adjusted under certain circumstances. The controller's preset parameters become invalid in the event that the system situations shift or become nonlinear. Overcompensation, oscillations, and even instabilities may occur. PID controllers are vulnerable to significant long-term or long-period accumulated errors, particularly if they are not properly tested, as discussed in the sections preceding this one about cumulative error [37]. As such, PID controllers are quicker than other controllers, but because of their limited flexibility, they are less useful in situations where system parameters change often.

- **MRC Controller**

Because of its ANN-based structure, the MRC controller offers better control accuracy over time by continually evaluating system dynamics and implementing the required modifications. This results in improved prediction and adaptive control capabilities. This is especially helpful in situations where precision is difficult for typical PID controllers, such as irregular or time-varying systems [60]. The enhanced long-term stability and error reduction of the MRC offset the modest delay in reaction time caused by its greater processing burden. Applications like robots, which need extreme accuracy and control flexibility, greatly benefit from MRC's versatility. Trajectory tracking and cumulative error graphs show that the MRC works better than PID controllers overall, providing more efficient adaptive control in scenarios where reliability and accuracy are more important than speed.

3.2.4 Response Versus Reference

- **Comparison of Tuned PID and MRC Response**

An optimized PID controller works fairly well, but the MRC works better in terms of precision and reliability over the course of time. This further illustrates the benefits of ANN-based controllers for long-term activities demanding high accuracy, since even small departures from the reference path may lead to an error accumulation of some kind [56].

The near-total overlay of the blue line (an MRC output) and the green dashed line (reference) indicates superior tracking capability from the MRC controller. The adaptive learning capability of the ANN, which allows constant modifications in the control behaviors of the MRC based on real-time system input, is responsible for the documented improved efficiency. Because of MRC's versatility, the controller may detect complex system variables that PID controllers can find challenging to manage [37].

- **Evaluation of MRC Response and Untuned PID**

In general, we can see that in the initial phases of the system reaction, the untuned PID response differs significantly from the original trajectory. The PID results, shown by the red line, oscillated significantly more than the reference trajectory and continued to oscillate for the duration of the measurement; this indicates that the system is volatile and that the precision of the PID output does not match sufficient tuning requirements [55]. These are signs of an improperly tuned PID controller; self-correcting, it overshoots the desired response during long oscillations.

Here, on the other hand, the MRC controller achieves nearly ideal tracking of the reference trajectory once more. The blue line shows the MRC controller's output as time passes, staying in close range to the reference's green dashed line. This demonstrates the robustness of the ANN-based control technique. This implies that, in comparison with PID, which counts so heavily on exact tuning to turn it in the best possible manner, MRC is inevitably far less susceptible to things like system modification or other tuning needs [68].

- **Concluding Remark**

Although the efficiency of the adjusted PID controller is acceptable, it is not quite as good as that of the MRC, which provides significantly more precise and reliable tracking in contrast. A significant decrease in the performance of the untuned PID controller appears because of the strong fluctuations and variations from the original trajectory. These findings show that while PID controllers perform very well when tuned to perfection, they are sensitive to tuning and unable to manage changing or nonlinear systems. In contrast, the MRC controller performs more effectively, requiring less effort to tune and exhibiting flexibility in responding to changes in the system's conditions. Thus, when complex system dynamics arise, the MRC stands out as a more reliable and effective substitute for applications requiring a high degree of accuracy and dependability [5].

3.2.5 RMS Error

- **Comparison of MRC RMS Error and Tuned PID**

The tuned PID controller's RMS error is 0.012, but the MRC's is almost nonexistent. Because of this, even a well-adjusted PID cannot surpass MRC's accuracy. The PID controller attempts to utilize fixed gains for the proportional, integral, and derivative actions in order to minimize the disparity between the system's output and its intended target [64]. Only at times when the system is linear and relatively simple does the control mechanism work well; in other cases, it struggles to handle nonlinear dynamics, which increases the RMS error.

It is obvious that the ANN MRC controller function has a much-reduced RMS error due to its adaptive learning capabilities. In relation to the system's behavior, the ANN changes the model on a regular basis. As a result, it helps the MRC to govern behaviors and predict future states. As a result, the system achieves more precise control as it mostly stays on the intended trajectory [5].

The ANN's ability for learning and adaptation increases the sustainability of the MRC for systems with complicated and complex circumstances, particularly those that are not constrained by set gain values as PID controllers are. This contrast indisputably shows that the PID controller will function significantly less well than the MRC, even with more precise settings. It would thus imply that MRC provides a superior solution due to its flexibility and forecasting characteristics in applications needing high accuracy and minimal error, particularly for complex systems [40].

- **Comparison of MRC RMS Error and Untuned PID**

The RMS error for the untuned PID controller and the MRC is compared in the second chart. The RMS error of the untuned PID controller is 0.35, far greater than the MRC values of 0.08. This indicates that, in the absence of tuning, the MRC's effectiveness maintains a respectable low error value while the PID's efficiency rapidly decreases [62]. This is to be anticipated, given that the success of PID control depends heavily on the proportional, integral, and derivative gains being properly tuned. The PID controller's gains are not ideal in such an untuned state, which results in inferior control quality. As a consequence, the controller may overcompensate or undercompensate for faults that cause fluctuations or even unstable conditions, which causes greater departures from the original trajectory [40].

An MRC, on the other hand, performs quite consistently both with and without tweaking. This is so that a system with nonlinear and dynamic characteristics may be handled by the MRC itself via its ANN without the need for any adjustment at all. It eliminates a great deal of the tedious error and trial tweaking procedures, which is a major benefit over the PID controller. In the majority of real-world situations, the MRC offers more resilience and, thus, greater dependability for a solution when the system variables happen to be time-varying or naturally nonlinear [40].

- **To Summarize**

While the adjusted PID works rather well, it is still far from the precision and adaptability offered by the model reference. Larger error values are shown by the untuned PID controller, highlighting the need for accurate tuning for effective control [64]. The MRC, on the other hand, produces a much-reduced RMS error for both configurations, demonstrating its advantages in applications requiring high accuracy and system change resilience. The ANN's adaptive learning capacity makes the MRC a dependable solution for nonlinear and time-varying systems, where a standard PID controller's performance might quickly deteriorate [5].

3.2.6 Control Input Comparison

- **Comparison of Tuned PID and MRC Control Input**

In terms of control inputs vs. time, both controllers are quite similar. The blue dashed line, which represents the MRC's control input, is typically below the red line, which represents the PID control input. The tuned PID controller shows very little oscillation in the corresponding control signal and tracks the intended trajectory with high precision. This shows that the PID's gains were properly adjusted, and the controller maintained an appropriate error-compensation level without encountering severe instabilities or overshoots [31].

In addition, the MRC provides a controlled control input that closely resembles the output of the tuned PID. MRC, on the other hand, offers an additional benefit of adaptability and self-tuning, allowing control behaviors to be dynamically changed to match the system's real-time behavior. The MRC's ability to adapt makes it especially helpful in systems where tight control must be maintained over an extended period of time or when dynamics are time-dependent [43].

- **Evaluation Between MRC Control Input and Untuned PID**

The control signal difference between the MRC and an untuned PID controller is shown. For the two controllers, the control input characteristics can vary significantly. The extensively bigger and more variable control input (the red line) is provided by the untuned PID controller. Not only do the fluctuations in the control input have a greater intensity, but they also continue throughout time [67]. This type of behavior suggests that the PID controller has been improperly tuned, which may result in unstable and poor control performance.

The blue dashed line, nevertheless, shows that the control input from the MRC controller is, in fact, considerably smoother and more accurate. The MRC exhibits fewer oscillations and a less volatile control input than the untuned PID. It makes sense since, in MRC, control actions are updated continually based on actual system behavior to achieve the intended system output with the least amount of fluctuation in control input [41].

- **Concluding**

The tuned PID controller works well in general, and its control input is quite similar to the MRCs. The untuned PID controller, on the other hand, demonstrates a significant decrease in performance. Its control input is noticeably unreliable and unpredictable [67]. The MRC controller has consistently provided more steady and effective control input, even in the face of changes in the entire system dynamics. It also suggests that an MRC will perform better, in particular, in systems where great accuracy and flexibility are practical requirements. In contrast to traditional PID controllers, which rely heavily on human tuning, the MRC continually updates its real-time corrective operations to be resilient and dependable because of the adaptive learning properties of ANN [16].

3.2.7 Settling Time

- **Comparison of Tuned PID and MRC Settling Times**

Both the MRC and the tuned PID behave badly, fluctuating constantly in an attempt to hit the intended targets. It is clear that the periodic behavior of both PID and MRC controllers is present. But in this case, the settling time—which is shown by markers on the graph—is a crucial finding. The settling period is significantly quicker from the tuned PID's blue circle than from the MRC's red circle [18]. This indicates that the PID controller stabilizes more quickly than the MRC on average.

This indicates that the tuned PID has a shorter settling time, indicating it is more precisely adjusted for the unique dynamics of the system and hence achieves the intended output more quickly. This time, it might be a sign that the PID, even while it becomes stable more quickly, may need an additional effort and cause stronger fluctuations, which would result in increased energy usage. According to Zhao et al. [75], these resulting downsides may lead to greater wear and tear of system components.

In contrast, an MRC may be useful in altering the behavior of a system that may be nonlinear or time-varying since it adjusts itself to resemble the reference model. Even while the MRC could take a little longer to get used to, it offers smooth control and is thus better able to adjust to the changing environment [23, 24].

- **Comparing Settling Times for MRC and Untuned PID**

The untuned PID is plotted against the MRC in the second figure. One additional marker is situated at the reference settling time. The blue and red markers represent the PID and MRC controllers' respective settling time locations. Based on the settling time markers, we observe that the untuned PID controller in this particular case significantly falls behind the MRC in terms of stabilization time [74]. These periodic fluctuations are substantially more noticeable and have greater amplitudes that last longer when the PID controller is out of tune. This indicates that the controller performs much worse when it comes to stabilizing the system. In comparison, MRC exhibits much more stability with less oscillation, leading to a quicker settling period.

This is a result of its adaptive neural network-based technique, which continually modifies the variables in real time to enhance the dynamics of the system. On the other hand, the disappointing results of an untuned PID controller are exactly what highlight the significance of tuning, among other factors, in obtaining the best possible control performance [75]. An extremely lengthy transient response, a large number of oscillations, and reduced system stability might result from an inefficiently tailored PID.

- **Final Thoughts**

The MRC and tuned and untuned PID controller comparison shown here highlights the need for responsive control in contemporary systems. Fast settling times can be achieved with a well-tuned PID controller, but doing so requires methodical, cautious tuning, which may potentially lead to forceful control actions that worsen the stability of the system [18]. Conversely, in the event that the system is functioning under operational circumstances or is exponentially dynamic, the model reference control offers the most stable and dependable solution [74]. Because of this, MRC is unquestionably a superior option for a precision application that needs long-term stability, even in situations where the settling period is greater.

3.2.8 Steady-State Error

- **Steady-State Error in Tune PID**

To provide an overview of how the error alters over the duration of the simulation, the graph presents the steady-state error for the same tweaked PID on a bigger scale. Although the error seems to be stable in this zoomed view, a broader perspective reveals that the PID controller experiences slight error variations as it attempts to stabilize at the intended state [73].

These fluctuations indicate that even while the PID often reduces errors rather well, it does not eliminate discrepancies entirely over an extended period of time. This would highlight the trade-off between reliability and response time: although a tailored PID can rectify small faults rapidly, it might not be able to eliminate them entirely over time [65].

- **Steady-State Inaccuracy in Untuned PID**

On the other hand, an untuned PID controller's outcome is seen in the third plot. With high peaks right when it started and high oscillations throughout time, the error is now much larger. This shows that the PID has a considerably bigger and louder error and is unable to stabilize the system properly if it is not tweaked [3]. Since the performance of PID controllers is mostly dependent on proportional, integral, and derivative gains, tuning becomes very important. The graph of the untuned PID amply illustrates the severe oscillations or extremely slow behavior that poor tuning causes in a PID. Additionally, the untuned controller shows excessive oscillations in a steady-state error, which substantially impairs control performance.

- **MRC Steady-State Error Comparison**

The steady-state error is significantly less than that of the tuned and untuned PID controllers. As shown, the steady-state error for MRC-based controlling has stayed close to zero for the majority of the simulations, confirming its superior ability to adjust to the dynamics of the system for accurate control. Because of the ANN's improved modeling precision, the MRC can change its control settings in real time and efficiently minimize errors that could have remained in conventional PID controllers [14]. The accuracy and flexibility of MRC are said to be some of its main benefits over traditional control methods.

Furthermore, the MRC controller eliminates these spikes and variations, in contrast to the PID controller. This alone means the intended state will not only be attained quicker but will also be maintained with very little inaccuracy in the event of external disruptions.

- **Discussion and Implications**

The noteworthy gap in steady-state error between the MRC and PID controllers emphasizes how crucial it is to choose a controller in accordance with the needs of the system. Although PID controllers are easier to set up and adjust, their limitations start from the fact that they cannot completely remove steady-state faults, particularly in higher-level systems. However, MRC using ANN performs far better in terms of stability and reduction of errors, making it an excellent choice for systems that need a high degree of accuracy and flexibility [65, 73].

4 Conclusions and Future Work

4.1 Conclusion

In particular, for a two-degree-of-freedom (2DOF) robotic arm, this research shows that model reference controllers (MRCs) based on artificial neural networks (ANNs) provide significant benefits over conventional PID controllers in complicated control system management. Comparing and evaluating the tracking precision, stability, and reaction time of the two controllers was the main goal. The results show over and over again that the ANN-based MRC controller performs better in terms of lowering steady-state errors, obtaining greater trajectory accuracy, and enhancing the system's capacity for dynamic situation adaptation. PID controllers struggle with nonlinear behaviors or shifting system dynamics, as seen by their increased cumulative error and restricted flexibility. Nevertheless, they may provide excellent results for basic, linear systems. In contrast, the ANN-based MRC makes use of its intelligence for learning to dynamically modify control techniques, which reduces errors and enhances trajectory tracking in challenging real-world situations [39, 47].

Because of its adaptive characteristics, which enable ongoing learning and self-adjustment, the MRC is a great option for applications requiring accuracy and dependability over long periods of time. The robotic arm can sustain precise motions even in the face of changes in external circumstances or system parameters because of its flexibility and adaptability, which also improve control performance and provide resilience against interruptions [65]. As a result, the results highlight how ANNs have the ability to take the place of conventional PID controllers in a variety of intricate and dynamic contexts, offering more precise and reliable control solutions for contemporary robotics applications.

4.2 Future Work

Subsequent investigations have to concentrate on enhancing the ANN-based MRC's computing efficiency so that it may be used in real-time applications without causing appreciable delay. Even though the present research has shown the MRC's better performance, there is still an opportunity to improve by lowering the amount of time and computing power needed for training in order to get optimum control. Furthermore, investigating the use of ANNs in combination with other control approaches, such as fuzzy logic control (FLC) or model predictive control (MPC), may result in hybrid control techniques that include the most advantageous features of each technique. This may improve the system's capacity to handle more complicated dynamics and disruptions [14].

The assessment of the ANN-based MRC's performance in systems with increased complexity and degrees of freedom is another topic for future study. More thorough testing and analysis of MRCs might be beneficial for real-world applications, such as multi-joint robotic arms and autonomous vehicles. Moreover, while the present research concentrated on circular tracks, investigating a broader spectrum of motion trajectories, such as parabolic or obstacle-avoidance pathways, may provide more profound understandings of the controller's potential and flexibility in real-world situations. Finally, the performance of the MRC in managing complicated spatiotemporal dynamics may be enhanced by the creation of more proficient neural network topologies, such as recurring or convolutional neural networks [14].

The study's overall findings support the use of ANN-based MRCs in modern control systems by highlighting their adaptability and effectiveness in managing environments that are nonlinear and spontaneously changing. They also offer clear directions for future research and improvement.

Appendices

Methodology

The First Strategy

Three percent of the data were set aside for testing,15% for validation, and 70% for training. Identifying this distinction is crucial for recognizing the model's capacity for adaptation. It is possible to assess the network's performance on unseen data both during and after training by assigning distinct validation and test sets.

- Training observations: 11,393
- Validation observations: 2,441
- Test observations: 2,441

Network Architecture: A two-layer feedforward network with a linear output layer and 1,000 neurons in the hidden layer, activated by sigmoid activation. An effort to simulate complicated nonlinear relationships in the data is shown by the large number of neurons (Fig. 23).

The Second Strategy

It employed no test set or validation, using all of the data for training. This makes sure that as much data as possible is used for learning, but it also makes it impossible to verify adaptation efficiency, which leaves it open to overfitting.

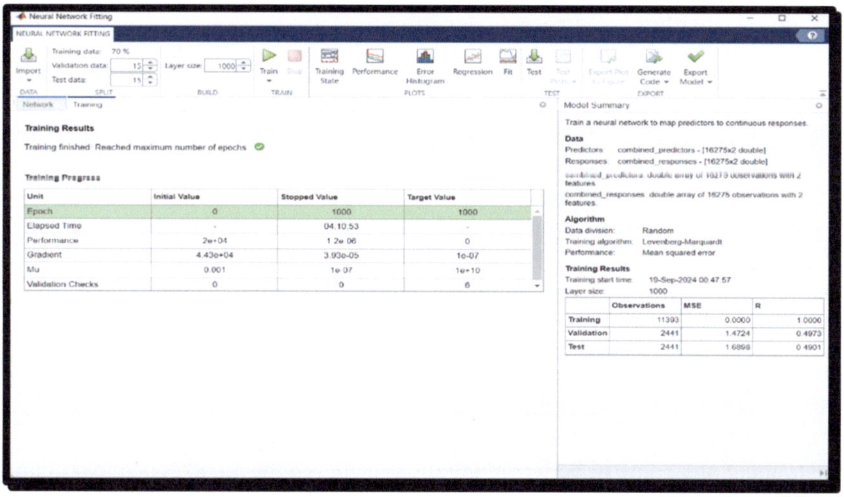

Fig. 23 Neural fitting setting of approach 1

- Training observations: 16,275

No Validation or Test Sets:

The model was trained only using the available data, and as a result, its fit and accuracy were evaluated only using that data and not on any other, previously untested samples (Fig. 24).

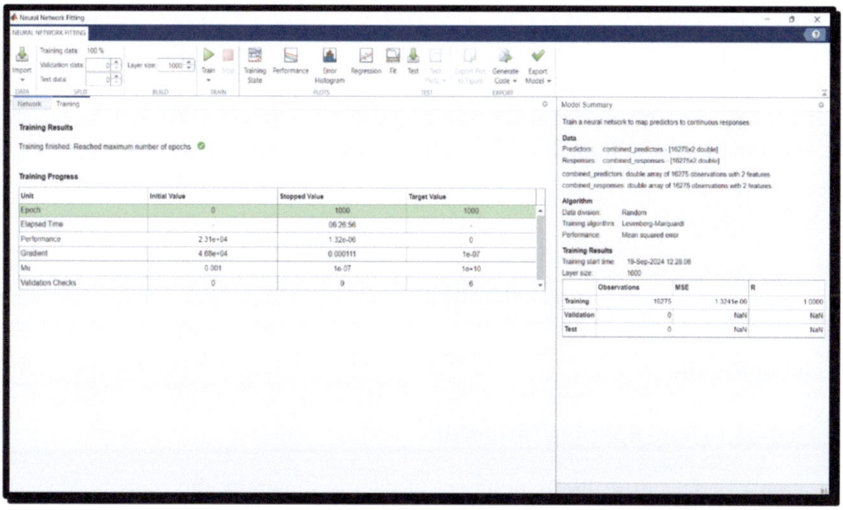

Fig. 24 Neural fitting setting of approach 2

Training Algorithm and Process Algorithm

- Both strategies make use of the Levenberg–Marquardt algorithm, which is well-liked for neural network training because of how well it converges quickly, particularly for smaller networks. It aids in rapidly lowering the error gradient and works well for nonlinear least squares situations.
- Performance measure: Model performance is assessed using mean squared error (MSE), and a better network fit is sought by lowering this parameter.

Training Outcomes and Reflections

Consistency with Gradient Patterns

The training procedure in both configurations achieved the 1,000 epoch maximum.

- **Strategy 1 (divided data):**

The slope decreased to around 3.98e–05, suggesting convergence but maybe leaving space for more reduction.

- **Strategy 2 (no split)**:

The gradient reduced to 0.000111, showing almost perfect convergence on the training set, although perhaps with the drawback of overfitting because of insufficient validation.

Validation and Termination Standards

- **Approach 1:**

It shows the training process was monitored for validation performance but hit the maximum epoch limit. It additionally includes validation checks (none were triggered).

- **Approach 2:**

Since no validation data was allocated, no validation tests could be performed.

Training Results

- **Method 1 (Data Splitting Included)**

An almost perfect fit to the training set of data is shown by the training MSE of 0.0000 and R value of 1.0000.

Validation MSE: 1.4724, R = 0.4973, indicating possible overfitting and a decline in performance.

Test MSE: 1.6898, identical to validation with R = 0.4901, confirming concerns about overfitting.

- **Method 2 (No Data Splitting)**

Near-perfect fit across the whole dataset is shown by the training MSE of 1.3241e–06 and R value of 1.0000. However, it is difficult to evaluate generalization in the absence of test or validation data.

Approach-Based Learnings

- **Impact of Data Splitting**

To appropriately assess model performance on unseen data, the first option exemplifies the methodologically sound practice of dividing data into training, validation, and test sets. It ensures that the model is routinely verified, which helps to minimize overfitting and permits monitoring throughout training.

- **Overfitting Risk in the Absence of Validation**:

The second method maximizes the fitting accuracy on all available data for training, without validating or testing for overfitting. Although this approach may provide good training results, it is neither very predictive nor generalizable in real-world scenarios.

- **Selecting the Training Algorithm and Layer Dimensions**:

To fit complicated data effectively, a large hidden layer (1,000 neurons) and the Levenberg–Marquardt algorithm are used. But doing so also raises the possibility of overfitting, particularly if validation data is not used to control the model's complexity.

Results

Approach 1

The distribution of errors for the test, validation, and training sets is shown in the histogram. Once again indicating overfitting, the errors for the test sets (in red) and validation (in green) vary more widely than the errors for the training set (in blue), which is extremely close to zero.

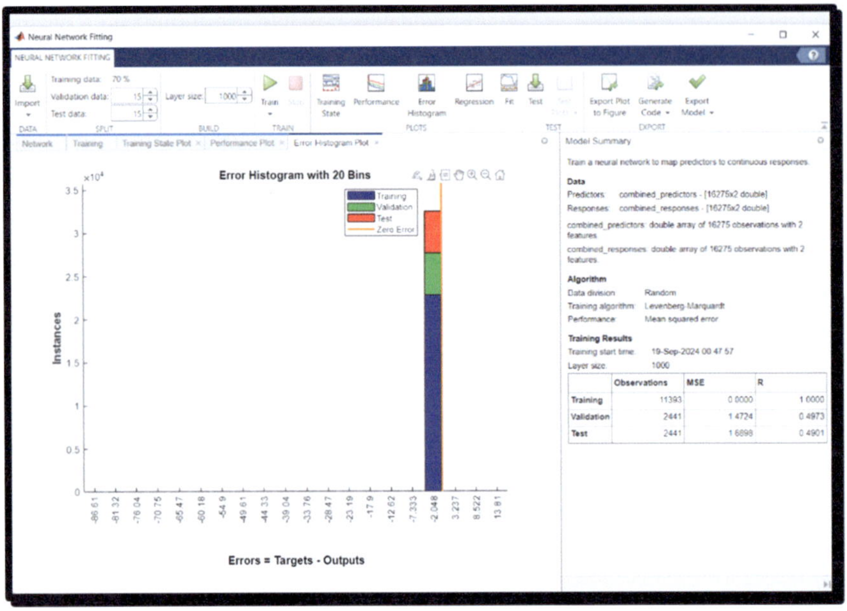

Fig. 25 Error histogram

The model's poor predictive ability is shown by the large range of errors in the validation and test sets compared to the narrow error distribution in the training set (Figs. 25 and 26).

The MSE for the test, validation, and training sets is shown in the graphic. The network seems to be overfitting because, while the training MSE hits zero, the validation and test MSEs stop at higher values (1.4724 and 1.6898, respectively), suggesting that the network fits the training data very well but does not generalize well to new data.

Overfitting is shown by the noticeable difference between the training curve (blue) and the validation/test curves (red and green). This difference suggests that normalization is required or that the level of detail in the model is too high for the supplied data (Fig. 27).

Gradient plot: The slope shows that the network is converging but may not have completely reduced the error as training goes on, since it does not reach a very low value. Although the ultimate gradient value is just 3.9255e–05, it indicates that if additional epochs were permitted, there may be an opportunity for further convergence.

Mu plot: As training progresses, the damping factor Mu stays steady at 1e–07, suggesting that the training procedure followed a reliable route of optimization. A constant Mu value, which is often a good indicator, indicates that the Levenberg–Marquardt method did not need a substantial amount of adaptation.

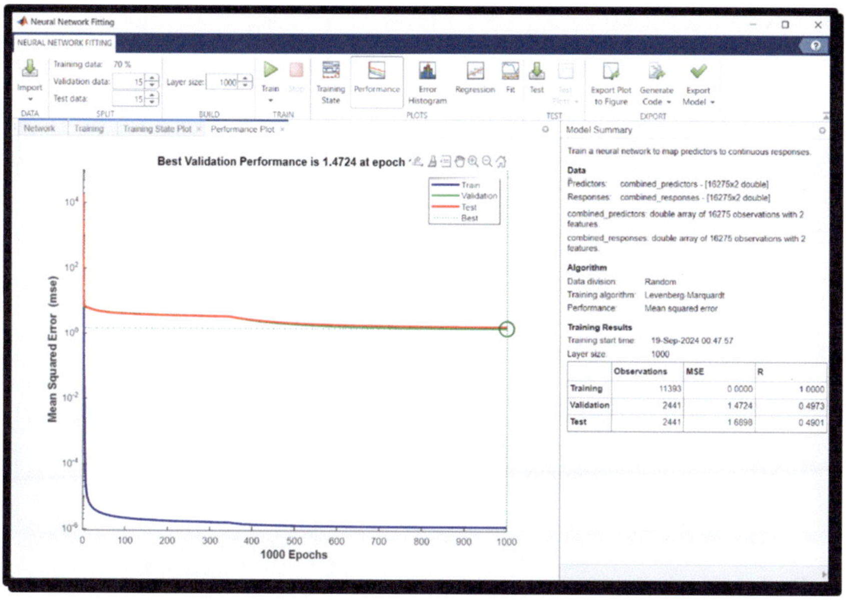

Fig. 26 Validation performance

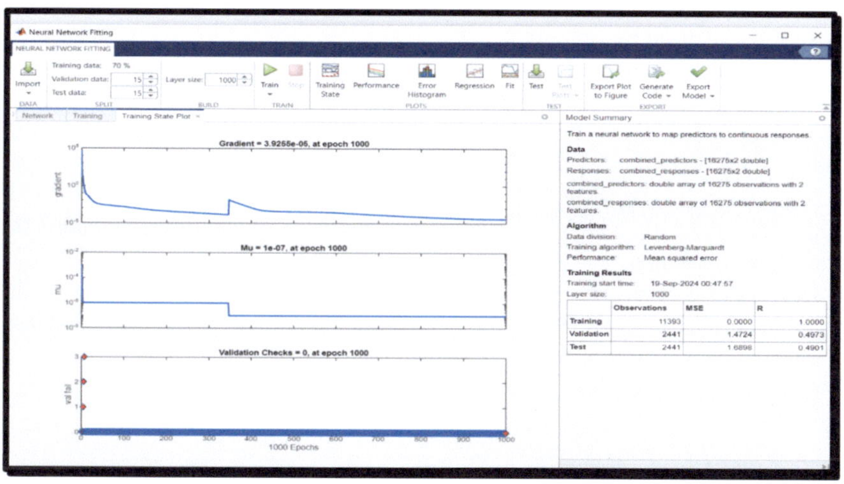

Fig. 27 Training state plot

Validation checks: According to the validation check plot, there was no overfitting of the neural network. Nevertheless, the network may not have completely used the validation data to determine the optimal moment to end, as it reached the maximum number of epochs without improving validation performance (validation checks = 0) (Fig. 28).

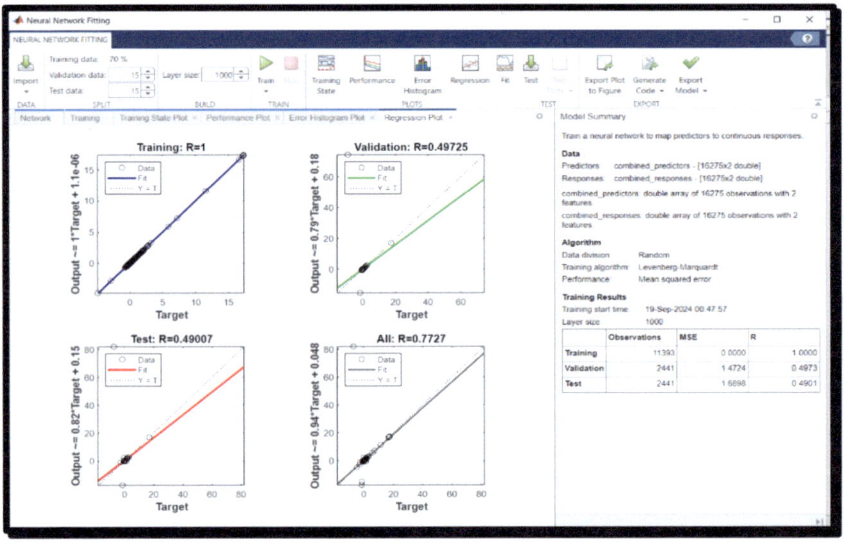

Fig. 28 Regression plot

Training regression (R = 1.0000): The network correctly remembers the training data, validating overfitting. It shows an ideal linear connection between output and goal for the set used for training.

Regression validation (R = 0.4973): A much lower correlation for the validation set, indicating a poor forecasting performance of the network on the validation data. The data points' increased deviation from the Y = T line suggests that the prediction accuracy was low.

Regression testing (R ≈ 0.4901): The test set exhibits poor accuracy with an R value of around 0.49, which is similar to the validation set and further supports the overfitting conclusion.

Regression overall (R = 0.7727): Since the high training accuracy skews the combined result, the combined regression for all data (training, validation, and test) is around 0.77, which is better but still suggestive of overfitting.

Approach 2

The difference between the goals and the output errors for the training set is compared via the error histogram. The narrow distribution and apparent zero center of the histogram indicate that the forecasts are quite accurate (Fig. 29).

Given that it is symmetric around zero and limited, suggesting no apparent bias or deviation toward the network's predictions, this distribution style would be perfect. The fact that most mistakes fall within a narrow range and that the distances to the

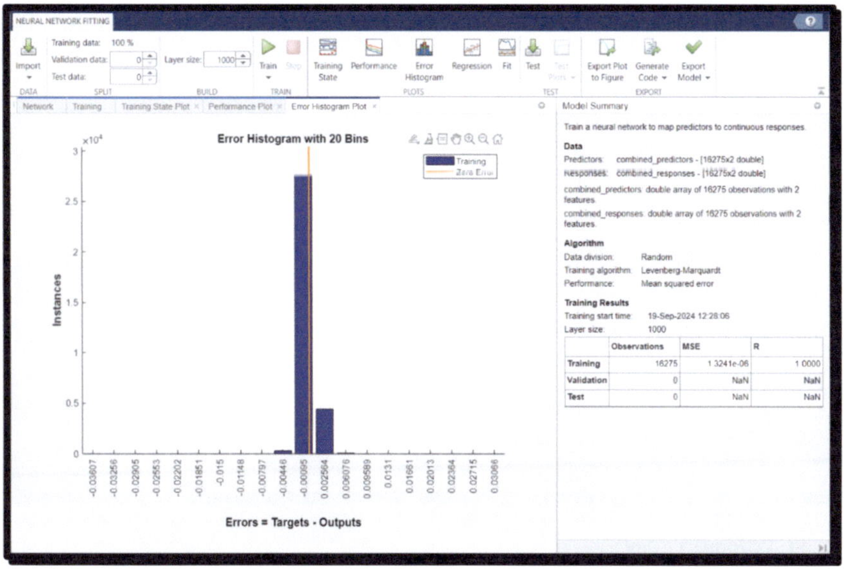

Fig. 29 Error histogram

target values are short supports the network's capacity to provide estimates that are quite near the target function.

The mean squared error vs. epoch graph is shown. A trained network usually has an abrupt decrease at first, followed by stability of the MSE. With a relatively low inaccuracy of 1.3241e–06, the "Best Training Performance" is highly promising. Stabilization, which occurs after around 100 epochs, shows that the network trained rapidly and was able to optimize the weights in the first training phase (Fig. 30).

In addition, the constant smoothing of MSE indicates training that is free of needless obstacles like overfitting and underfitting (Fig. 31).

Gradient plot: The gradient plot illustrates how the result function's gradient varies over the course of the training epochs. As in this particular case, a falling gradient shows that the network converges successfully toward a minimal error. In this instance, the gradient's smooth decline results in a steady value, which suggests substantial learning without any hint of deviation or instability.

Levenberg–Marquardt parameter, or Mu: This figure shows how much the damping factor Mu changes during the duration of the algorithm. When the Mu value approaches zero, it indicates that the training process had little to no adjustments to make along the way, leading to a smooth convergence of the error. Generally speaking, a stable Mu shows that the model did not need any major modifications during training.

Validation plot: Since there are no epochs shown on this plot, no independent validation set was employed during this training session. This might lessen the potentially important findings about how well the model applies to unobserved data (Fig. 32).

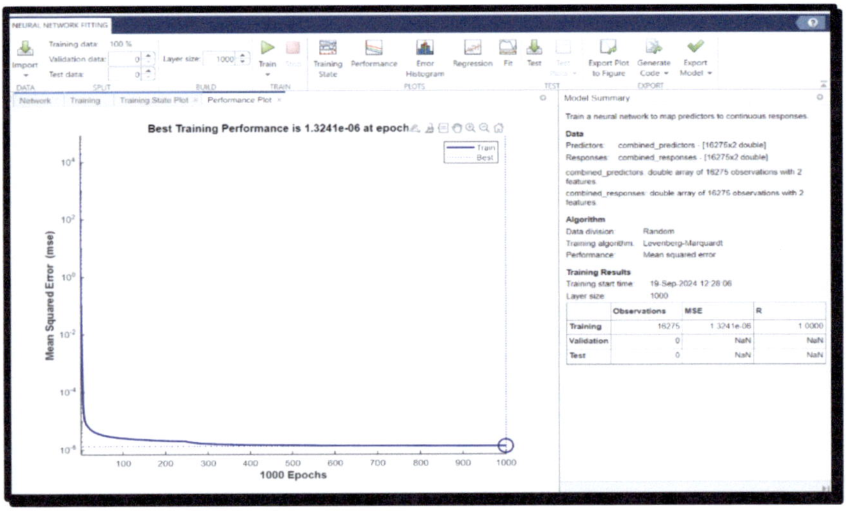

Fig. 30 Performance plot

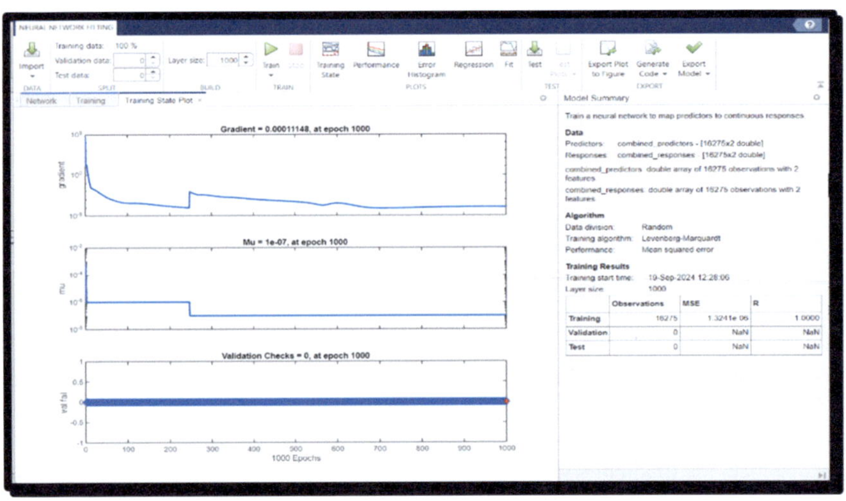

Fig. 31 Training state plot

The relationship between goal values and researched outputs has been graphically represented by the regression plot. A perfect match, or all points falling on the line Output = Target, is shown by a perfect linear connection of + 1. The training regression coefficient in this instance is R = 1.0000, indicating that it is really close to perfection. As an outcome, this proves the model's excellent accuracy using the training set.

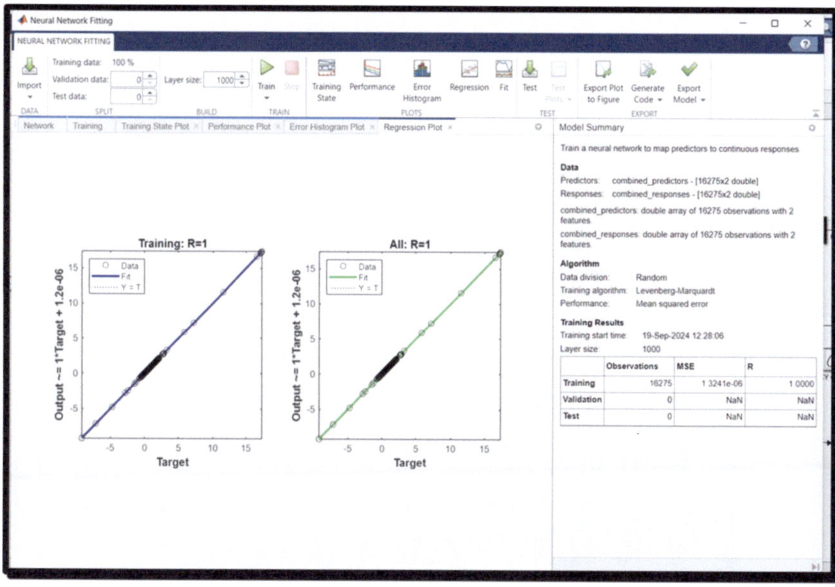

Fig. 32 Regression plot

The fact that the data points for the training and overall datasets are aligned along the line indicates that the system is predictively reliable and can thus map inputs to the appropriate targets.

Discussion on Neural Fitting

When the graphs are combined, they show that the neural network model has been trained successfully; there are relatively few mistakes and a strong correlation between the forecasts and the goals. A low mean square error and an R-squared of 1.0000 indicate a very efficient model in learning the underlying mapping. However, as the model's capacity to generalize to new data was not explicitly examined, over-fitting may be assumed. All data, without validation or test split, was utilized for training.

Improvement Consideration

One of the next essential stages is to separate the data into training, validation, and test sets. In a similar vein, this will provide one chance to assess model performance

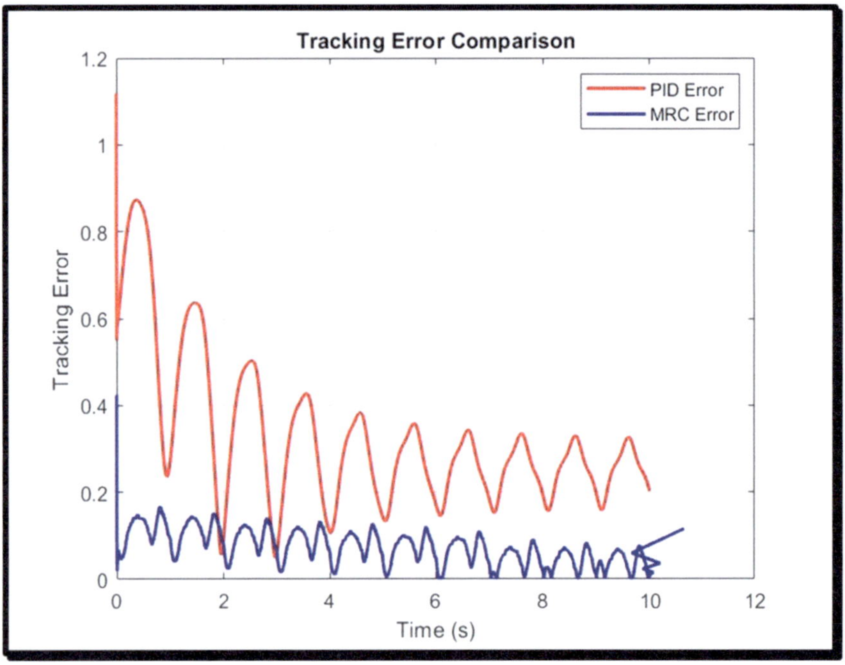

Fig. 33 Tracking error graph of untuned PID versus MRC

on data that has not been examined before and assist in fine-tuning the model to prevent overfitting.

Concerns about overfitting: A model with 1,000 hidden neurons may be too sophisticated and lead to overfitting. Enhancing generality may be accomplished by using regularization techniques or by shrinking a hidden layer's size.

In addition to using other training techniques than Levenberg–Marquardt, more fine-tuning might be carried out on the variety of network designs and layer sizes to provide insights into the optimal configuration for this issue.

Trajectory Error Comparison

See Figs. 33, 34 and 35.

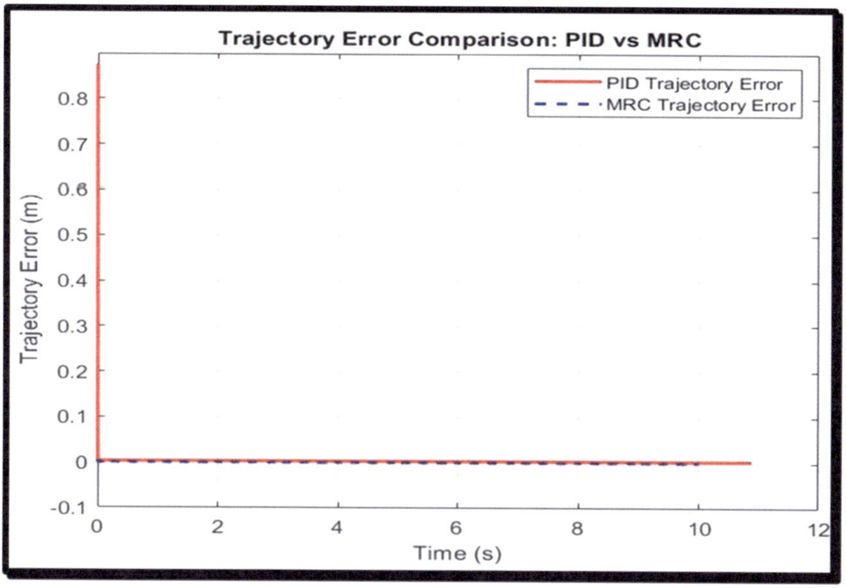

Fig. 34 Tracking error graph of tuned PID versus MRC

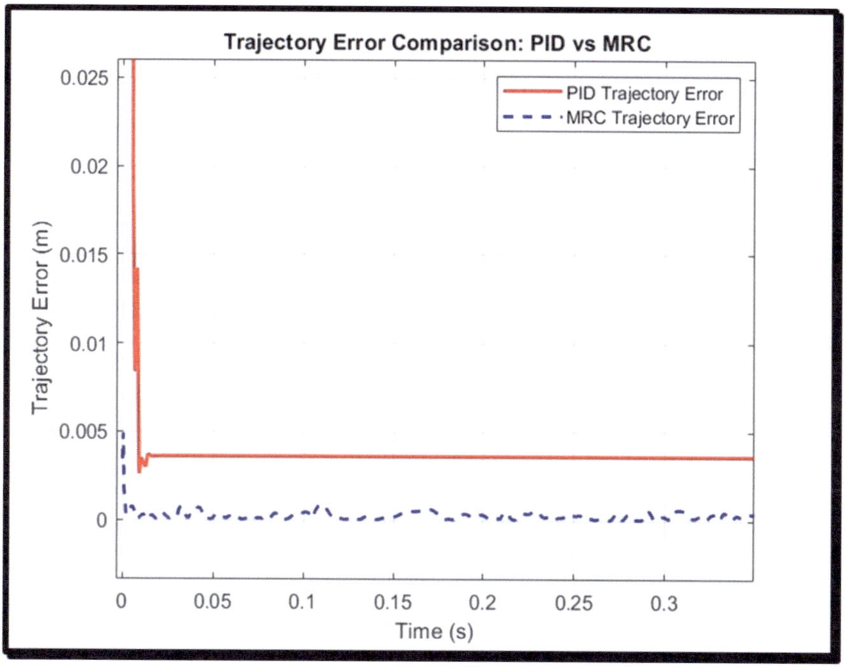

Fig. 35 Zoomed image of tracking error graph of untuned PID versus MRC

References

1. Abbas M, Narayan J, Dwivedy SK (2022) Event-triggered adaptive control for upper-limb robot-assisted passive rehabilitation exercises with input delay. Proc Inst Mech Eng Part I J Sys Control Eng 236(4):832–845
2. Abdusamad A, Aburakhis M (2021) Adaptive control of nonlinear systems represented by Extreme Learning Machine (ELM) and the Fuzzy Logic Control (FLC). In: 2021 IEEE 1st international Maghreb meeting of the conference on sciences and techniques of automatic control and computer engineering MI-STA. IEEE, pp 92–97
3. Ali I, Askir AN, Nazlibilek S (2023) Design of an adaptive PID controller based linear neural network for two-area interconnected power system. IEEE, pp 43–49
4. Alsaied A, Li K (2023) An intelligent PID controller based on ultra-local model for trajectory tracking in lower limb exoskeleton robot. IEEE, pp 31–35
5. Andersen AHD et al (2022) Model-based control algorithms for the quadruple tank system: an experimental comparison. arXiv:2212.04798
6. Aravkin AY, Baraldi R, Orban D (2024) A levenberg–marquardt method for nonsmooth regularized least squares. SIAM J Sci Comput 46(4):A2557–A2581
7. Aslan MF, Sabanci K, Durdu A, Unlersen MF (2022) COVID-19 diagnosis using state-of-the-art CNN architecture features and bayesian optimization. Comp Biol Med 142:105244
8. Berger T, Dennstädt D (2024) Funnel MPC for nonlinear systems with arbitrary relative degree. Automatica 167:111759
9. Bilski J et al (2023) Fast computational approach to the Levenberg-Marquardt algorithm for training feedforward neural networks. J Artif Intell Soft Comput Res 13(2):45–61
10. Bingi K et al (2018) A comparative study of 2DOF PID and 2DOF fractional order PID controllers on a class of unstable systems. Arch Control Sci 635–682
11. Chai J et al (2024) Concurrent learning-based trajectory tracking control of a novel multi-node flexible lander for asteroid missions. Aircraft Eng Aerosp Technol
12. Cherivirala YK, Wentzloff DD (2023) A capacitor-less digital LDO regulator with synthesizable PID controller achieving 99.75% efficiency and 93.3-ps response time in 65 nm. IEEE Trans Circ Sys II Express Briefs 70(5):1769–1773
13. Choe J et al (2023) Seamless reaction strategy for bipedal locomotion exploiting real-time nonlinear model predictive control. IEEE Robot Autom Lett 1–8
14. Choudhari P, Kulkarni NR, Bakshi M (2023) A system theoretic-based optimum controller for single-tank system and its performance comparison with PID controller. J Inst Eng (India): Ser B 104(3):551–561
15. Crnokic B, Peko I, Grubisic M (2023) Artificial neural networks-based simulation of obstacle detection with a mobile robot in a virtual environment. Int Robot Autom J 9:62–67
16. Cruz FT et al (2021) A comparison of gain scheduling PID and μ-synthesis robust level control for a conical tank system. IEEE, pp 1–4
17. Divakar K, Kumar MP (2022) A novel efficacious PID controller for processes with inverse response and time delay. IEEE Access 10:63626–63639
18. Eltayeb A et al (2024) Comparative analysis: fractional PID versus PID controllers for robotic arm using genetic algorithm optimization. Automation 5(3):230–245
19. Eslami A, Abdollahi F (2018) Simultaneous stochastic fault detection and consensus control in multi-agent systems. In 2018 6th RSI international conference on robotics and mechatronics (IcRoM) (pp. 189-196). IEEE
20. Forouzesh M, Thiran P (2021) Disparity between batches as a signal for early stopping. Springer, pp 217–232
21. Glushchenko A, Lastochkin K (2023) Quadrotor trajectory tracking using model reference adaptive control, neural network-based parameter uncertainty compensator, and different plant parameterizations. Computation 11(8):163
22. Gotov B-E et al (2022) Quadcopter stabilization using neural network model from collected data of PID controller. ICT Focus 1(1):10–21

23. Guo C, Hu J (2023) Fixed-time stabilization of high-order uncertain nonlinear systems: output feedback control design and settling time analysis. J Syst Sci Complexity 36(4):1351–1372
24. Guo R et al (2023) A robotic arm control system with simultaneous and sequential modes combining eye-tracking with steady-state visual evoked potential in virtual reality environment. Front Neurorobot 17:1146415
25. Guras R, Strambersky R, Mahdal M (2022) The PID and 2DOF control of the integral system-influence of the 2DOF parameters and practical implementation. Meas Control 55(1–2):94–101
26. Heydari A (2018) Optimal codesign of control input and triggering instants for networked control systems using adaptive dynamic programming. IEEE Trans Ind Electron 66(1):482–490
27. Hsieh WW (2022) Evolution of machine learning in environmental science—a perspective. Environ Data Sci 1:e3
28. Jarosz M, Sniezynski B (2024) Personalization of robot behavior using approach based on model predictive control. Appl Sci 14(24):11805
29. Kannan G, Saravanakumar G, Saraswathi M (2018) Two-degree of freedom PID controller in time delay system using hybrid controller model. Int J Autom Control 12(3):399–426
30. Kasi VR, Majhi S, Gogoi AK (2018) Identification and PID tuning techniques for stable and unstable SOPDT processes. Int J Dyn Control 6(3):1194–1206
31. Khan H et al (2022) Speed control comparison of wheeled mobile robot by ANFIS, Fuzzy and PID controllers. Int J Inf Technol 14(4):1893–1899
32. Kim H, Chang H (2023) Model reference Gaussian process regression: data-driven state feedback controller. IEEE Access 11:134374–134381
33. Kubatko M et al (2024) Sensorless in duction motor drive based on reactive power mras estimator using artificial neural network. IEEE, pp 1–4
34. Lakhdar S et al (2023) Enhanced sensor-less control of DFIG-based generators using hybrid MRAS-ANN observer and PSO parameter optimization. IEEE, pp 1–6
35. Lee D et al (2022) Comparison of deep reinforcement learning and PID controllers for automatic cold shutdown operation. Energies 15(8):2834
36. Liang Z, Zhou Y, Sesia M (2023) Conformal inference is (almost) free for neural networks trained with early stopping. PMLR, pp 20810–20851
37. Liu J, Fang H, Xu J (2021) Online adaptive PID control for a multi-joint lower extremity exoskeleton system using improved particle swarm optimization. Machines 10(1):21
38. Liu Z et al (2022) Adaptive neural network-based fixed-time control for trajectory tracking of robotic systems. IEEE Trans Circuits Syst II Express Briefs 70(1):241–245
39. Manna S, Akella AK, Singh DK (2023) Implementation of a novel robust model reference adaptive controller-based MPPT for stand-alone and grid-connected photovoltaic system. Energy Sour Part A: Recov Utilization Environ Effects 45(1):1321–1345
40. Milijas R et al (2020) A comprehensive lidar-based slam comparison for control of unmanned aerial vehicles
41. Mitov A et al (2023) Variable displacement axial-piston pump control by genetic algorithm tuned PID controller. IOP Publishing, p 012013
42. Mohammed RH et al (2018) Trajectory tracking control and robustness analysis of a robotic manipulator using advanced control techniques. Int J Eng Manuf (IJEM) 8(6):42–54
43. Mohd Tumari MZ, Ahmad MA, Suid MH, Hao MR (2023) An improved marine predators algorithm-tuned fractional-order PID controller for automatic voltage regulator system. Fractal Fractional 7(7):561
44. Mousa MAA, Elgohr AT, Khater HA (2024) Trajectory optimization for a 6 DOF robotic arm based on reachability time. Ann Emerg Technol Comput 8(1):22–35
45. Müller A (2020) An O (n)-algorithm for the higher-order kinematics and inverse dynamics of serial manipulators using spatial representation of twists. IEEE Robot Autom Lett 6(2):397–404
46. Nagado T (2022) Feed-back controller reduction considering the variation of the plant in 2DOF control system. IEEJ Trans Electron Inf Syst 142(11):1229–1234
47. Nayak N, Khan S, Bansfore SK (2020) Inter-area and intra-area oscillation damping for a multi-machine power system integrated with UPFC using DIW-PSO based PID controller. IEEE, pp 1–6

48. Nouri N, Zamani AA, Barakati SM (2018) Load frequency control with considering impact of wind turbine using a Fractional PID controller. IEEE, pp 1–6
49. Palomino-Jaime AJ et al (2022) Design and simulation of a 2 DoF robotic arm for TIG welding. IEEE, pp 287–291
50. Pascher M et al (2023) In time and space: towards usable adaptive control for assistive robotic arms. IEEE, pp 2300–2307
51. Protonotarios NE et al (2022) Sigmoid and beyond: algebraic activation functions for artificial neural networks based on solutions of a riccati equation. IT Prof 24(5):30–36
52. Qi R et al (2023) Design of the PID temperature controller for an alkaline electrolysis system with time delays. Int J Hydrogen Energy 48(50):19008–19021
53. Qi X, Ma H, Jing Y (2023) Arbitrary settling time control with prescribed performance for high-order nonlinear systems. Asian J Control 25(3):2424–2432
54. Rahmati V, Ghorbani A (2019) A novel low complexity fast response time fuzzy PID controller for antenna adjusting using two direct current motors. Indian J Sci Technol 11(13):8
55. Rodriguez EX, Lu Q (2024) Tuning PID controller for quadrotor using particle swarm optimization. IEEE, pp 168–175
56. Sarma H, Bardalai A (2023) Tuning of PID controller using driving training-based optimization for speed control of DC motor. IEEE, pp 1–8
57. Schmidgall S et al (2024) Brain-inspired learning in artificial neural networks: a review. APL Mach Learn 2(2)
58. Shamseldin MA (2022) Adaptive controller with PID, FOPID, and NPID compensators for tracking control of electric–wind vehicle. J Robot Control (JRC) 3(5):546–565
59. Sharkawy A-N, Khairullah SS (2023) Forward and inverse kinematics solution of A 3-DOF articulated robotic manipulator using artificial neural network. Int J Robot Control Syst 3(2)
60. Sharma AB, Tiwari S, Singh B (2020) Intelligent speed estimation in induction motor drive control using feed-forward neural network assisted model reference adaptive system, IEEE, pp 1–6
61. Silvestrini S, Lavagna M (2022) Deep learning and artificial neural networks for spacecraft dynamics, navigation and control. Drones 6(10):270
62. Simón P, Hällström J, Bergman A (2022) Supplementary comparison for the traceability of AC high voltage reference measuring systems up to 200 kV (EURAMET. EM-S33). Metrologia 59(1A):01008
63. Soni SK, Khemariya M, Singh A (2022) Model reference adaptive system based sensor-less induction motor with Fuzzy Interface. IEEE, pp 1–6
64. Urrea C, Saa D (2023) Design, simulation, implementation, and comparison of advanced control strategies applied to a 6-DoF planar robot. Symmetry 2023(15):1070
65. Veerendar T, Kumar D (2023) AVOA-based PID + IDF controller for frequency control of isolated hybrid thermal power system. IEEE, pp 1–5
66. Villa-Tiburcio JF et al (2024) ANN enhanced hybrid force/position controller of robot manipulators for fiber placement. Robotics 13(7)
67. Widhiada IW et al (2019) Comparison of PID and fuzzy logic to control the motions of robotic prosthetic limbs. IOP Publishing, p 012109
68. Wisotzki S (2019) The incorruptible integrator: a streamlined approach to IMC-PID controller tuning. The University of Vermont and State Agricultural College
69. Xie H et al (2021) Disturbance rejection MPC framework for input-affine nonlinear systems. IEEE Trans Autom Control 67(12):6595–6610
70. Xu Q et al (2023) Constructing deep spiking neural networks from artificial neural networks with knowledge distillation, pp 7886–7895
71. Yanarates C, Zhou Z (2022) Design and cascade PI controller-based robust model reference adaptive control of DC-DC boost converter. IEEE Access 10:44909–44922
72. Yu P, Tan N, Mao M (2023) Position-based visual servo control of dual robotic arms with unknown kinematic models: a cerebellum-inspired approach. IEEE/ASME Trans Mechatron 28(4):2328–2339

73. Zafar MH et al (2023) Optimal tuning of pid controller for boost converter using meta-heuristic algorithm for renewable energy applications
74. Zhang X, Liu X, Wang P (2023) Distributed fixed-time secondary control for MTDC systems using event-triggered communication scheme. Processes 11(8):2329
75. Zhao ZQ, Liu SJ, Pan JS (2021) A PID parameter tuning method based on the improved QUATRE algorithm. Algorithms 14(6):173

Study and Optimization of Electric Torque Vectoring System in an FSAE Race Car

Arthur Julve

Abstract This dissertation presents the development and optimization of a torque vectoring system for a Formula Student vehicle, aimed at enhancing dynamic stability and handling performance. The research addresses four key objectives: creating a realistic vehicle model, identifying the optimal control strategy, ensuring a high level of safety, and enabling future improvements for the vehicle and the team. Simulink, in conjunction with data from VSM laptime simulation software, was used to model the vehicle and integrate three different torque distribution controllers: a proportional-integral-derivative (PID) controller, a sliding mode controller, and a fuzzy logic controller. The performance of these controllers was evaluated by comparing the vehicle's yaw acceleration to an ideal reference value. Among the tested controllers, the fuzzy logic controller demonstrated the best performance in terms of response speed, accuracy, and reduced oscillations. To ensure safety, a failsafe mechanism and dashboard alerts were implemented, mitigating risks associated with torque vectoring malfunctions. This study provides a foundation for improving vehicle dynamics in Formula Student competition cars and offers insights into future developments for the team. **Highlights** • Development and optimization of a torque vectoring system to significantly enhance dynamic stability and handling performance for a Formula Student vehicle. • Test and comparison of three torque distribution controllers—PID, sliding mode, and fuzzy logic—identifying the fuzzy logic controller as the best performer in response speed, accuracy, and reduced oscillations. • Implementation of a failsafe mechanism and dashboard alerts to ensure high safety levels, protecting against potential torque vectoring malfunctions. • Foundation for improving vehicle dynamics and offering valuable insights for future advancements in Formula Student competition cars.

A. Julve (✉)
School of Engineering, Computing and Mathematics, Oxford Brookes University, Oxford, UK
e-mail: 19276215@brookes.ac.uk

A. Azizi (ed.), *Intelligent Mechatronic Control Systems*, Emerging Trends in Mechatronics, https://doi.org/10.1007/978-981-95-3925-3_3

List of Symbols and Abbreviations

Symbols

a	Acceleration
C	Tyre stiffness
CG	Centre of gravity
δ	Steering angle
F	Force
h	Height
K	Understeer gradient
l	Wheelbase
m	Metre
%	Percentage
rad	Radians
s	Seconds
t	Track

Abbreviations

ABS	Anti-lock braking system
ADAS	Advanced driver assistance systems
ASOSM	Automated surface observing system
BMS	Battery management system
DYC	Direct yaw control
ECU	Electronic control unit
ETAS	Empowering Tomorrow's Automotive Software
ESP	Electronic stability programmes
FL	Front left
FLC	Fuzzy logic controller
FPGA	Field-programmable gate array
FR	Front right
FS	Formula Student
FSAE	Formula of the Society of Automotive Engineers
FSUK	Formula Student United Kingdom
GPU	Graphics processing unit
GPS	Global positioning system
H	High
L	Low
M	Medium
MPC	Model predictive controller
NH	Negative high

NL	Negative low
OBR	Oxford Brookes Racing
PH	Positive high
PID	Proportional integral derivative
PL	Positive low
RL	Rear left
RWS	Rear-wheel steering
RR	Rear right
SMC	Sliding mode controller
TV	Torque vectoring
VSM	Vehicle simulation model
XH	Extra high
XL	Extra low
Z	Zero

1 Introduction

In Formula SAE (FSAE), as in any other motorsport competition, the entirety of a project is focused on one clear goal: going faster than the opponents. Teams invest significant hours and resources in incorporating innovative ideas every season to gain a competitive edge.

1.1 Background

Within this context, this study is being undertaken as part of that incessant pursuit of performance improvement within Oxford Brookes Racing. More specifically, the aim lies in the integration of torque vectoring (TV) techniques within the Vehicle Controls system of the car.

The torque vectoring approach refers to the precise manipulation of torque delivery to each wheel, optimizing the dynamics of the car and consequently reducing lap times. It aids the car in maintaining an optimal yaw rate through turns, while also reinforcing longitudinal motion by giving authority to the most loaded wheels. This concept is being investigated by major manufacturers regardless of the propulsion type with a view to enhancing drivability.

Torque vectoring is a multifaceted concept that intersects various disciplines within motorsport engineering. It draws upon expertise in dynamics, power electronics, control engineering, and data analysis. When executed correctly, it has demonstrated remarkable effectiveness in enhancing performance on track.

1.2 *Objectives*

The next step in the process is to define the main objectives of the project. Once the overall aim is established, more specific goals can be set to aid in planning the work, determining the scope, and establishing criteria for measuring the successfulness of the project. For this application and the context of the study, the project is guided by four main objectives.

1.2.1 Truthful Car Modelling

The output of the torque vectoring system combines information from multiple sensors, car characteristics, and the integration of a control model through dynamic equations. Given that the validation of this project is entirely virtual, it is crucial to ensure the car model is accurately replicated. While some assumptions are made to simplify calculations, the goal is to incorporate as much known information about the car as possible to achieve reliable results. A robust data acquisition system, paired with an accurate car model, is essential for producing sensible and valid outcomes.

1.2.2 Optimum Control Technique

Another objective of this study is to identify the most suitable control technique. Since the system is Simulink-based, three different control strategies within this software are compared: proportional-integral-derivative (PID) controller, sliding mode controller, and fuzzy logic controller. From this set of results, the most effective alternative will be chosen for the final car design. Effectiveness is measured by considering response time, aggressiveness, and the remaining error.

1.2.3 Safety

In every automotive advancement, safety must be a top priority, and this project is no exception. By incorporating torque vectoring to assist driving, the driver's actions may no longer directly influence the behaviour of the car. Therefore, it is crucial to implement strategies and measures to ensure safety, avoiding any dangerous situations for the driver or others around the car. This aspect is considered paramount when developing the system.

1.2.4 Documentation

Finally, considering that this study is part of a bigger project, which is Oxford Brookes Racing, a certain interest arises to keep information transparent and understandable.

One of the objectives is to establish a foundation for future improvements, either focused on the torque vectoring itself or other control systems. To facilitate the work of prospective members, all steps, the functioning, and the results must be extensively documented.

1.3 Literature Review

1.3.1 Introduction to Torque Vectoring

Torque vectoring is the name given to a computer-controlled system capable of selecting left-to-right and front-to-rear torques transmitted from the powertrain to the wheels [12]. In general, the torque vectoring control impacts longitudinal, lateral, yaw, roll, and pitch dynamics of the vehicle. When driving, situations can arise where one or several tyres reach the friction limit, as can be seen in Fig. 1. To avoid unstable vehicle behaviour under these conditions, the distribution of the individual wheel torques has to be tuned [10].

Torque vectoring is nothing but the natural evolution of the differential concept with the aid of electronic controls. Since the relevance of torque distribution in the yaw moment and the stability of the car were evidenced, several systems have been implemented into road cars [8]. The first one was the mechanical differential, which ensures that the inner tractive wheels rotate at a lower speed than the outer wheels to avoid slipping. The next advancement was the anti-lock braking system (ABS), which electronically regulates torque under braking to improve deceleration effects. Later, traction control systems appeared, balancing the excess of torque under acceleration. Last, electronic stability programmes (ESP) detect and reduce loss of traction, preventing the tyres from skidding uncontrollably [4]. Torque vectoring inspires from all its predecessor techniques to give a more complete system with high potential in electric mobility.

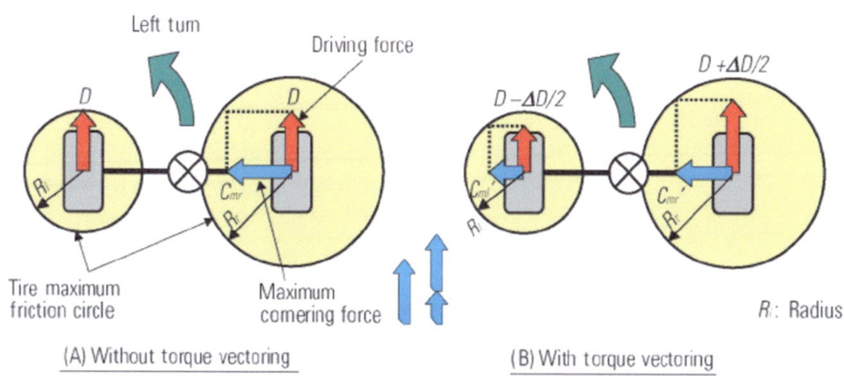

Fig. 1 Effect of torque vectoring [8]

The approach of torque vectoring in a road car compared to a racing car is very distinct. Commercial vehicles are developed with an eye on enhancing safety, ride comfort, stability, and energy consumption [25]. Although these are also strengths of a torque vectoring system, racing cars focus on the improvement of vehicle dynamics to reduce lap time and torque vectoring is leveraged with that idea in mind.

1.3.2 Vehicle Dynamics Principles

The successful development of a torque vectoring system relies on a good understanding and implementation of basic vehicle dynamic principles. The first concept to be considered is yaw rate. It is the magnitude at which the vehicle turns around its own centre of gravity [7], and it is the value to be optimized with torque vectoring. In each corner, there is an ideal yaw rate, the closer the car is to that optimum value, the more stable it will be and the faster it will manage to turn [18]. This velocity is determined by the force contribution of each of the four wheels on the car. The distribution of force is the output of the system, and it is dictated by the tyre's capability to generate force in each driving scenario [20]. When developing the system, it is essential to inherently consider all of these concepts.

The inputs of the system are all the different magnitudes that act on the tyre, so that a precise estimation of its force contribution is performed. For that, a second key concept of vehicle dynamics is introduced: tyre load. The greater the tyre load, the more longitudinal and lateral force the tyre will be able to generate [3]. This is ruled by acceleration, load transfers as depicted in Fig. 2, suspension characteristics, and aerodynamics [11]. Everything that happens inside and outside the vehicle is to be modelled in torque vectoring for a more realistic result [21]. The detail of the modelling and the number of considerations will have a direct effect on how complex the control and the system in general will end up becoming.

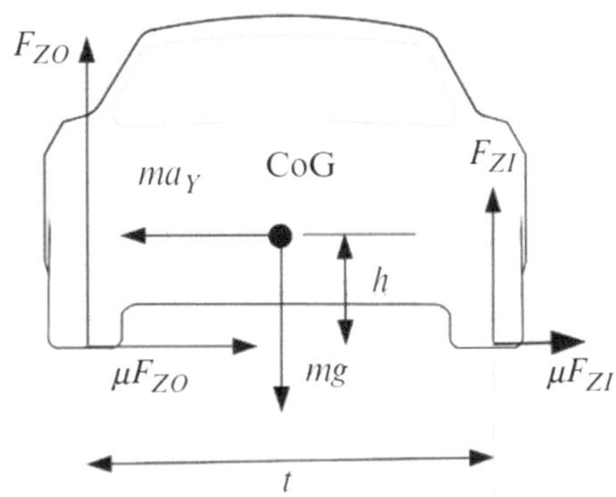

Fig. 2 Lateral weight transfer in a rigid chassis model [3]

1.3.3 Powertrain Topology Effects

The application of torque vectoring is viable and beneficial regardless of the geometry and topology of the powertrain. Internal combustion engine cars and electric vehicles can be divided into rear-wheel drive, front-wheel drive, and four-wheel drive [19]. In two-wheel drive cars, which is a common topology of internal combustion engine cars, the ability to act on each of the wheels independently is difficult. However, with the disruption of electric vehicles, two- and four-in-wheel motor cars are possible, giving much more room for a proper torque vectoring application [24]. In short, the configuration of the end car has a significant impact on the design of the system.

Oxford Brookes Racing's powertrain design accounts for an electric motor in each of the wheels, as showcased in Fig. 3. This allows for better control of the yaw moment with precise optimization of the torque in all four wheels and not just two of them [24]. Referring to vehicle dynamics, if the four forces creating the yaw moment are carefully controlled, the yaw rate achieved will be closer to the ideal yaw acceleration and yaw rate [22] and therefore the performance of the system will experience an improvement. Apart from that, the control complexity is reduced since each engine has one inverter and the requested torque, being the magnitude that indicates how much current is flowing to the motors, is easily modifiable. That part of the powertrain chain is where the torque vectoring system comes into play. Otherwise, a further calculation of how the torque is distributed must be carried out, or new mechanical components must be included.

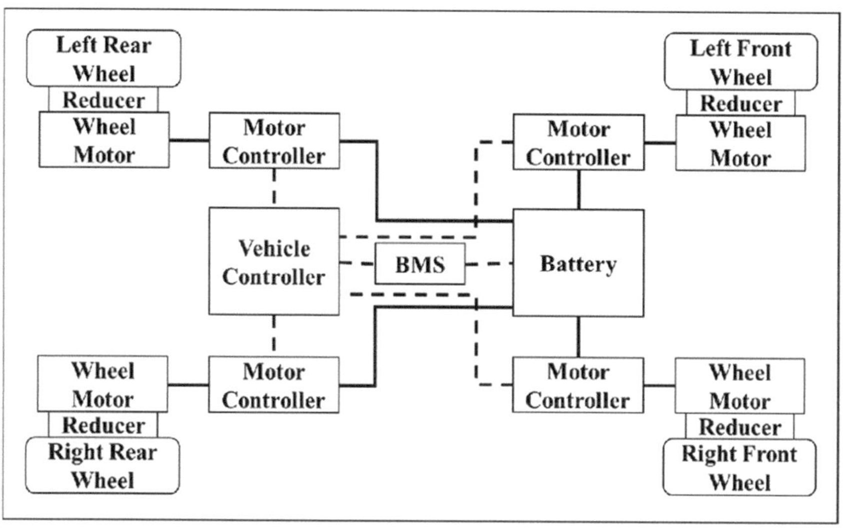

Fig. 3 Oxford brookes racing powertrain type [19]

1.3.4 Control Strategies

Previous research evaluates the impact of different control techniques on vehicle dynamics and their inherent downsides. Although all of them can be refined to achieve a proper result that significantly reduces lap time, they showcase differences.

The first study introduces a hardware-in-the-loop torque vectoring control for a fully electric vehicle with four independently controlled motors. The controller design incorporates loop shaping, a state observer, and gain scheduling based on vehicle speed [16]. It compares the robustness of the proposed controller to a proportional integral controller, with simulations addressing various tyre and vehicle parameters. Experimental results on a four-wheel-drive electric vehicle show that the control provides effective cornering response without needing feedforward control.

Another paper published by the Politecnico di Milano explores coordinating active rear-wheel steering (RWS) and torque vectoring (TV) with independent electric motors to enhance vehicle handling and stability during cornering [29]. The controller adjusts RWS and TV contributions based on their impact, analysed using phase portraits. A performance index prioritizes the most effective control action. Simulations with a nonlinear vehicle model show the controller's performance compared to using RWS and TV individually.

Zhejiang University presents a fuzzy torque vectoring differential controller aimed at improving the lateral stability of electric buses with two motor-driven wheels. By using an unscented Kalman filter to estimate the sideslip angle from measured yaw rate and lateral acceleration, the controller applies fuzzy logic rules to adjust torque distribution, enhancing stability and handling [32]. Real-time hardware-in-the-loop tests demonstrate the effectiveness of the controller in various driving conditions, showing significant improvements in bus handling and path accuracy.

Kwang-Ki Kim proposes two model predictive control methods to enhance stability and energy efficiency in electric vehicles with four in-wheel motors [14]. These methods, linear parameter-varying model predictive controller (MPC) and nonlinear MPC, use adaptive weighting to balance stability and energy savings. Simulations show that both methods improve fuel efficiency by 2–3% and ensure yaw stability compared to standard controls.

Research in the University of the Basque Country in Spain explored combining machine learning with heterogeneous embedded platforms like field-programmable gate arrays (FPGAs) and graphics processing units (GPUs) to develop advanced control concepts for vehicle dynamics and advanced driver assistance systems (ADAS). It focuses on implementing neural networks for fast predictions to enhance torque vectoring in multi-electric-motor vehicles. The study discusses technological constraints in the automotive field and proposes efficient neural network implementations on GPUs and FPGAs [6]. The results demonstrate the strengths and weaknesses of these platforms, providing insights and valuable enablers for future developments in vehicle control systems.

Other studies also implement an adaptive second-order sliding mode (ASOSM) controller combined with the backstepping method to enhance vehicle manoeuvrability and stability [31]. Unlike traditional controllers, ASOSM does not need prior knowledge of uncertainties and mitigates chattering issues. Simulations show that ASOSM improves vehicle response bandwidth by 21%, is robust to changes in vehicle mass and tyre type and requires less control effort compared to conventional first-order and second-order sliding mode controllers.

1.3.5 Results Analysis

In order to evaluate the functionality of the design, torque vectoring systems are tested in a series of phases before they are fully operational and approved. The first phase checks the results that are being obtained as the model is still being developed. This uses recorded data from the vehicle to feed the model and confirm that the results are as expected [15]. The second phase accounts for a full emulation of the vehicle in a simulation software such as AVL VSM, CarSim, or IPG CarMaker, as illustrated in Fig. 4 [9]. When the tests are completed successfully and the proper modifications are implemented, the model can be part of the Vehicle Controls system of the car. Prior to making use of it in an official event, the system is evaluated under real track conditions [5].

Apart from the process to approve the system, data analysis is also paramount. It determines if the system is performing as expected and to what degree. In motorsport, data logging systems and telemetry provide teams with information to carry on those evaluations [27]. The signal of hundreds of sensors in the car is stored and plotted in different channels. From there, engineers can post-process the information and extract conclusions about the car's behaviour [13]. The same techniques apply to software in the loop and driver in the loop scenarios. In this particular project, the focus lies on the yaw acceleration error and the reduction of lap time that this can entail.

Fig. 4 FSAE car testing simulation [9]

1.4 Original Contribution

In the course of reviewing relevant literature pertaining to the implementation of torque vectoring in Formula SAE cars, examples of its application by other teams have been encountered. Additionally, it is noteworthy that torque vectoring systems have already been successfully integrated into some road cars, further underscoring their potential benefits and feasibility.

However, a notable gap emerges within our specific team context, which, despite possessing enough expertise and resources, torque vectoring has not yet been integrated into the car. This presents an opportunity for the study to fill this gap and gain an advantage.

At the same time, the car characteristics, the project limitations, the mathematical assumptions, the software used, and the validation plan make it a unique control systems project. It is inspired by previous work, but adapting it to the team's needs; for that reason, unpublished ideas were required.

1.5 Structure of the Report

After establishing the context and the background of the project, the report is divided into four main sections. The first one encompasses the methodology, including planning, workflow, methods, and resources. Next, the system design is thoroughly explained, detailing the entire development process up to validation. In third place, a results evaluation is undertaken, contrasting various outcomes obtained after completing the system design. Last, following that discussion, the conclusions together with a series of future improvements for the project are listed in two final chapters.

2 Methodology

Before starting the work, a planning phase is conducted to organize the various steps and ensure they are completed on time and as required. This plan considers several factors, including assumptions, limitations, resources, deadlines, and cost.

2.1 Assumptions

The dynamics of a car represent a highly complex field of study, encompassing numerous variables and factors. Furthermore, this research is undertaken by a Formula Student (FS) team with limited knowledge and resources. Consequently, the team must make a series of assumptions which, while less precise, facilitate a better understanding of the system and enable a quicker resolution of the primary equations.

The first assumption involves calculating the yaw rate and yaw acceleration reference values. This setpoint, derived from the steering radius formula, corresponds to the widest turning radius and is assumed to be the most stable and fastest way to navigate any corner. While this is generally true, other factors must be considered. For instance, the driver plays a crucial role; some drivers may feel more comfortable with a less stable car that they can rotate faster, resulting in better lap times. Additionally, the widest turning radius may not always be the fastest path through a series of corners. Although it may be optimal for a single corner, drivers often use strategies that sacrifice the ideal racing line in consecutive turns to maintain higher speeds leading into a long straight. Having said that, this will be disregarded during the development of the project.

To simplify vehicle dynamics calculations, a second widely accepted assumption, known as the 'rigid chassis approach', is considered valid. This approach involves treating the mass of the vehicle as a whole, thereby ignoring the effects of the suspension system on tyre loads and not differentiating between unsprung and sprung mass. While this is not an entirely accurate method, it significantly reduces the complexity of calculations and saves development time, making it easier to understand and implement.

A third assumption is helpful when distributing torque based on the estimated tyre load. In this study, torque is allocated as if the grip were directly proportional to the load. However, not only is this relationship nonlinear, but load is also not the only factor influencing grip. Temperature, wear, and compound type also play significant roles, though they are challenging to estimate without additional information. Consequently, the linear relationship is accepted as true for the purposes of designing the torque allocation strategy.

Finally, the simulation conducted with the lap time simulation tool assumes ideal data acquisition. This means that all necessary sensors are correctly installed, the information is complete, and there is no extraneous noise interfering with the signals. In fact, this is unrealistic for a car with numerous internal movements and varied signal types operating on diverse terrains. However, minimizing these additional issues is beyond the scope of this project, so they are neglected.

2.2 Limitations

To create an effective plan and understand the series of decisions made, it is essential to recognize the limitations surrounding the project.

Firstly, the mathematical assumptions involved inevitably introduce some limitations to the accuracy. Certain equations used are approximations of reality, and some values obtained from the various components are estimations. Therefore, it is understandable that the final validation results may slightly differ from real-world values.

Another limitation to be addressed during the project is the limited access to physical testing. Conducting a testing day requires not only the readiness of the torque vectoring system but also the preparation of the car, a suitable location, and sufficient personnel to manage the event. Consequently, teams often consolidate multiple upgrades to test on a single day to maximize efficiency. Given these constraints, the project and development of the torque vectoring system cannot wait for all these factors to align. The frequency of required tests exceeds what the team can realistically manage. Therefore, the decision has been made to simulate all testing using laptime simulation tools.

The fact that the project is part of an existing or developing car introduces additional limitations regarding components and procedures. Firstly, the powertrain topology is restricted to four electric in-wheel motors, limiting the potential impact and testing of torque vectoring with different topologies. Secondly, there is a rigid path to follow for the control software. The entire vehicle control system is designed in Simulink and then translated into C language for the electronic control unit (ECU). This process restricts the exploration of other programming languages that could be more optimal for development and testing.

2.3 *Development Workflow*

The flow chart that rules the progress of the project is depicted in Fig. 5. Each of the steps is further described as follows.

- **Vehicle study**: Prior to any modelling activity, the behaviour of the car and the input parameters that have an effect on yaw velocity are to be internalized. This involves reviewing component documentation, studying the vehicle dynamics, and contacting any team member with relevant advice.
- **Vehicle modelling**: Once the different parts of the car are studied, they can be modelled in the system using Simulink. These account for the longitudinal and lateral load transfer, the steering ratio, and the tyre behaviour in essence. From there, enough information is obtained to proceed with the subsequent steps.

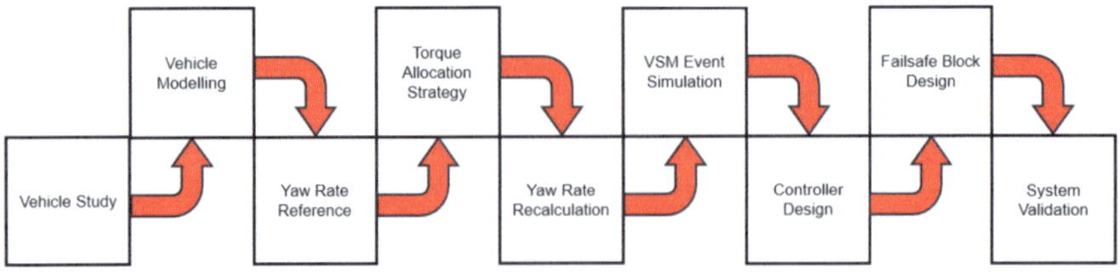

Fig. 5 Development flow chart (own elaboration)

- **Yaw acceleration reference**: Third, the yaw acceleration setpoint can be estimated. The desired yaw acceleration is a function of the linear velocity, steering, and characteristics of the car. It defines a yaw speed which allows for the smallest turning radius corresponding to optimal performance.
- **Torque allocation strategy**: In order to reach the desired yaw acceleration in a more sensible way, a torque allocation strategy is designed. This considers the tyre load distribution, the throttle value, and the steering input from the driver.
- **Yaw acceleration recalculation**: From the torque distribution and the characteristics of the tyre curves modelled into the system, the reaction force that the tarmac is exerting onto the vehicle and the accelerations that it generates can be calculated. The new yaw acceleration can then be compared to the setpoint and used to implement control techniques upon that difference.
- **VSM event simulation**: Since Formula Student testing is quite limited due to the preparation it requires and the great time investment it represents, the initial data for the development of this system has been extracted from digital lap simulations. Using AVL VSM software, an event has been created where the car and the Formula Student United Kingdom (FSUK) endurance track have been modelled. Running the simulation allowed the team to export data and fit it into Simulink to start working with truthful signals from the car.
- **Controller design**: With the implementation of the signals, the first conclusions for an open-loop system can be extracted. However, if the goal is to close the system and generate feedback where the system reads the value in the output, the controller needs to be designed. The goal is to test three different controllers and assess the performance of each of them.
- **Failsafe block design**: To meet the safety requirements, a failsafe block is designed. This component detects potentially dangerous situations and attempts to correct them or disable the torque vectoring system if it might increase the problem. By shutting down the system, full control is returned to the drivers, allowing them to address the issue effectively.
- **System validation**: At the end of the development phase, a crucial part of the project is dedicated to evaluating the data obtained from the simulations and assessing its validity. This step is essential for identifying any necessary adjustments within the system and ensuring that the results and conclusions are interpreted accurately and meaningfully.

2.4 Cost

Since most of the tools and components were already acquired for previous versions of the Oxford Brookes Racing (OBR) car, it is challenging to provide an accurate estimation of the total cost. The minimum expenditure necessary to develop a torque vectoring strategy involves purchasing the powertrain parts. Assuming that the hardware and software are exclusively procured for this project, the total cost amounts to £70,900.

Table 1 Cost breakdown (own elaboration)

Item	Price
AMKmotion racing kit	£ 25,000
Vehicle control unit	£ 10,000
Power distributor	£ 4,000
Accumulator pack	£ 7,000
BMS circuit	£ 2,000
Wiring	£ 3,000
Simulink licence	£ 1,500
AVL VSM licence	£ 10,000
Engineering hours	£ 8,400
Total	£ 70,900

This price encompasses nine key assets. Firstly, the AMKmotion racing kit, which includes four in-wheel engines and their inverters. Secondly, the cost for the Empowering Tomorrow's Automotive Software (ETAS) vehicle control unit is included. The power distribution circuit is also accounted for. Regarding the accumulator, this includes the cells, their housings, the wiring, and the battery management system (BMS) circuit. Additionally, the annual subscriptions for the two main software applications used are factored in. Finally, the cost of 600 engineering hours anticipated for a project of this scale is included. Table 1 provides a detailed breakdown of the costs for each item.

3 System Design

The schematic of the torque vectoring control system is formed by six different blocks, as can be observed in Fig. 6. The functioning of each of the parts is explained in depth, together with their relevance in the system.

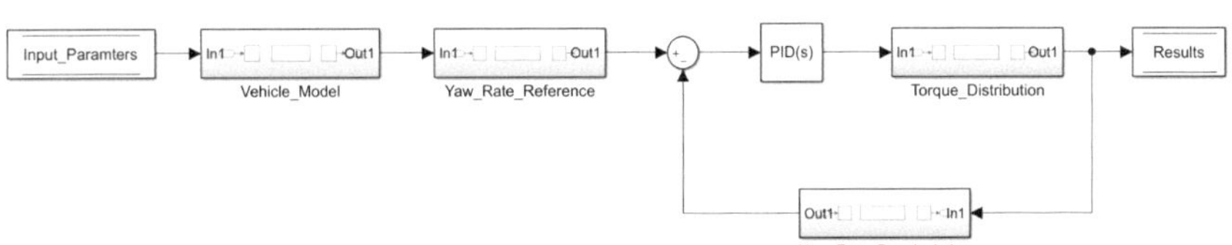

Fig. 6 Schematic of the control system (own elaboration)

3.1 *Input Parameters*

Despite being one of the last parts to be developed in the process flowchart, the importation of data logged from the sensors in the car is where the system starts. Since physical testing is limited in Formula Student due to the time restrictions and resources, a decision was taken to obtain the required parameters from a laptime simulation software.

For that part, AVL VSM, which is the software chosen, allows us to configure a car from scratch and simulate it around a track. The signal from all the sensors of the car is recorded and can be exported into an Excel file. This matches the fact that the signal builder in Simulink can read Excel files in order to generate the inputs of the system. The use of this block can be observed in Fig. 7.

Once in VSM, all the parameters of the car have been inserted to replicate the same exact behaviour the car would have on track. The suspension and vehicle dynamics section in the team has helped in modelling the suspension characteristics, while the vehicle geometry, the powertrain architecture, and even the tyre curves were also added, as represented in Fig. 8.

The track in which the simulation is performed has also been modelled following the Formula Student UK competition track profile from the previous edition. It is the circuit in which the autocross and the endurance events are held, making it a really good representation of the vehicle dynamics and the impact of torque vectoring.

After the simulation is run, more than 725 channels can be read, each of which focuses on a specific measurement of the car. However, in this project, most of them must be filtered to stick to the ones that are relevant to the torque vectoring and can actually be measured with the sensors that are embedded in the current version of the single seater. The main parameters to be considered are six, and they are listed in Table 2.

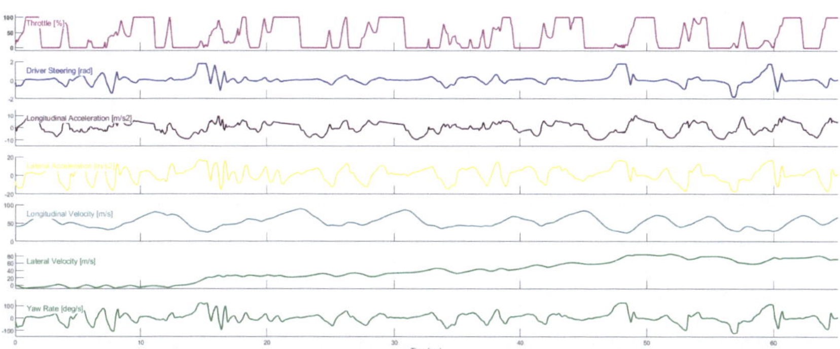

Fig. 7 Signal builder layout (own elaboration)

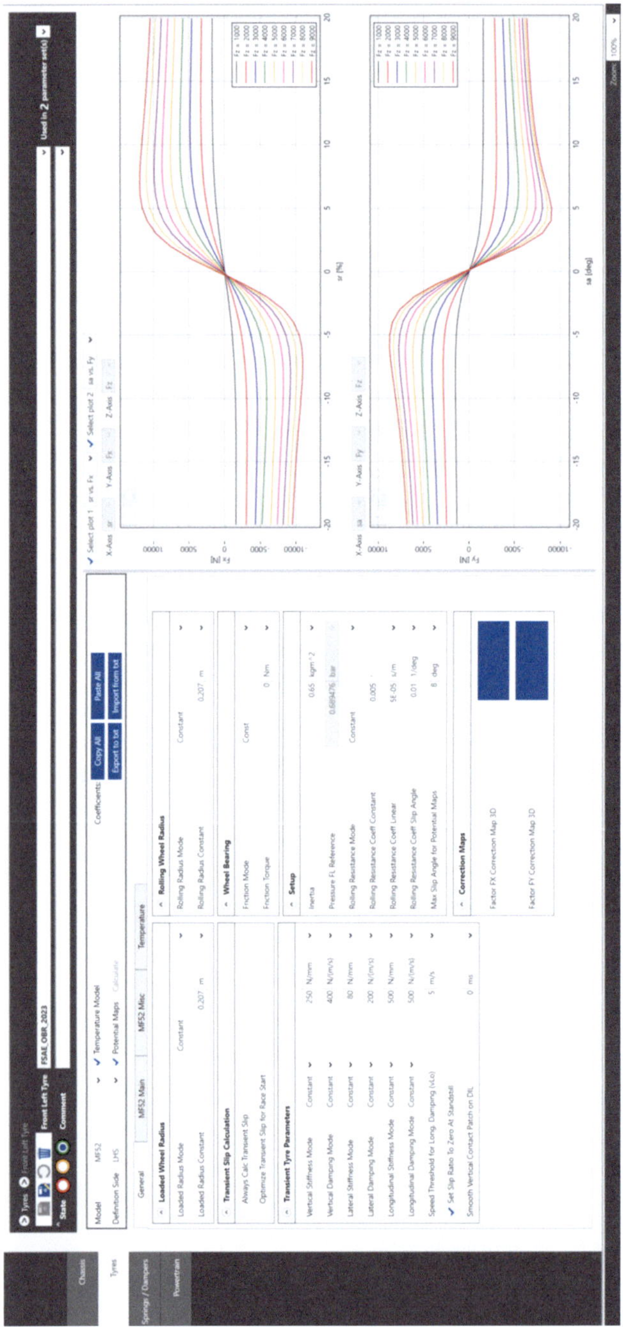

Fig. 8 Tyre model implemented in VSM simulation (own elaboration)

Table 2 Torque vectoring input parameters (own elaboration)

Magnitude	Sensor	Application
Throttle pedal percentage [%]	It is measured with a potentiometer attached to the throttle pedal, and it converts a voltage signal ruled by the pedal position into a percentage	It gives a reference to the amount of acceleration the driver is requesting, and it is used in the torque allocation block to set the maximum torque value
Driver steering angle [rad]	It is measured by a sensor located in the steering column that relates an angle value to a voltage pulse	It gives a reference of the amount of rotation the driver is expecting, and it is employed to calculate the yaw acceleration reference and the direction of the tyre forces
Longitudinal velocity [m/s]	It is registered by the GPS transponder integrated in the car. It can also be estimated by the rotation of the four wheels	It is useful when calculating the yaw reference and is key when finding the slip ratio of the tyres
Lateral velocity [m/s]	It is registered by the GPS transponder integrated in the car	It is especially meaningful in order to predict the slip angle and the lateral forces exerted by the tyres
Longitudinal acceleration [m/s^2]	It is sensed by the gyroscope installed inside the vehicle	It is the main parameter to be considered when estimating the load transferred from one axle to the other. For that reason, it is also paramount in the torque allocation phase
Lateral acceleration [m/s^2]	It is again registered by the gyroscope installed inside the vehicle	It is key when estimating the load transferred from one side of the car to the other. It is also critical in the torque allocation phase

3.2 Vehicle Model

From the signals imported, some other parameters are derived, which are not directly measured by the sensors and here is where the vehicle model created in Simulink comes into play. This part incorporates many different values of the car in order to estimate two things: the vertical load that is being put in each of the tyres at each moment, and the steered angle of the front wheels. These values are later used in the control part to define which is the torque requested at each of the engines.

For the load transfer formulation, the rigid chassis assumption is made, which means that the car mass is taken as a whole, disregarding unsprung and sprung mass differentiation effects. The value of the load transferred is computed by the formulas of longitudinal load transfer and lateral load transfer illustrated in Eqs. 1 and 2.

Load transfer formulation [3]

$$LongitudinalLoadTransfer = \frac{h}{l} \cdot m \cdot a_x \tag{1}$$

$$LateralLoadTransfer = \frac{h}{t} \cdot m \cdot a_y \tag{2}$$

The longitudinal load transfer is the weight that is moved from the front axle to the rear axle when accelerating and all the way around under braking. It is due to the moment generated by the longitudinal tyre forces around the centre of gravity of the car. As illustrated in the formula, it depends on the height of the centre of gravity (h), the wheelbase of the car (l), the mass (m), and the value of the longitudinal acceleration (a_x).

As for the lateral load transfer, it follows the same principle, but in this case, considering the left and right sides of the car under cornering. It is proportional to the height of the centre of gravity (h), the track of the vehicle (t), the mass (m), and the lateral acceleration (a_y).

These load transfers are added and subtracted from the static load of the tyres when no acceleration is being applied to the car. These generate a different load value for each of the wheels at any given moment, which is considered by the torque vectoring system in order to allocate the correct torque depending on the grip level they can provide.

The equations have been transcribed in Simulink blocks as showcased in Figs. 31 and 32 in the Appendices. In the end, the value for the load affecting each of the tyres is obtained, which allows us to carry on to the next step of the process.

Inside the vehicle model section, it has been also included some characteristics of the steering system, more precisely, the conversion from radians to degrees and the steering ratio. The signal obtained from the software is in radians and is multiplied by a gain of 57.3 to adapt it to the rest of the calculations. The steering ratio is the relationship between the angle steered in the steering wheel by the driver and the angle that the wheels turn around their own axis. It is not only a main parameter in any car, but it is paramount when predicting the trajectory that the driver is expecting from their input in the torque vectoring system.

In this case, the steering ratio is 5.57:1, which means that for every 5.57 degrees applied to the steering wheel, the front wheels rotate 1°. This has been modelled in Simulink, allowing for the estimation of the direction of the wheels at each time and the corresponding slip angles. The block in charge of this function is presented in Fig. 33 in the Appendices.

3.3 Yaw Acceleration Setpoint

With the relevant parameters of the car introduced, a yaw acceleration reference is calculated. The yaw rate is the speed at which the car changes direction around its z-axis. This value is to be optimized in order to create a neutral steering car, which means that the smallest possible turning radius is needed for a given velocity, which corresponds to optimal performance. This desired behaviour can be later tuned depending on the preferences of the driver. To be able to match this with the torque and forces exerted by the tyres, the value of the yaw rate is derived to obtain the angular acceleration that will be aimed.

The optimal rate of direction change is calculated by Eq. 3, which is defined as the longitudinal velocity of the vehicle divided by the corner radius. However, the system cannot know the radius of the corner unless it is explicitly defined; for that reason, it is estimated with the perception of the driver and the steering input. The resultant formula is shown in Eq. 4, which also takes into consideration the steering angle (δ), the understeer gradient (K), and the wheelbase (l).

Desired yaw rate calculation [3]

$$\dot{\gamma}_{desired} = \frac{v_{CG}}{R} \tag{3}$$

Desired yaw rate estimation [3]

$$\dot{\gamma}_{desired} = tan\delta \cdot \frac{v_{CG}}{l \cdot (1 + K \cdot v_{CG}{}^2)} \tag{4}$$

The understeer gradient is a value derived from the load in the different axles and the cornering stiffness of the tyres, as illustrated in Eq. 5, which helps quantify the amount by which the steering response deviates from Ackermann [3]. If K is positive, then the car would require more steering input than the Ackermann car to achieve the same radius. Alternatively, if the same steering input is maintained, then the car will corner with a larger radius. If K is negative, the opposite situation is encountered.

Understeer gradient expression [3]

$$K = \left(\frac{F_{ZF}}{C_F} - \frac{F_{ZR}}{C_R} \right) \tag{5}$$

All these relationships have been replicated as a separate block in Simulink, illustrated in Fig. 9. For each time step, the steering angle is recorded along with the longitudinal velocity of the car and the understeer gradient, and then the ideal yaw rate is determined.

3.4 Torque Allocation Strategy

The next step is to design the torque allocation strategy, defining the parameters that will be considered in order to distribute the torque in all four wheels differently. The values to be considered are the amount of throttle from the driver, the maximum power allowed by the engine and the FS regulations and the vertical load at each of the corners as a representation of the available grip at each tyre.

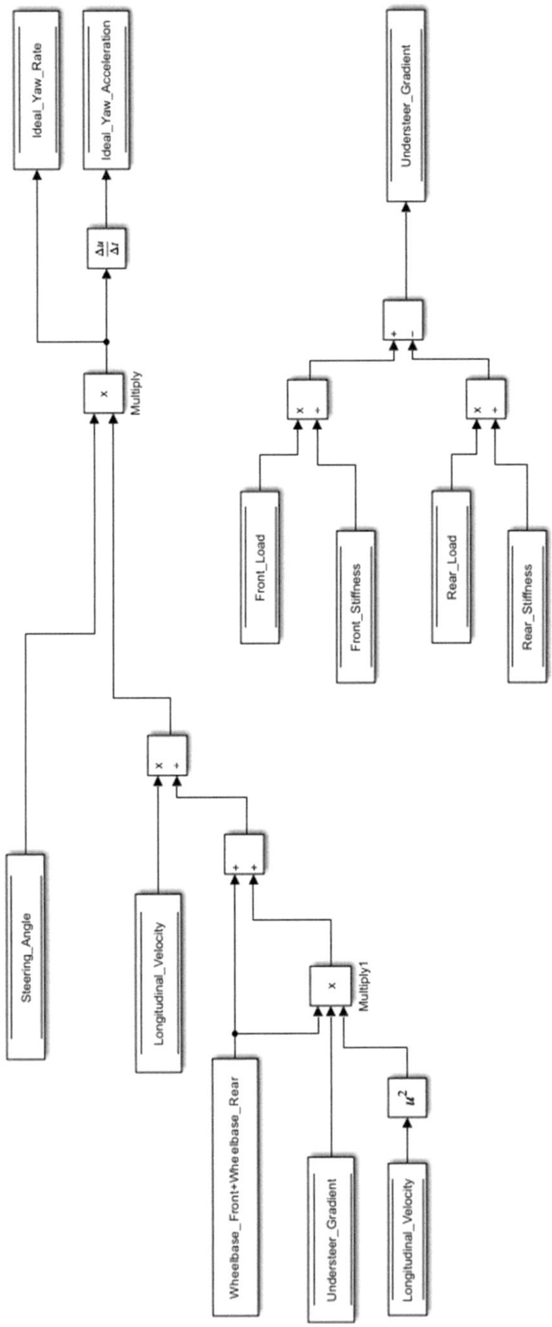

Fig. 9 Yaw acceleration reference block (own elaboration)

In a first stage, the system has been configured as an open loop, where the yaw setpoint has been disregarded. The equation that calculates the final torque for each wheel is defined in Eq. 6, and it is applicable for any of the four wheels.

Requested torque formula for open loop (own elaboration)

$$\textit{Maximum torque} \cdot \textit{Throttle pedal value} \cdot \frac{\textit{Tyre load}}{\textit{Most loaded tyre}} = \textit{Requested torque} \tag{6}$$

The first component of the formula, the maximum torque, is used to determine a value of torque that you do not want to surpass at any time. Since the rest of the components are percentages, the maximum value of the requested torque will match the maximum torque defined. In Formula Student, this value is set as 21 Nm per wheel and so is reflected in the model.

The second component accounts for the amount of throttle the driver is applying. It is a value compared to full throttle, which is 1. Any value lower than that means that the driver is not in a situation where the maximum torque from the wheels is needed; therefore, only a fraction of that is used.

Last, the tyre load contribution can be found. In this case, the tyre load from each wheel is compared to the most loaded tyre in the shape of a fraction. This is carried out that way so the most loaded tyre will obtain a value of 1 in this part, meaning that the maximum value of torque will be transmitted. However, if the tyre is not the one with the greatest load, it will only receive a value of torque proportional to the tyre load it is being applied. The most loaded wheel is determined by a series of comparisons as displayed in Fig. 10.

In short, by using this expression, each of the engines in the hubs will request a different value of torque proportional to the maximum torque, the amount of throttle, and the grip capability. This is showcased in Fig. 11 as a Simulink block diagram.

The upcoming stage of the development accounts for a closed-loop configuration in which the yaw reference is taken into account. The error obtained when comparing the actual yaw acceleration of the car with the ideal yaw acceleration previously calculated and processed by the corresponding controller is added to the torque allocation formula. Depending on whether the wheel forms part of the right side or the left side, the controller output is the maximum torque or the remaining value. The expressions can be observed in Eq. 7.

Requested torque formula for closed loop (own elaboration)

$$\begin{gathered} \textit{Controller Output} \cdot \textit{Throttle pedal value} \cdot \frac{\textit{Tyre load}}{\textit{Mos loaded tyre}} \\ = \textit{Requested Torque Right Wheels} \\ (\textit{Maximum torque} - \textit{Controller Output}) \cdot \textit{Throttle pedal value} \cdot \frac{\textit{Tyre load}}{\textit{Most loaded tyre}} \\ = \textit{Requested Torque Left Wheels} \end{gathered} \tag{7}$$

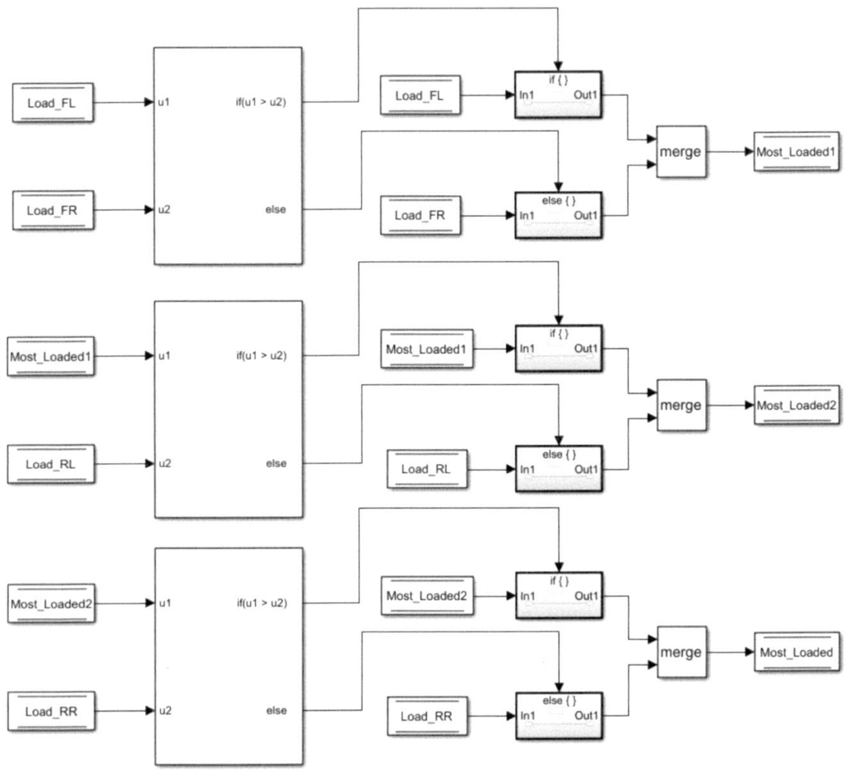

Fig. 10 Most loaded tyre identification (own elaboration)

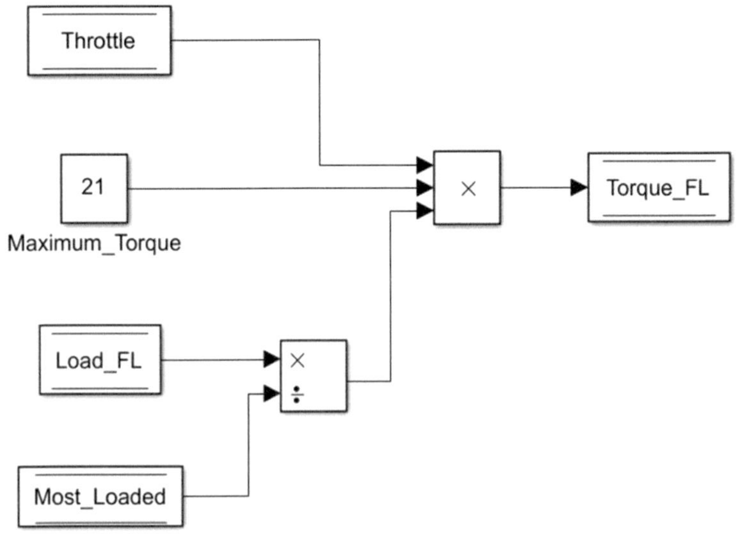

Fig. 11 Torque allocation for open loop (own elaboration)

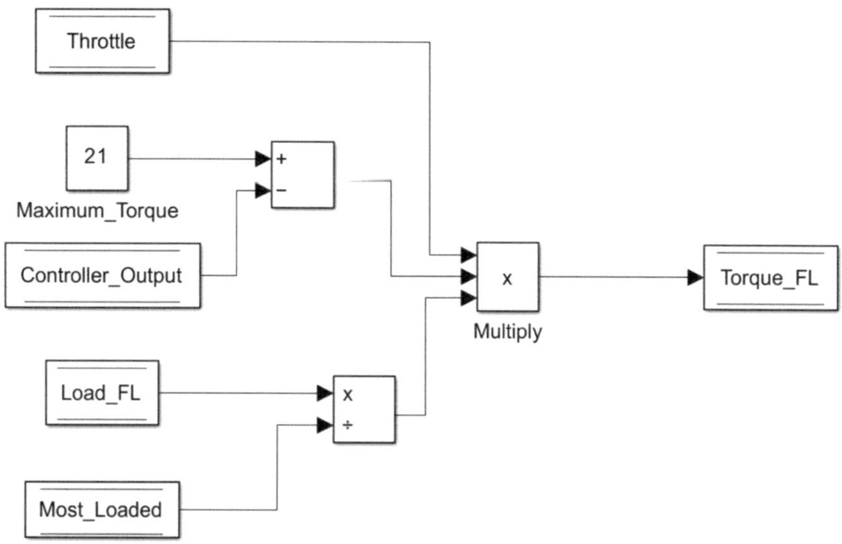

Fig. 12 Torque allocation for closed loop (own elaboration)

This entails a restructuring of the torque allocation block as depicted in Fig. 12.

3.5 *Yaw Acceleration Recalculation*

As one of the final components of the system, the estimation of the actual yaw acceleration is performed. In a real-world scenario, this data would be provided by an onboard sensor. However, due to limited testing opportunities, a dedicated block has been created to simulate this function. Equation 8 illustrates the formula that is being used for the block designed in Simulink.

Yaw acceleration equation [18]

$$\begin{aligned} I_z\ddot{\psi} &= I_f\left(\cos\delta\left(F_{yFL} + F_{yFR}\right) + \sin\delta(F_{xFL} + F_{xFR})\right) - I_r\left(F_{yRL} + F_{yRR}\right) \\ &\quad + 0.5t_f\left(\cos\delta(F_{xFL} + F_{xFR}) - \sin\delta\left(F_{yFL} + F_{yFR}\right)\right) \\ &\quad + 0.5t_r(F_{xRL} + F_{xRR}) - M_{zFL} - M_{zFR} - M_{zRL} - M_{zRR} \end{aligned} \tag{8}$$

The principle behind the block is to replicate the calculations of the yaw acceleration formula for any given car. This formula begins by considering the longitudinal and lateral forces exerted by the tyres at any given moment. Using trigonometry, the yaw moment around the centre of gravity is then determined. The relationship between the yaw moment and the vehicle's inertia around the z-axis yields the yaw acceleration.

Additionally, the complexity of the formula is increased by the need to estimate the forces exerted by the tyres. These forces are not fixed values; they depend on the type of tyre, the tyre model used, the load, and the slip ratio or angle. To address this, tyre information and curves have been replicated in Simulink as lookup tables. This data, obtained from prior testing and research within the team, uses slip ratio or angle and tyre load as inputs and produces longitudinal or lateral force as outputs. The resulting values are then fed into the yaw recalculation formula. Figures 34 and 35 in the Appendices illustrate the two blocks that perform this task.

3.6 Feedback Control

The next phase of development, briefly introduced in previous sections, focuses on designing and calibrating the control strategy. The primary objective is to minimize the error between the actual yaw rate and the ideal yaw rate calculated using a formula. There are various approaches to achieving this goal.

The first strategy to be tested is the PID controller. This involves not only implementing the PID block but also fine-tuning its coefficients for optimal performance. Additionally, a sliding mode controller and a fuzzy logic controller are incorporated. The results from these three feedback strategies are evaluated and compared to determine the most effective one for integration into the control system.

3.6.1 PID Controller Approach

The proportional-integral-derivative controller manages to minimize the difference between the reference signal and the real signal by producing an output, which is the sum of the contributions from the P, I, and D components. The proportional part produces an output that is proportional to the current error. The integral component sums the error over time, considering the accumulated error to reduce any residual error that the proportional part leaves behind. The derivative component reacts to the rate of change of the error, providing a prediction of future errors based on current trends [1].

Having the basic concepts clear, the implementation of this controller is straightforward thanks to Simulink. The controller output is generated as illustrated in Fig. 13, where the difference between the signals is the input of the controller, and the output is the result of the coefficients and the calculations. The output is strategically added to the torque allocation formula to obtain the desired behaviour as seen in Fig. 12.

Precise tuning of PID parameters typically requires a deep understanding of the plant, the governing equations, and control engineering theory. Knowledge of the poles and zeros of the system allows for a more informed and theoretically grounded tuning approach, often leading to better or quicker results. However, it is also possible to manually tune a PID controller effectively by observing and adjusting the behaviour, without explicitly calculating poles and zeros.

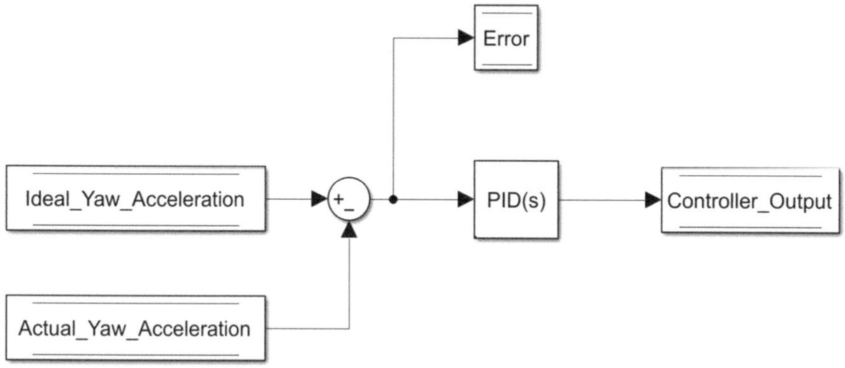

Fig. 13 PID controller implementation (own elaboration)

To simplify the tuning process, Simulink includes the Control Design toolbox, which uses advanced algorithms for the automatic identification of optimal PID coefficients. The toolbox also provides sliders to adjust the desired aggressiveness and robustness of the response, giving users intuitive control over system performance. The PID tuner interface is shown in Fig. 36 in the Appendices.

3.6.2 Sliding Mode Controller Approach

The second controller tested is the sliding mode controller, which is available as a block within the Simulink environment, similar to the PID controller. Its implementation involves two inputs, one for the reference value and another for the actual value, together with the calculated output, as it is shown in Fig. 37 in the Appendices. This output is fed into the torque allocation block to adjust the torque applied to the wheels minimizing the error in yaw acceleration.

Sliding mode controllers work by guiding a system to a predefined sliding surface that represents the desired behaviour of the system. The sliding surface is carefully designed based on the desired dynamics, and the control strategy focuses on driving the system state onto this surface. Once the state reaches the sliding surface, the system exhibits the intended behaviour, effectively reducing the impact of uncertainties or disturbances.

The control law in sliding mode controller (SMC) has two components: equivalent control and switching control [30]. Equivalent control maintains the state on the sliding surface, ensuring the system follows the desired dynamics. Switching control, which is typically discontinuous, forces the state toward the sliding surface when it deviates. This dual approach provides high robustness, though it can introduce 'chattering', a high-frequency oscillation around the sliding surface, which may require additional techniques to be mitigated.

Unlike the PID controller and its tuner, Simulink offers limited options for modifying the sliding mode controller, which restricts the potential for improvements.

Enhancements would require creating additional blocks to supplement the controller, rather than adjusting directly.

3.6.3 Fuzzy Logic Controller Approach

The last control strategy to be studied is fuzzy logic. It is a form of logic that allows for reasoning with imprecise or uncertain information, mimicking human decision-making [2]. Unlike traditional binary logic, where variables are either true or false, fuzzy logic deals with degrees of truth, enabling a more flexible approach to problem-solving.

Such systems have a fuzzy logic controller that, just like other controllers, receives a series of signals and produces a corrected output. The set-up process begins with fuzzification, where crisp input values are converted into fuzzy sets using membership functions that represent degrees of truth [28]. The next step involves creating a rule base consisting of if–then rules that define the relationship between input fuzzy sets and output fuzzy sets, based on expert knowledge or empirical data. During inference, these rules are applied to the fuzzy input sets to determine the corresponding fuzzy output sets. The process concludes with defuzzification, which translates the fuzzy output sets back into a crisp value, producing the final output of the controller.

In this particular case study, the only input of the system is the error between the ideal yaw acceleration value and the actual yaw acceleration value. Taking this into consideration, five triangular membership functions have been defined to encompass an error range between –10 and 10: the PH membership function, which stands for positive high and represents the higher part of the positive values. The PL which stands for positive low and represents the lower part of the positive values. The closest one to the zero-error value is the membership function named Z for zero. The next one is the NL which stands for negative low and represents the low values of the negative side of the spectrum. And last, there is the NH representing the high negative values. The visual representation of this relationships is illustrated in Fig. 14.

The same process is carried out to determine what the output of the block will be. In this case, the output is the value of torque that is added and subtracted from the different wheels, but it is known that it cannot be a value lower than 0 Nm or greater than 21 Nm. For that reason, the five membership functions are spread out along these values as can be seen in Fig. 15: XL for very low torque, L for low torque values, M for middle values, H for high values, and XH for the highest values of torque.

To create an adequate relationship between the error values and the torque generated at each of the wheels, a set of rules is introduced. These rules depend directly on how the system is configured and how the left or right-side wheels affect the yaw acceleration value. By analysing the behaviour of the system, it is noticeable that a negative error means that the car is not rotating to the left as much as it is desired, or it is rotating too much to the right. Therefore, the right-side wheels experience an increase in torque to regulate that. Conversely, when the error is very high on the positive side, it means it is rotating too much to the left or it is resisting turning right.

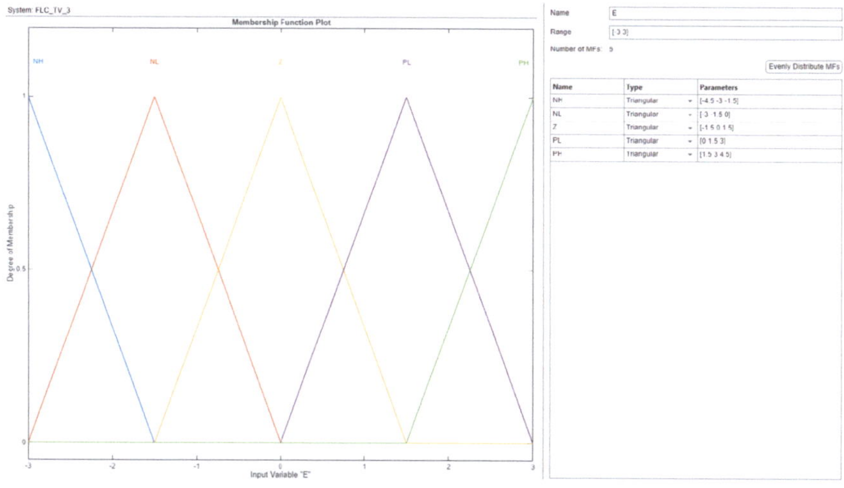

Fig. 14 Error membership functions (own elaboration)

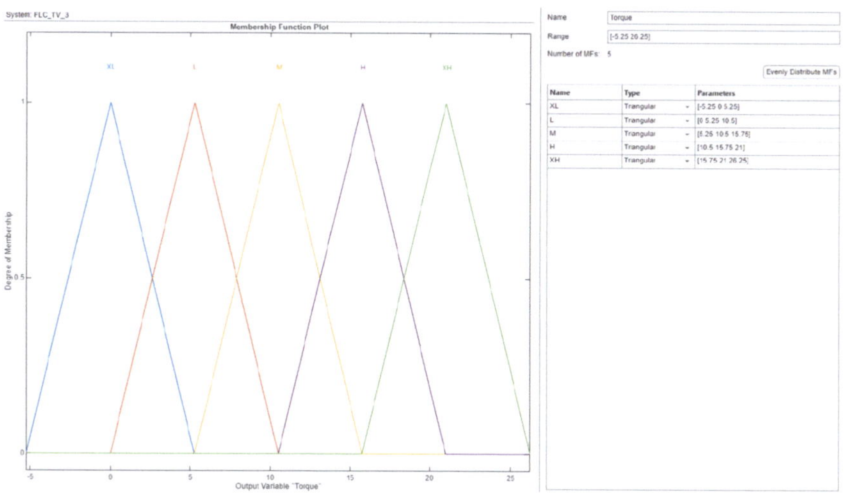

Fig. 15 Torque membership functions (own elaboration)

In this case, the torque at the left-hand wheels is increased. This basic behaviour is inserted in the system with the set of tools presented in Fig. 38 in the Appendices.

Last, the fuzzy logic designer interface allows for the visualization of the system inference and the control surface. These display the response of the system to the different expected inputs to spot any misbehaviour prior to implementing the controller in the final system. In this project, the pursued result is a linear proportion between the error and the demanded torque, as illustrated in Fig. 39 in the Appendices.

The final implementation of the fuzzy logic block into the feedback control section can be observed in Fig. 40 in the Appendices.

3.7 Failsafe Block

The objective of this final block is to establish a cyclic monitoring process that continuously assesses the state of the system. In the event of any detected issue, the system undergoes an automatic shutdown, effectively pre-empting further drivability concerns or potential complications.

The code is designed to establish distinct conditions that yield an output of one when the system operates correctly and zero when it encounters a failure. Multiplying this resultant value by the total gain or torque serves as the determinant for the automatic operation or shutdown.

3.7.1 Type of Errors

The failsafe block is designed to detect three types of errors or misbehaviours during system operation.

The first category involves sensor failures or out-of-range values, encompassing any malfunction of the system's sensors, such as those monitoring steering angle, linear velocity, and throttle position. If a sensor signal is lost or an unexpected value is received, the block immediately shuts down the system to prevent further issues.

The second category addresses contradictory sensor readings. The block cross-validates data from multiple sensors to identify conflicting information, such as a steering input that contradicts the yaw acceleration or a situation where longitudinal acceleration is positive while the brakes are engaged. These inconsistencies indicate a malfunction in the vehicle operation; for that reason, the best solution is to report the issue and disengage torque vectoring to prevent interference with vehicle handling.

The third category focuses on dynamic misbehaviours, which are anomalies in the physical behaviour rather than logical inconsistencies. Examples include detecting wheel slippage or a yaw rate that does not align with the direction of the wheels. These issues suggest that the vehicle is not adhering to its intended dynamics, such as the Ackermann steering geometry. In such cases, the system deactivates torque vectoring to avoid further complications and allows for an assessment of the underlying cause.

3.7.2 Block Description

Three levels within the failsafe block have been created to organize the types of errors that can cause the torque vectoring to stop running. The upper level classifies the different groups of errors that have been found relevant and showcases the final

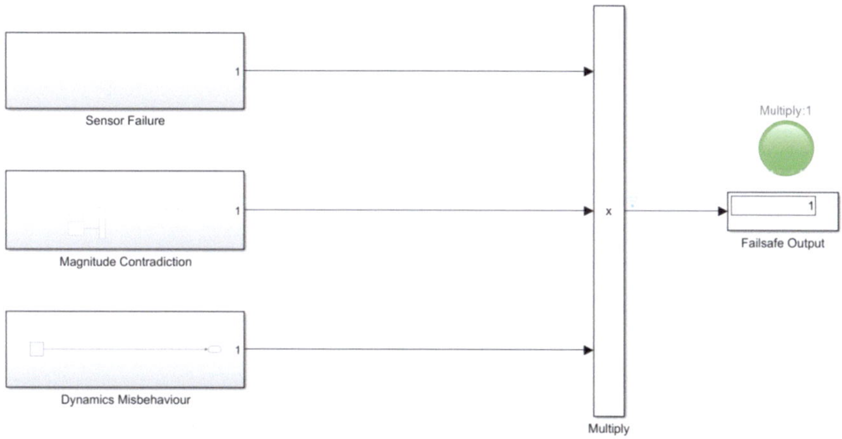

Fig. 16 Failsafe block upper level (own elaboration)

output as presented in Fig. 16. The final output is also represented with a green or red light for better visualization during development.

The middle level differentiates the failures inside the same group, as illustrated in Fig. 17.

Finally, the lower level contains the conditions that detect the errors and make the output 0 or 1 accordingly. In this part, the real values from the sensors are considered, and they are compared to specific values in an if structure, as shown in Fig. 18.

Ultimately, the idea is that all these outputs are multiplied, meaning that even if only one of them is true, representing that there is one failure, the final output will be zero, and the torque vectoring will not affect the behaviour of the car.

3.7.3 Further Concerns

During the development of the failsafe block, two additional concerns emerged: how to communicate the status of torque vectoring to the driver and how aggressively the system should shut down in the event of a fault.

To address driver communication, the solution involves adding an indicator on the dashboard that shows the current status of torque vectoring, along with a warning message if the system becomes ineffective. This ensures that the driver is aware of the level of assistance available, particularly when approaching a turn, helping to prevent overconfidence and reduce the risk of an accident.

As for the shutdown response, various approaches were considered, ranging from a gradual deactivation to an immediate forced stop. Ultimately, the decision was made to implement an instant and complete shutdown. This choice was guided by the need to handle the most dangerous scenarios, such as a sudden spin on a high-speed straight caused by torque vectoring. In such cases, the quickest possible shutdown is essential to prevent a far more serious accident.

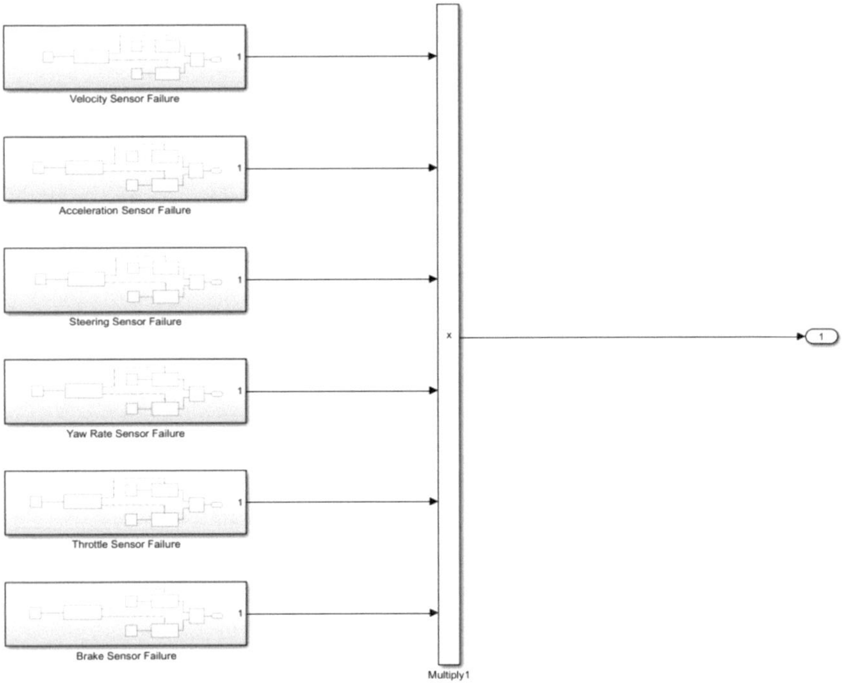

Fig. 17 Failsafe block middle level (own elaboration)

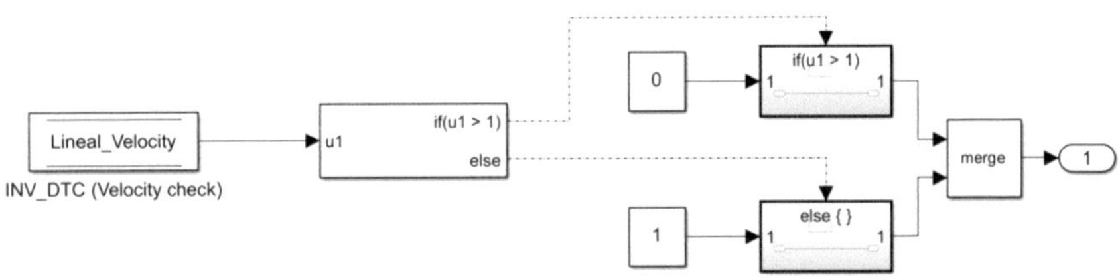

Fig. 18 Failsafe block lower level (own elaboration)

4 Results

After developing the system and implementing various control techniques, it is essential to validate the performance and effectiveness of each case study. With all the information gathered, it is then possible to go on and discuss the differences between each of them in order to find the most suitable solution.

4.1 Results Obtained

In the following section, the four strategies are evaluated and fine-tuned to achieve the best possible outcomes within the constraints of the available tools and time. A panel such as the one presented in Fig. 41 in the Appendices has been prepared in Simulink to give access to any meaningful piece of data in the form of graphs.

4.1.1 Feedforward Model Results

The first configuration to be tested is the feedforward model. It refers to the simulation with an open-loop control with no reading of the final output. Although they are also applicable to real jobs, these draft configurations are firstly tested to ensure that the system is following the desired behaviour. Confirming this allows for a much easier definition of the controller afterwards.

Once the model is simulated, the first information to be reviewed is the graph with all four tyre loads over time. The results appear reasonable, as each tyre supports a different amount of load. In sections influenced only by longitudinal load distribution, the front and rear tyres form pairs, showing symmetry. In contrast, during sections with lateral acceleration, there are greater differences between the tyre loads, although a certain degree of symmetry is always maintained. This behaviour is expected, as load transfer occurs between tyres, meaning that an increase in load on one tyre corresponds to a decrease on another. This can be observed in more detail in the graph displayed in Fig. 19. Simulation time has been shortened to have more precise data and point out interesting aspects.

Another great indicator of the good functioning of the model is the torque value at each of the wheels, which allows for the verification of the torque allocation formula. The first 10 s of the simulation can be observed in Fig. 20. On one hand, it is evident that the torque limit incorporated into the formula is functioning correctly, as none of the curves exceeds this value under any circumstances. Additionally, the maximum torque values coincide with the throttle position, differentiated as a green line on top, being at its peak. This confirms that the torque values are always proportional to the throttle input given by the driver. On the other hand, the behaviour observed in the tyre load results is also reflected here, with each tyre having a different torque value depending on the driving scenario and always proportional to the tyre experiencing the greatest load.

However, although the previous requisites were met satisfactorily, it is observed that there is no approximation obtained between the desired yaw acceleration and the actual one. In fact, any similarity seen in Fig. 21 is due to a coincidence between the ideal torque and the tyre loads, which leads to a close value of yaw acceleration. An accommodation of the curves is what is sought when implementing a controller that feeds back the output of the system.

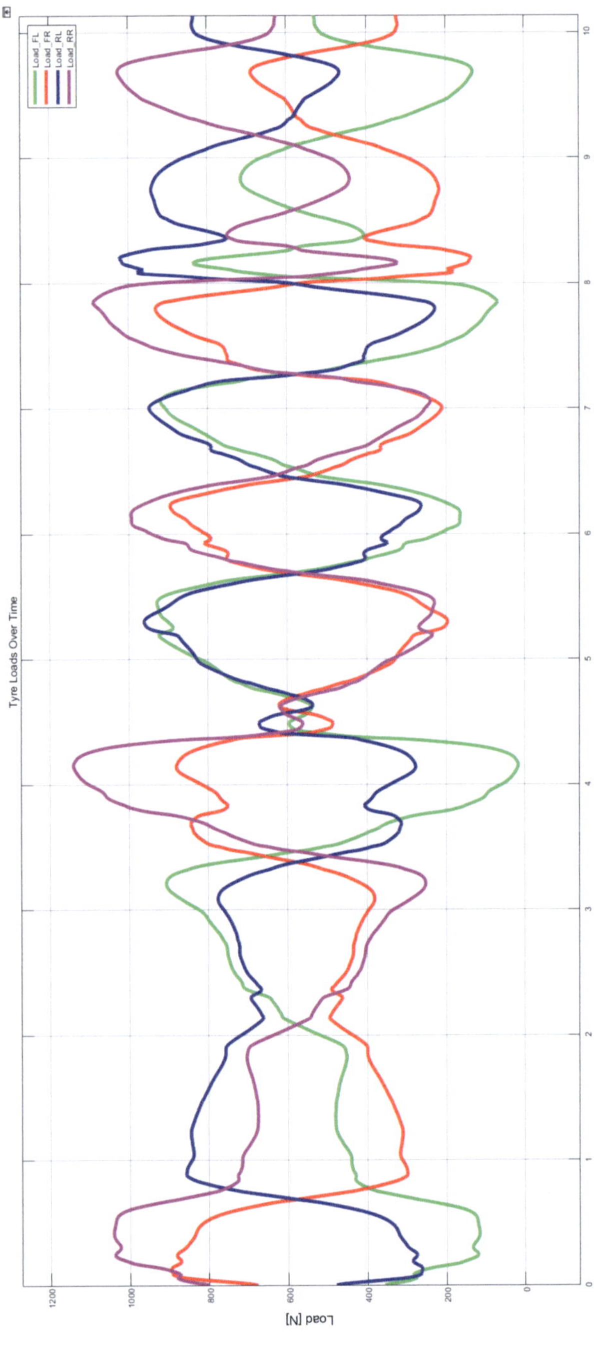

Fig. 19 Tyre loads for an open-loop configuration (own elaboration)

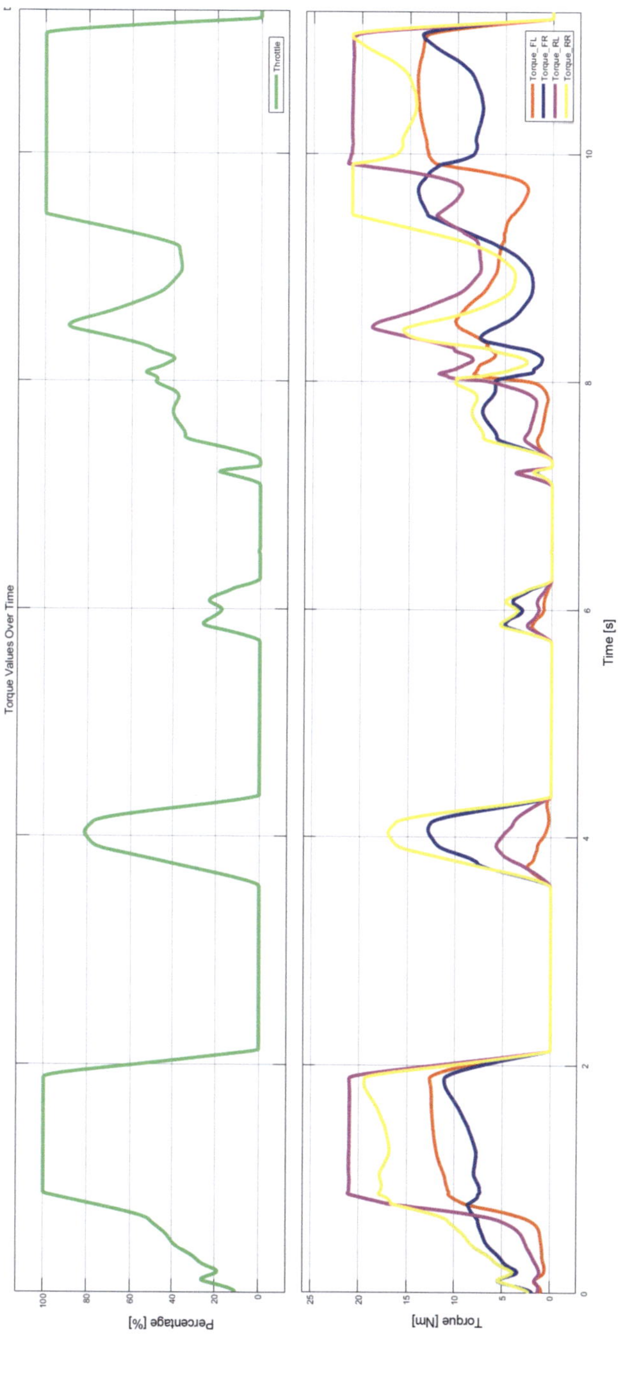

Fig. 20 Torque values for an open-loop configuration (own elaboration)

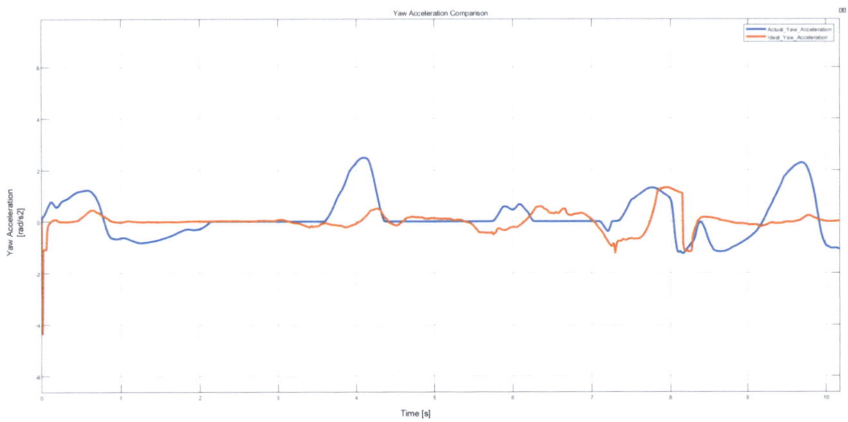

Fig. 21 Yaw acceleration for an open-loop configuration (own elaboration)

4.1.2 PID Controller Results

The first controller to be tested is the proportional derivative controller. This encompasses the assessment of the new torque values and the effectiveness in reaching the ideal yaw acceleration, which considers response time, aggressiveness, and the remaining error. The tyre loads are no longer taken into account since they are not expected to have a significant change.

Having said that, a PID does not provide a singular outcome, but it gives different results depending on how the controller is tuned. Aggressiveness and robustness are balanced to obtain the best result possible, and the process to get there is explained step by step.

The tuning process starts with the automatic coefficients given by the toolbox, which provide the results of torque and yaw acceleration showcased in Fig. 22. These represent a baseline of 2 s of response time, which is right in the middle between a fast and a slow response, and 0.6 of transient behaviour, which slightly tends to be robust.

By analysing the two graphs, two main conclusions can be drawn. First, the torque values appear accurate, as they remain within the expected ranges and adhere to the programmed rules. Additionally, the disparity in torque between the sides is evident in the yaw acceleration graph. However, the relationship between the ideal and actual yaw acceleration is unclear. Despite attempts to adjust, the proportionality and response speed are far from meeting expectations. For that reason, the model is tested again with the lowest response time of 0.2 s.

As illustrated in Fig. 23, with a much faster response, the yaw acceleration follows closely the reference value. Nonetheless, the result is not ideal. There is still a certain delay perceptible, and in some cases, where the frequency at which the yaw acceleration varies is too high, the controller ignores it and completes the curve with a straight line. To see if these traits could be refined, the controller has been tested

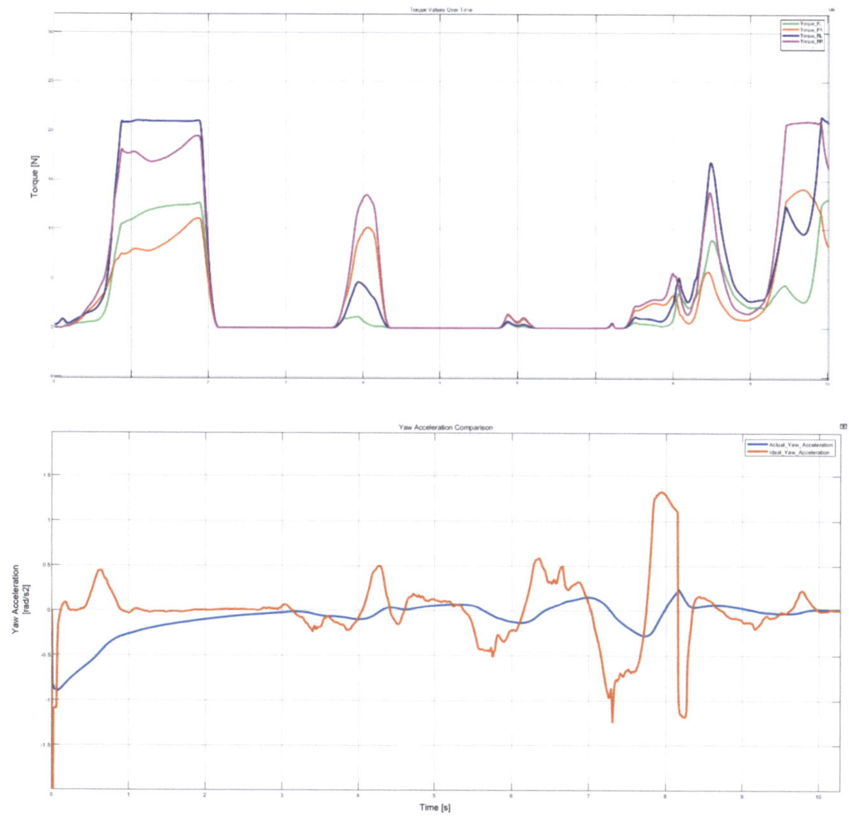

Fig. 22 Results for an automatically tuned PID (own elaboration)

with different values of transient behaviour, more precisely, a robust value of 2 and an aggressiveness value of 0.

After analysing the outputs of both experiments, represented in Figs. 24 and 25, no variation was observed in the yaw acceleration graph. Independently of the transient behaviour, the system keeps performing the same way, and the error can not be reduced. The best result available using a PID is achieved when the time response is reduced to the minimum value of 0.2 s.

4.1.3 Sliding Mode Controller Results

The second controller that has been fitted into the model is the sliding mode controller. The parameters to play with in this type of controller are the hysteresis band, the control action upper limit, and the control action lower limit. The hysteresis band is the range of the error signal where the inverter output current is controlled. The control limits, however, are the maximum and minimum values available for the

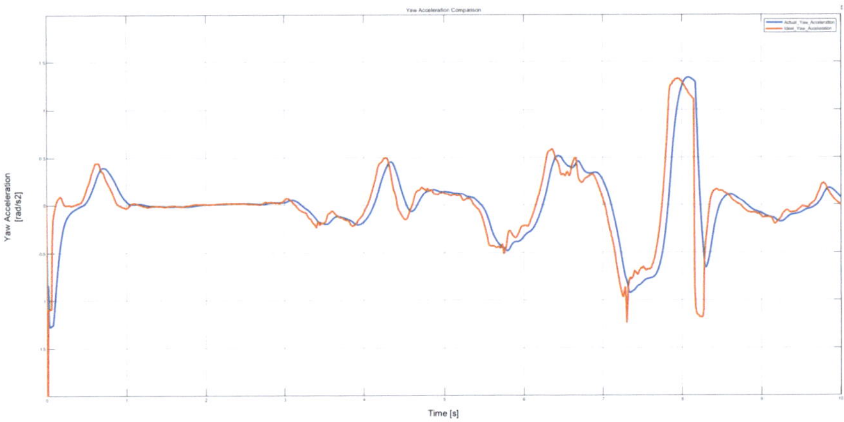

Fig. 23 Yaw acceleration for a fast PID (own elaboration)

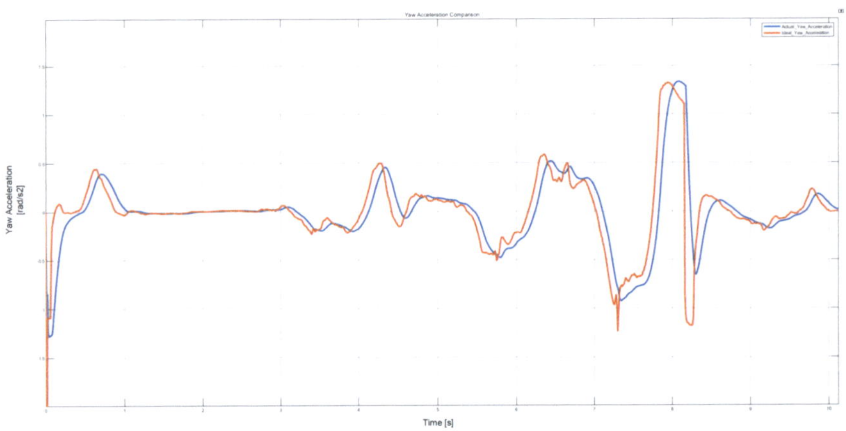

Fig. 24 Yaw acceleration for a fast and robust PID (own elaboration)

output. Since it is known that the torque values must remain between 0 and 21 Nm, this is constant throughout all the different tries. Different values of the hysteresis band are tested and evaluated. The Simulink interface to control all these parameters is presented in Fig. 42 in the Appendices.

When analysing the results for the sliding mode controller, a new phenomenon arose; this effect is called chattering. In sliding mode controllers, chattering occurs because the control signal switches rapidly to keep the system on track, but real-world limitations like actuator delays and imperfect switching make it hard to do this smoothly [31]. Instead of staying right on the target, the system ends up oscillating back and forth. This gets worse when the controller is tuned to handle uncertainties or disturbances, causing the control to switch even more aggressively, which leads to noticeable chattering.

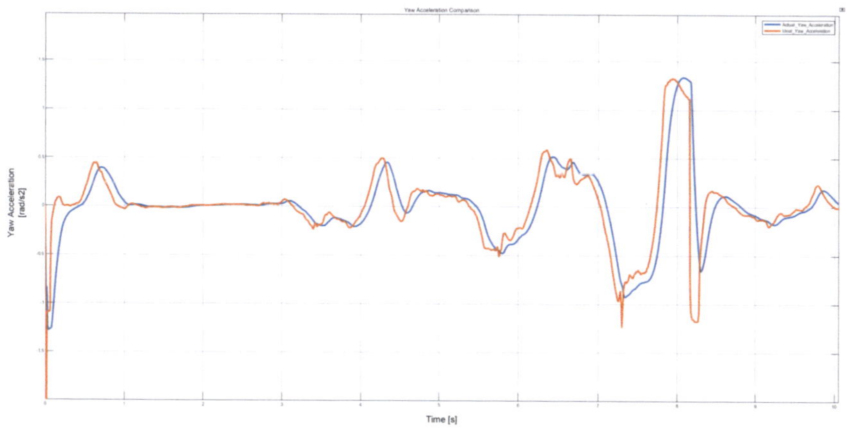

Fig. 25 Yaw acceleration for a fast and aggressive PID (own elaboration)

The behaviour of the controller varies significantly with different hysteresis band values, particularly in terms of chattering. Figure 26 illustrates the results for a controller with a very narrow hysteresis band of 0.1. During the first 10 s of the simulation, chattering is pervasive, except for the brief moment when the reference value turns negative and remains there for a few milliseconds.

This phenomenon affects the torque, which is depicted in the graph as fully coloured shapes, indicating the high frequency of oscillations. Despite the chattering, the values still adhere to the load and throttle criteria. In the real acceleration plot, the chattering hides the actual values. Notably, in the brief instance where chattering comes to a rest, the yaw acceleration does not match the desired value. Additionally, a graph combining the error value and the torque output sent by the controller has been created to visualize the relationship between the two. However, due to the overwhelming chattering, this relationship is barely appreciable.

Figure 27 shows the results for a simulation with an hysteresis band of 50. The results obtained in this run are a bit different to the previous test. This time, there are two very differentiated areas. During the first 7.3 s, the system is outputting the maximum torque value, meaning that only two of the wheels are requesting torque, as seen in the torque graph. When the error goes lower than -2, the output transitions to a chattering scenario as the one described in the previous simulation. The yaw acceleration graph shows a similar behaviour, with a first part not following the ideal value whatsoever and a second part with massive chattering.

The last trial for the sliding mode controller was made with an hysteresis band value of 100 to check what was occurring as this value was incremented. Nothing extremely surprising occurred as the system continued to behave similarly. In this case, the chattering stopped except for those moments where there was no chattering in the first place. The outcome was completely the opposite of the first simulation. Instead of having oscillatory values of torque in all four wheels, it ended up sending

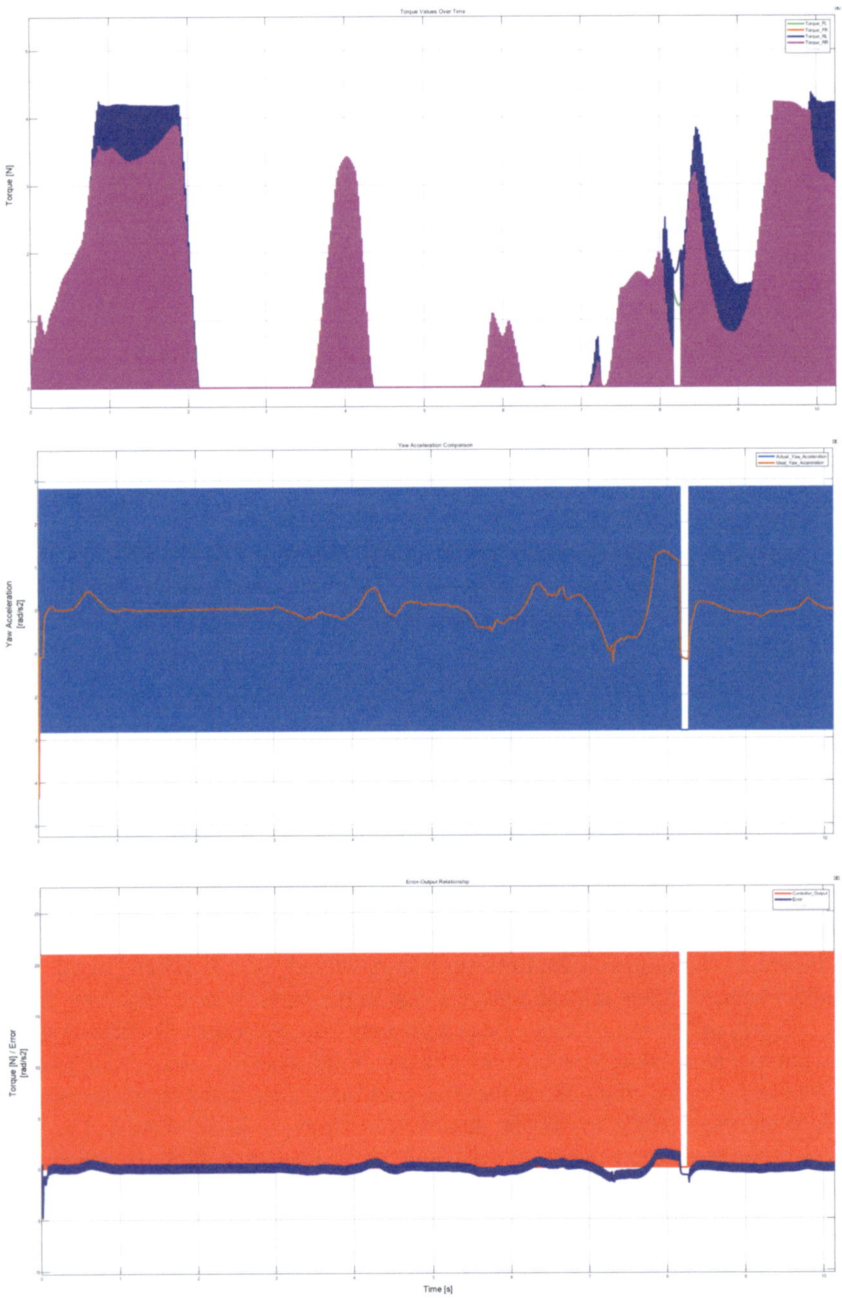

Fig. 26 Results for an SMC with an hysteresis of 0.1 (own elaboration)

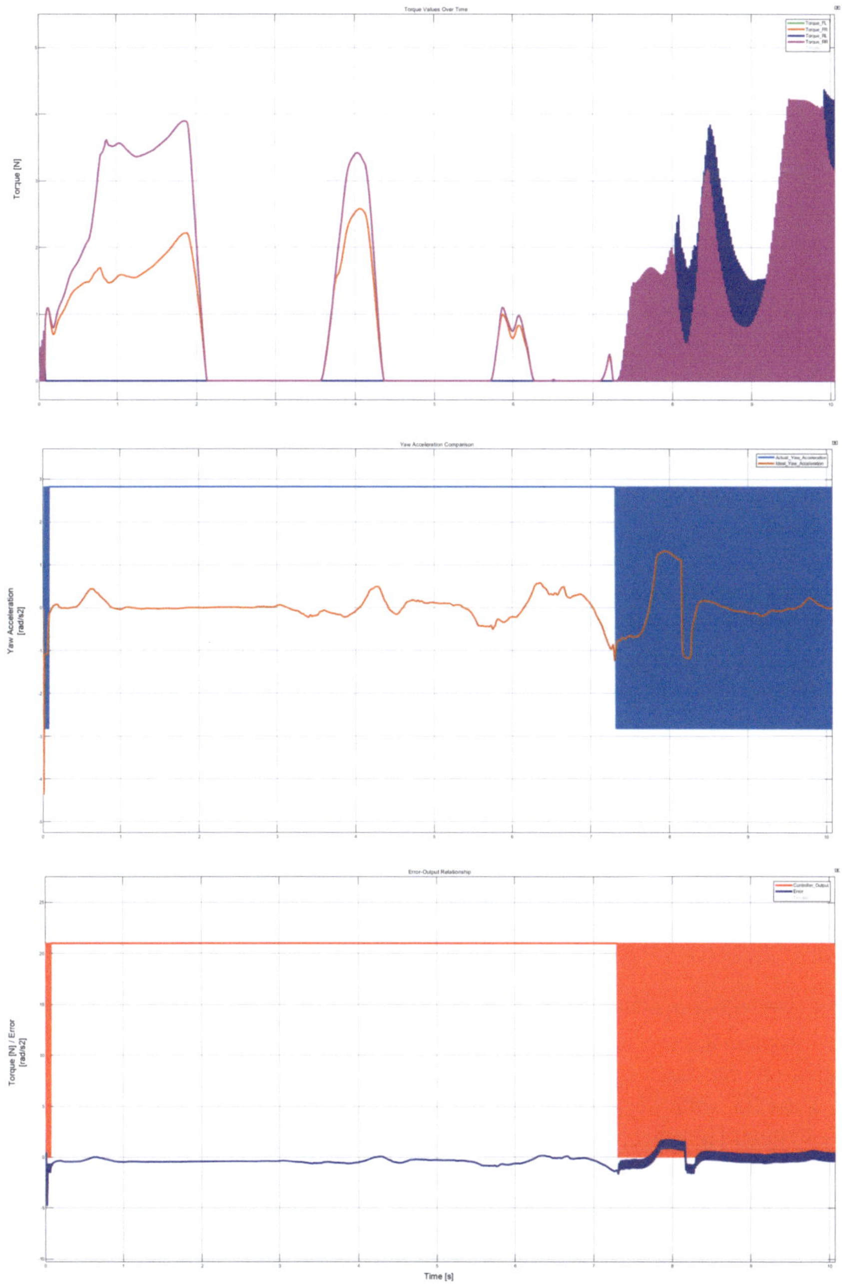

Fig. 27 Results for an SMC with an hysteresis of 50 (own elaboration)

torque only to one side of the car, which can be translated as a continuous spin. The results are presented in Fig. 28.

4.1.4 Fuzzy Logic Controller Results

Finally, the system is tested with a fuzzy logic controller. In an initial iteration, the controller is programmed with an error range going from –10 to 10. The results shown in Fig. 29 evidence that something is odd with the relationship between the error and the output. The performance seems to be adequate, but the torque and, therefore, the yaw adjustment are only triggered once the error goes above a certain level.

To solve the previous issue, the range of values has been evaluated and adjusted to ± 3. This way, the controller is working properly throughout the entire spectrum and not just when the error is too high. In Fig. 30, the results for the new test are presented, showing a much better performance.

Starting with the torque allocation, all four signals are consistent. There are different values for each corner depending on how the load is distributed, and the magnitudes fall between the specified range. The yaw acceleration graph shows impressive results, indicating a really close relationship between the ideal value and the actual value generated by the four torques in the wheels. There is almost constant overlap between the signals, meaning that there is no delay. As a remarking part, when the ideal yaw acceleration is very high, the vehicle falls short at around 80% of the peak values. As far as the error-output graph, there is a sensible relationship between the two signals, and it is easy to see how the torque is being accommodated depending on how much error is being developed.

4.2 Discussion

The performance of the control system and the different strategies were evaluated through a series of experiments that focused on ease of implementation, error, response time, and torque allocation.

The PID controller was the first one to be implemented since it was the easiest one to tune. The tuner interface allowed for an automatic calibration of the controller coefficients. After adjusting the transient behaviour and minimizing the response time, the best result was obtained. Anyways, this could not get rid of a permanent error due to a constant delay. This matches with what was commented by Medina et al. (2021) in previous research that 'Proportional Integral Derivative controllers are robust and easily implemented but lack complete knowledge of the vehicle and therefore are less precise'.

In the second place, the PID was replaced by a sliding mode controller to try improving the performance previously obtained. This step generated massive chattering at low values of the hysteresis band across all the main measurements: error,

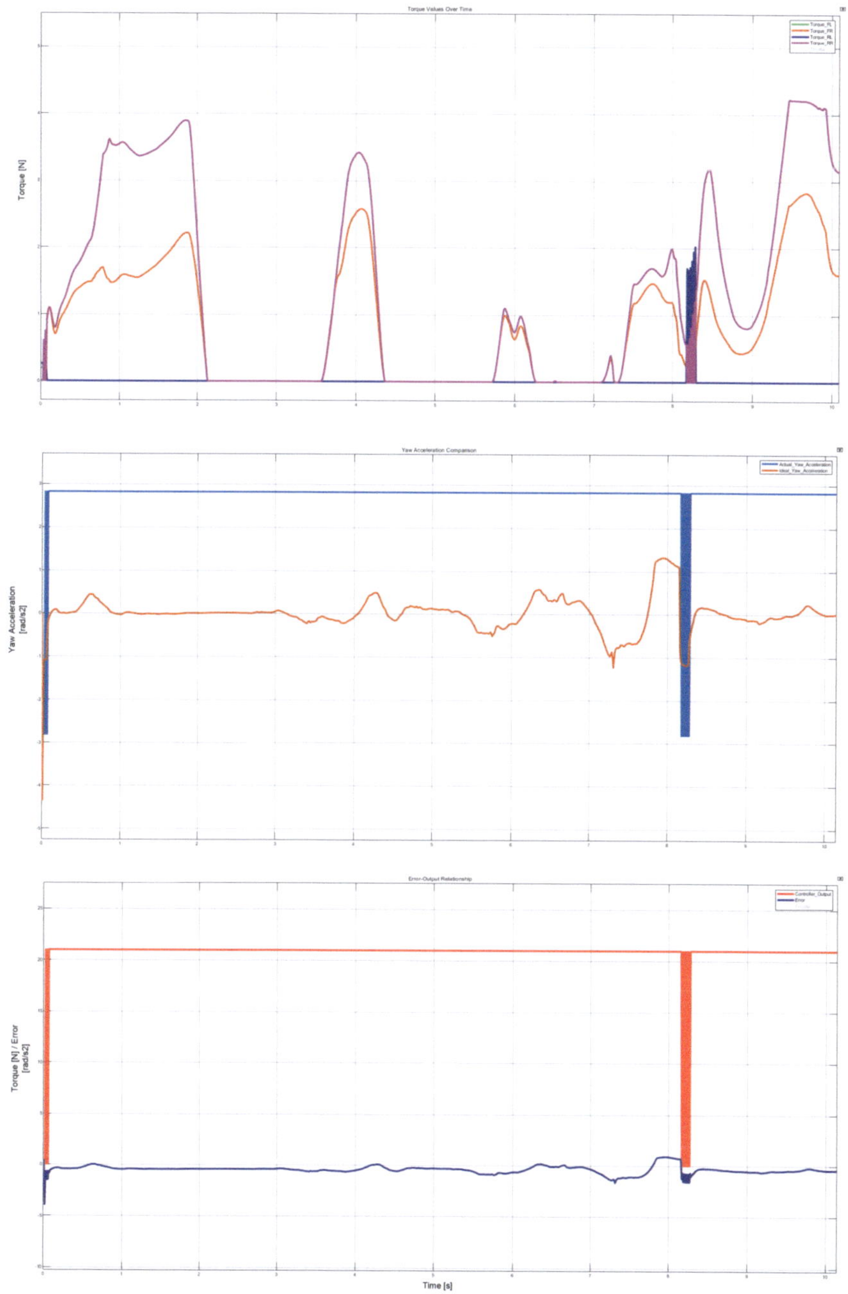

Fig. 28 Results for an SMC with an hysteresis of 100 (own elaboration)

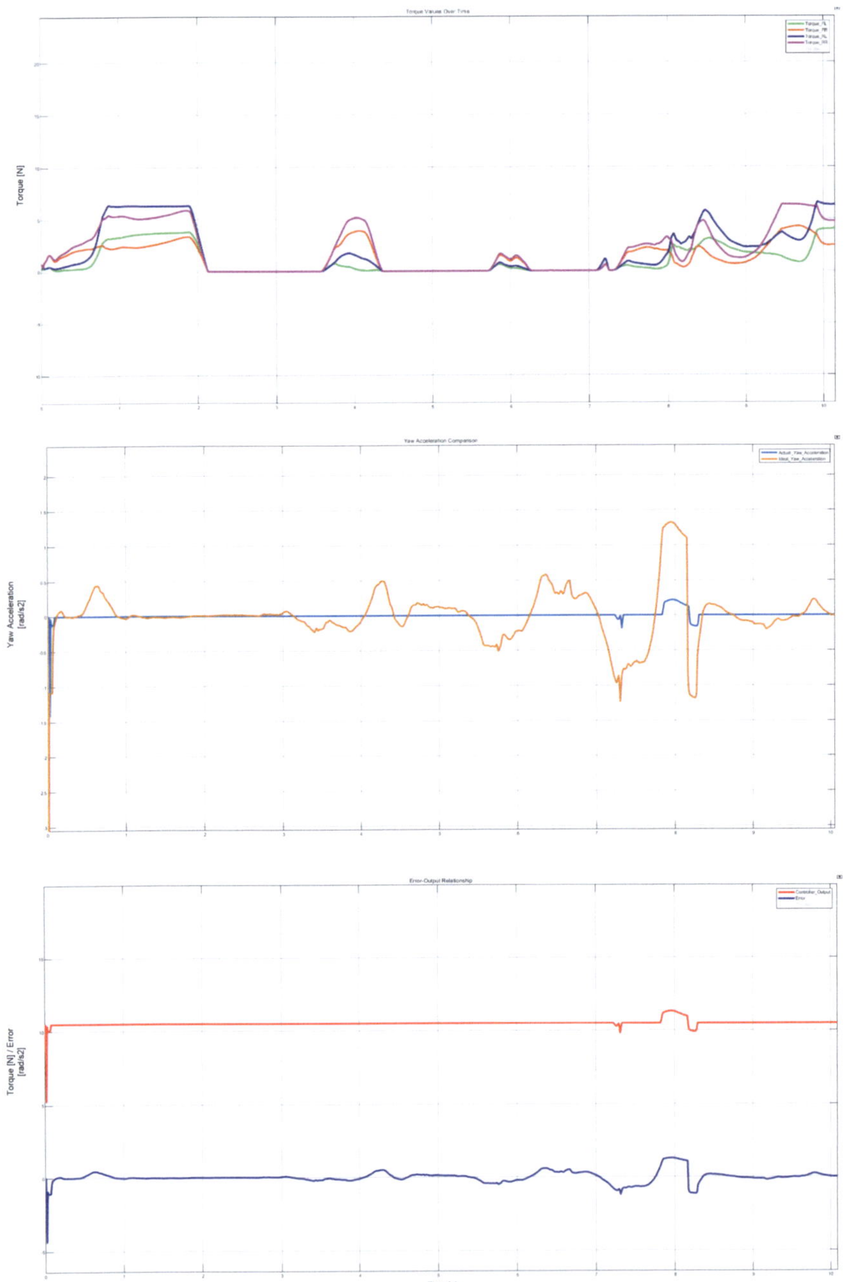

Fig. 29 Results for an FLC with an error range of ±10 (own elaboration)

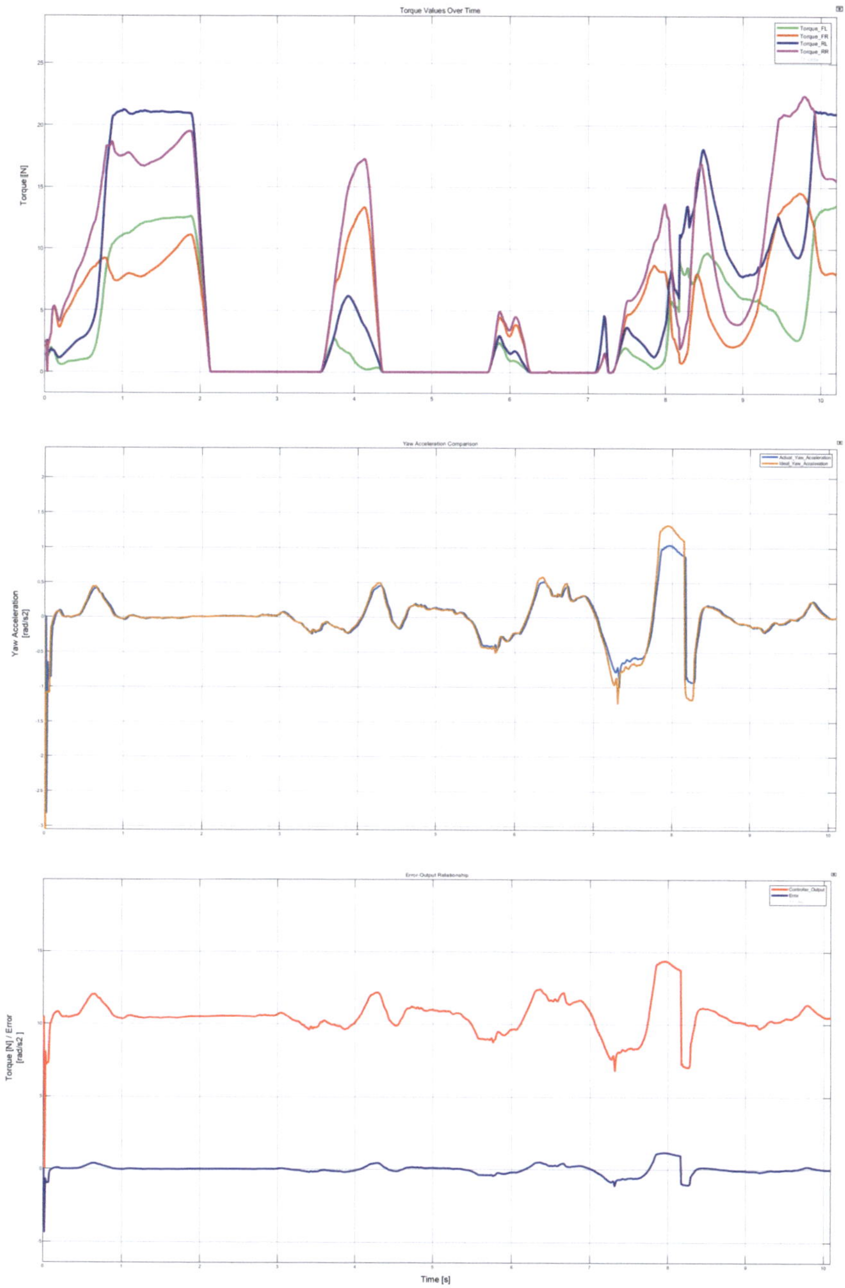

Fig. 30 Results for an FLC with an error range of ±3 (own elaboration)

torque, and yaw acceleration. At high values of the spectrum, the response was completely the opposite; no error value generated a sensible result in terms of torque and yaw acceleration. This piece of information is supported by results described in previous studies, where Palma et al. (2021) stated that 'the sliding mode controller saturated the actuators with an excess of information'.

In last place, it was given the chance for a fuzzy logic controller to take over the system. As Rastelli and Peñas (2015) confirmed, 'despite requiring an in-depth knowledge of the process under study, fuzzy logic controllers have also been applied to Direct Yaw Control successfully'. Out of the three integrated controllers, the fuzzy logic is the one that required the most development and a true meaning behind the parameters that were being inserted in the design. With careful adjustment, however, it showed the best result in terms of error magnitude and its proximity to the ideal value of yaw.

After evaluating all three cases and their corresponding results, a final decision must be made regarding the best strategy to implement in the car. Given the outstanding performance it delivered, the fuzzy logic controller is the top choice for integration into the torque vectoring model and, ultimately, the Oxford Brookes Racing car. The fuzzy logic controller not only produced the best results but also incorporates real physics in its configuration, making it more intuitive to understand. Additionally, its design allows for easy adjustments; the controller can be modified by opening the corresponding fuzzy logic designer file. As long as the file is saved under the same name that the model references, no further adjustments are required, minimizing the risk of unintended changes.

5 Conclusion

At the beginning of this project, four fundamental objectives were defined to establish a clear direction and align with the team's expectations for this study.

In the initial stages, the behaviour and dynamics of the FSAE vehicle were studied in detail to develop an accurate model in Simulink that closely represented the real car. This objective was successfully achieved and validated through simulation results.

The next key objective was to identify a suitable control technique for managing the pre-calculated torque allocation. This task was divided into three distinct designs, each focusing on the implementation of a different controller: a proportional-integral-derivative (PID) controller, a sliding mode controller, and a fuzzy logic controller. After carefully tuning each controller and assessing its performance through simulations, the fuzzy logic controller was found to be the most effective. It demonstrated the closest alignment with the target yaw acceleration, which was the main goal.

The third objective addressed the risk factor of integrating such a system into the vehicle, with a focus on ensuring safety during operation. A failsafe block was incorporated into the design to protect the driver and others around the car in the event of system failure. If any component of the car prevents the system from functioning properly, the system will automatically switch off, giving full control back to the driver. Additionally, various strategies were proposed to alert both the driver and the team if the system is malfunctioning.

Finally, the primary aim of this project was to lay the groundwork for a system that could be expanded with further complexities and to provide a learning platform for future members of Oxford Brookes Racing. To facilitate this, the report includes a step by step description of the process, alongside all relevant files from the development and simulation phases. Communication with the Vehicle Controls department was also maintained throughout to ensure consistency and collaboration.

In conclusion, this thesis outlines the conceptualization, development, and testing of a control system that meets the original objectives. However, it is important to note that no single control technique is universally the best fit for all systems. The ideal solution depends on factors such as the project architecture, available resources, tools, expertise, and time constraints. Balancing these factors is essential in determining the most suitable approach.

6 Future Work

Once the work is completed, new opportunities for further research and improvement emerge. The first potential enhancement, which could add significant value to the research, is to conduct a physical test of the system during a shakedown or an event. This would provide more accurate coefficient values, allowing the system to be better tailored to the specific track and driver, resulting in more meaningful data. Additionally, a more in-depth study could involve comparing the car's performance with the system active versus when it is disabled. This comparison could yield valuable insights into dynamic improvements and lap time performance.

As for the vehicle model formulation, it can also be expanded in many different directions. On the one side, the equations used to estimate the load in each of the four corners can be made more complex and accurate to account for suspension and aerodynamic effects. These are two things that have been disregarded in the first model, but depending on the vehicle and the speeds that are reached, they can be very relevant. On the other side, the brief tyre model that has been developed to calculate tyre forces and their impact on vehicle dynamics can be upgraded. For instance, it could consider variables, such as ambient and tyre temperature, wear, or compound type.

Similarly, the torque allocation strategy can also be investigated more in depth. So far, the strategy is only active as long as there is a throttle demand by the driver. However, if the system were also configured to work with the brake pedal, there would be an advantage at corner entry as well, giving the team a really good competitive edge.

Last, to elevate the system to the next level, artificial neural networks (ANNs) can be introduced. They can replace a vehicle model designed in Simulink by learning to approximate the behaviour of the original system through training on input–output data generated by the Simulink model. Once trained, the ANN can replicate the complex, nonlinear relationships inherent in vehicle dynamics without the need for detailed physical modelling or computationally expensive simulations. This allows for faster real-time predictions and control decisions, making the network a powerful alternative for certain tasks.

7 Appendices

Load Calculation

See Figs. 31 and 32.

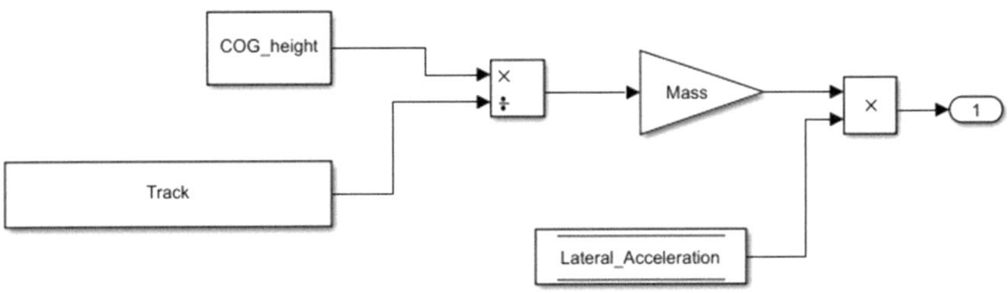

Fig. 31 Lateral load transfer block (own elaboration)

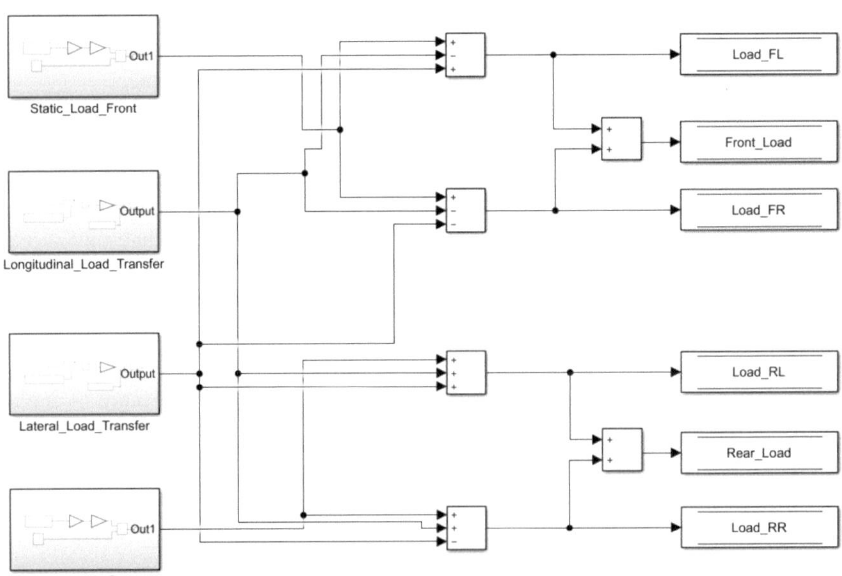

Fig. 32 Tyre load calculation block (own elaboration)

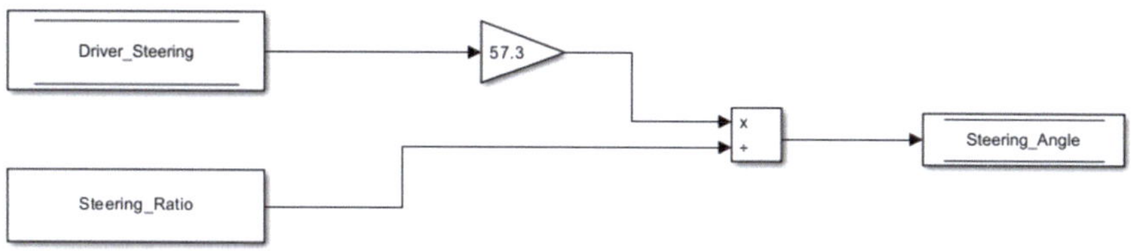

Fig. 33 Steering ratio representation (own elaboration)

Steering Ratio

See Fig. 33.

Tyre Force Calculation

See Figs. 34 and 35.

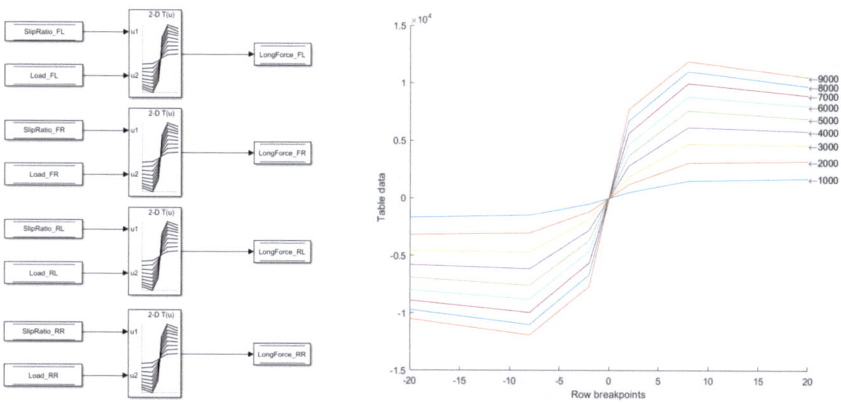

Fig. 34 Longitudinal tyre force calculation and lookup table (own elaboration)

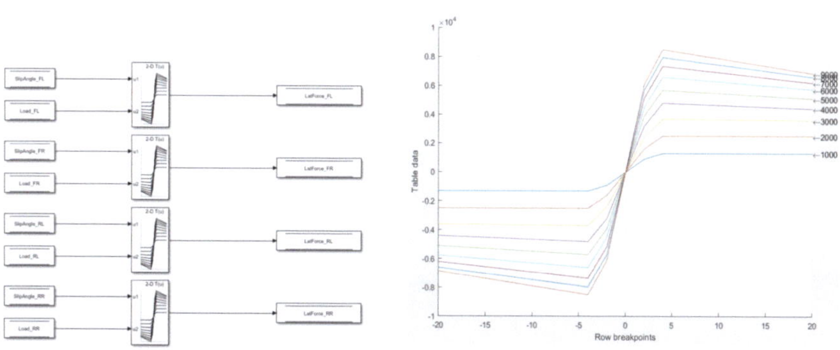

Fig. 35 Lateral tyre force calculation and lookup table (own elaboration)

PID Tuning

See Fig. 36.

Sliding Mode Controller

See Fig. 37.

Fuzzy Logic Controller

See Figs. 38, 39 and 40.

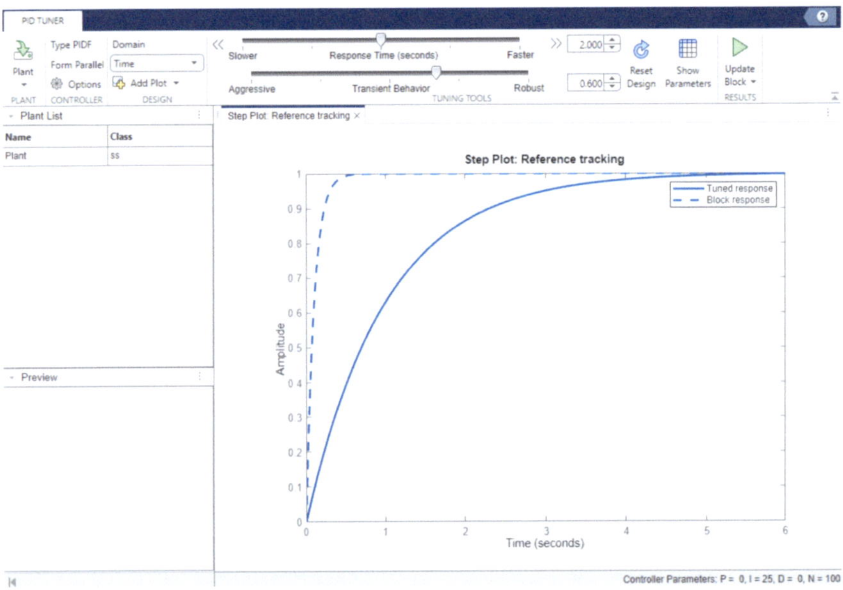

Fig. 36 PID tuner interface (own elaboration)

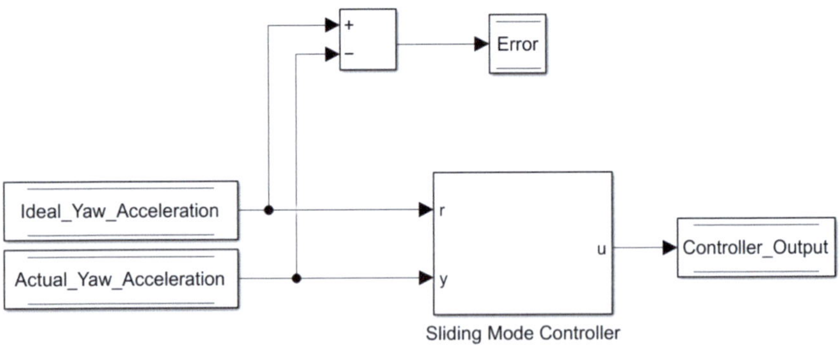

Fig. 37 Sliding mode controller implementation (own elaboration)

	Rule	Weight	Name
1	If E is NH then Torque is XL	1	rule1
2	If E is NL then Torque is L	1	rule2
3	If E is Z then Torque is M	1	rule3
4	If E is PL then Torque is H	1	rule4
5	If E is PH then Torque is XH	1	rule5

Fig. 38 System rules between input and output (own elaboration)

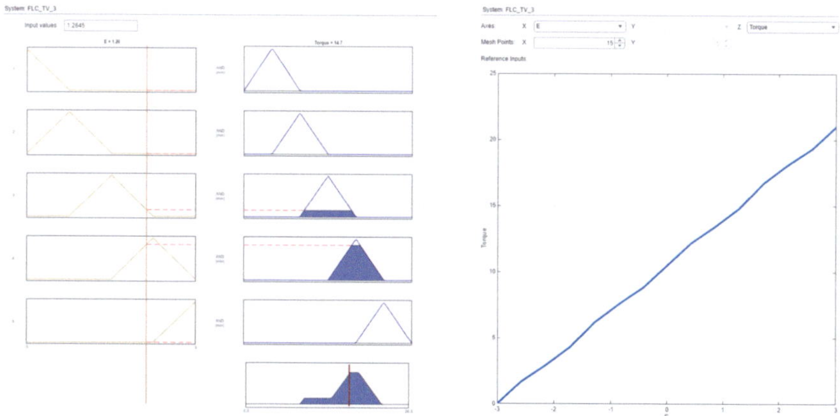

Fig. 39 System inference and control surface (own elaboration)

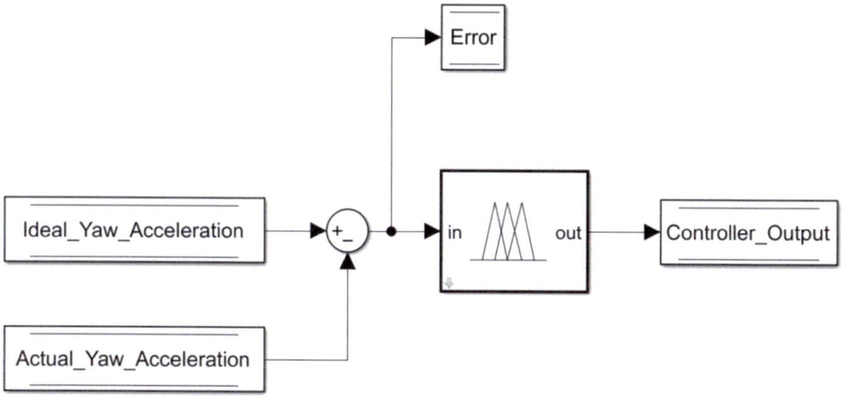

Fig. 40 Fuzzy logic controller implementation (own elaboration)

Results

See Figs. 41 and 42.

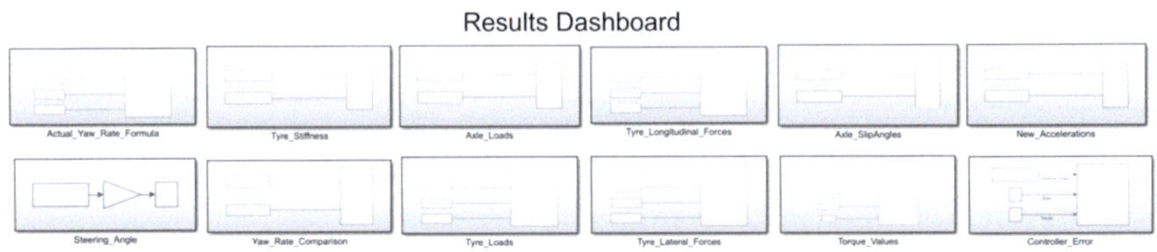

Fig. 41 Results dashboard prepared in Simulink (own elaboration)

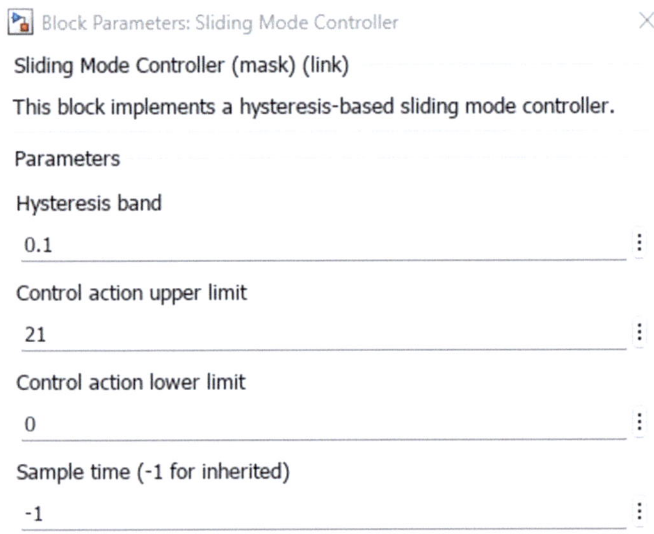

Fig. 42 Sliding mode controller parameters (own elaboration)

References

1. Antunes J, Antunes A, Outeiro P, Cardeira C, Oliveira P (2019) Testing of a torque vectoring controller for a Formula Student prototype. Robot Auton Syst 113:56–62
2. Bai Y, Wang D (2006) Fundamentals of fuzzy logic control—fuzzy sets, fuzzy rules and defuzzifications. In: Advanced fuzzy logic technologies in industrial applications, pp 17–36
3. Balkwill J (2018) Performance vehicle dynamics: engineering and applications. Butterworth-Heinemann, an imprint of Elsevier, Oxford
4. Bosch R (2007) Bosch handbook for automotive electrics-automotive electronics. Bentley Publishers
5. Choo CLW (2015) Real-time decision making in motorsports: analytics for improving professional car race strategy
6. Dendaluce Jahnke M, Cosco F, Novickis R, Pérez Rastelli J, Gomez-Garay V (2019) Efficient neural network implementations on parallel embedded platforms applied to real-time torque-vectoring optimization using predictions for multi-motor electric vehicles. Electronics 8:250
7. Gillespie TD, Taheri S, Sandu C, Duprey BL (2021) Fundamentals of vehicle dynamics. Revised, edition. SAE International, Warrendale, Pennsylvania

8. Herring J, Burnham K, Oleksowicz S (2012) Review and simulation of torque vectoring yaw rate control. In: Proceedings of the international conference on systems engineering (ICSE). Coventry, UK, pp 11–13
9. Hong CJ, Aparow VR (2021) System configuration of human-in-the-loop simulation for level 3 autonomous vehicle using IPG CarMaker. In: 2021 IEEE international conference on internet of things and intelligence systems (IoTaIS). IEEE, pp 215–221
10. Hu Z, Liao Y, Liu J, Xu H (2022) Investigation of vehicle stability by integration of active suspension, torque vectoring, and direct yaw control. SAE Int J Veh Dyn Stab NVH 6:441–459
11. Hucho W-H (1998) Aerodynamics of road vehicles: from fluid mechanics to vehicle engineering. Society of Automotive Engineers, Warrendale, PA
12. Ivanov V, Augsburg K, Savitski D (2012) Torque vectoring for improving the mobility of all-terrain electric vehicles
13. Kegelman JC, Harbott LK, Gerdes JC (2017) Insights into vehicle trajectories at the handling limits: analysing open data from race car drivers. Veh Syst Dyn 55:191–207
14. Kim SH, Kim KKK (2023) Model predictive control for energy-efficient yaw-stabilizing torque vectoring in electric vehicles with four in-wheel motors. IEEE Access 11:37665–37680
15. Lopez G, Seaber A (2009) The theory and practice of race vehicle data acquisition and analysis in motor sports engineering. Education 2009:14–1257
16. Lu Q, Sorniotti A, Gruber P, Theunissen J, de Smet J (2016) H∞ loop shaping for the torque-vectoring control of electric vehicles: theoretical design and experimental assessment. Mechatronics 35:32–43
17. Medina I, Santiago L, Hernández-Gómez JJ, Castillo R, Couder-Castañeda C (2021) Speed PID controller simulation of a reaction wheel for CubeSat orientation applications. J Phys Conf Series (IOP Publishing) 1723(1):012013
18. Milliken WF, Milliken DL (1994) Race car vehicle dynamics. PA, Society of Automotive Engineers, Warrendale
19. Nemeth T, Bubert A, Becker JN, de Doncker RW, Sauer DU (2018) A simulation platform for optimization of electric vehicles with modular drivetrain topologies. IEEE Trans Transp Electrification 4:888–900
20. de Novellis L, Sorniotti A, Gruber P (2013) Wheel torque distribution criteria for electric vehicles with torque-vectoring differentials. IEEE Trans Veh Technol 63:1593–1602
21. de Novellis L, Sorniotti A, Gruber P, Shead L, Ivanov V, Hoepping K (2012) Torque vectoring for electric vehicles with individually controlled motors: state-of-the-art and future developments. World Electr Veh J 5:617–628
22. Nowlan D (2010) The dynamics of the race car [Sydney], [ChassisSim]
23. Palma Á, Moreno J, Palma J, Domínguez R, Rohten J, Esparza V (2021) Comprehensive review of torque and vectoring control for electrical powertrains. In 2021 IEEE international conference on automation/XXIV congress of the chilean association of automatic control (ICA-ACCA), pp 1–7. IEEE
24. Parra A, Zubizarreta A, Pérez J (2019) A comparative study of the effect of intelligent control based torque vectoring systems on EVs with different powertrain architectures. In: 2019 IEEE intelligent transportation systems conference (ITSC). IEEE, pp 480–485
25. Piyabongkarn D, Rajamani R, Lew JY, Hai Y (2006) On the use of torque-biasing devices for vehicle stability control. In: 2006 American control conference, p 6
26. Rastelli JP, Peñas MS (2015). Fuzzy logic steering control of autonomous vehicles inside roundabouts. Appl Soft Comput 35:662–669
27. Sergers J (2014) Analysis techniques for racecar data acquisition. SAE International
28. Uzunsoy E (20180 A brief review on fuzzy logic used in vehicle dynamics control. J Innovative Sci Eng (JISE) 2:1–7
29. Vignati M, Sabbioni E (2022) A cooperative control strategy for yaw rate and sideslip angle control combining torque vectoring with rear wheel steering. Veh Syst Dyn 60:1668–1701
30. Young KD, Utkin VI, Ozguner U (1999) A control engineer's guide to sliding mode control. IEEE Trans Control Syst Technol 7:328–342

31. Zhang L, Ding H, Shi J, Huang Y, Chen H, Guo K, Li Q (2020) An adaptive backstepping sliding mode controller to improve vehicle maneuverability and stability via torque vectoring control. IEEE Trans Veh Technol 69:2598–2612
32. Zhu S, Huang X, Jiang D, Wu Z (2021) Low-cost electric bus stability enhancement scheme based on fuzzy torque vectoring differentials: design and hardware-in-the-loop test. IFAC-PapersOnLine 54:500–507

Design a Fuzzy Logic Controller for an Active Suspension System

Aydin Azizi

Abstract The suspension system's primary function is to provide passenger comfort by mitigating the effects of road roughness. Notably, minimising the effects of such vibrations also reduces the strain on mechanical vehicle components. The goal of this project is to create an artificial intelligence controller for a car's active suspension system in order to minimise the impact of imposed vibrations on the vehicle. In this study, the proposed controller has been designed using the fuzzy logic technique. Simulink environment has been used to model the dynamic system's controlled behaviour based on the imposed pavement condition. Gaussian white noise was used to depict the road surface condition. The performance of the fuzzy logic controller (FLC) has been compared to a proportional-integral-derivative (PID) controller. The results show that the developed controller outperforms the PID in minimising the effects of road roughness.

Keywords Active suspension system · Quarter-car model · Dynamic system · Controller design · Fuzzy logic · PID

1 Introduction

Many academics have been concentrating their efforts in recent years on developing effective suspension systems to mitigate the negative effects of vehicle pavement conditions. Rao et al. [1] evaluated the performance of a quarter-car semi-active suspension system using a proportional-integral-derivative (PID) controller in a MATLAB Simulink model in 2014, and Youn et al. [2] constructed an attitude motion control for an active suspension system with a tracking controller in 2015. Rao et al. [3] modelled and simulated the quarter-car suspension system in 2015 using the

A. Azizi (✉)
School of Engineering, Computing and Mathematics, Oxford Brookes University, Oxford, UK
e-mail: aydin.azizi@brookes.ac.uk

A. Azizi (ed.), *Intelligent Mechatronic Control Systems*, Emerging Trends in Mechatronics, https://doi.org/10.1007/978-981-95-3925-3_4

LQR control approach. Zhao et al. [4] suggested in 2015 an adaptive neural network control for camera stabilisation with an active suspension system. Bello et al. [5] have implemented the use of the whole state feedback controller as an active vehicle suspension control technology. Zhao et al. [6] suggested in 2016 an adaptive neural network controller for an active suspension system with actuator saturation. Deshpande et al. [7] studied nonlinear control for dual objective active suspension system approaches in 2016. In 2018, Azizi et al. [8] proposed a PID controller for regulating the rate of fuel consumption caused by changing pavement conditions. Later in 2020, they developed an AI-based controller based on RPLNN architecture to demonstrate the RPLNN controller's superior performance over the PID [9]. Azizi and Mobki designed a sliding mode controller for the active suspension of the model quarter vehicle in 2021 [10].

Many researchers utilised the fuzzy logic approach to address the problem, and the outcomes have reported that the fuzzy logic controllers (FLCs) are well suited to control the complex behaviour of dynamic systems such as active suspension systems, as they can handle imprecise or uncertain input variables [11–15].

Research on the design of a fuzzy logic controller for an active suspension system typically involves the use of simulation and experimentation to evaluate the performance of the controller and optimise its parameters. Here are a few key aspects that are often studied in such research [16–25]:

1. Fuzzy rule design: Researchers may investigate different methods for designing the fuzzy rules that govern the relationship between the input variables (e.g. road condition, vehicle speed) and the output variables (e.g. suspension stiffness and damping). The goal is to create rules that accurately reflect the relationship between the input and output variables and lead to optimal performance of the suspension system.
2. Performance evaluation: Researchers may use simulations and experiments to evaluate the performance of the fuzzy logic controller in terms of factors such as ride comfort, handling, and stability. The goal is to identify the optimal settings for the controller that lead to the best performance.
3. Comparison with other controllers: Research may also compare the performance of the fuzzy logic controller with other types of controllers, such as PID controllers. This allows researchers to determine the advantages and disadvantages of different types of controllers and identify when a fuzzy logic controller is the best choice.
4. Robustness analysis: Researchers may also conduct robustness analysis to evaluate the performance of the fuzzy logic controller under different road conditions and vehicle speeds to identify any possible limitations of the controller.
5. Real-time implementation: Researchers may also investigate the real-time implementation of the fuzzy logic controller in an active suspension system. This will

involve the use of embedded systems and microcontrollers to implement the controller in real time and the challenges that come with it.

Overall, research on the design of a fuzzy logic controller for an active suspension system aims to improve the performance of the suspension system, compare the performance of different types of controllers and identify the best controller for a specific application, as well as identify the limitations of the controller and propose solutions to overcome them.

In the current research, the focus of utilising the fuzzy logic approach has been set on the performance evaluation.

The current study's goal is to create an effective artificial intelligence (AI)-based controller to lessen the effect of the road profile on automobiles, which manifests as vehicle-imposed vibration. The road profile is modelled in this study using Gaussian white noise. The fuzzy logic (FL) paradigm is used in this case to build an effective controller. In this chapter, the MATLAB software, Simulink toolbox, and fuzzy logic toolbox were used to create a simulation-based model of an active suspension system on a quarter-car model in order to investigate and reduce the effect of imposed vibration from the road profile by designing an effective fuzzy logic controller. The proposed model may simulate the vertical displacement of the vehicle before and after the application of the fuzzy logic controller, as well as evaluate these vibrations by comparing the various parameter values. The suggested fuzzy logic controller outperforms the PID controller and successfully decreases road-profile-induced vibration, according to the results.

Following the brief introduction and literature review, the remainder of this chapter is divided into three sections. The Methodology section first delves into the mathematical modelling of the suggested active suspension of a quarter-car model, the fuzzy logic technique, and the simulation-based model setup. In the second phase, the findings section, the developed model was validated by assessing the effect of the proposed controller on the system's vertical displacement. A summary of the research and recommendations for future research are offered in the conclusion section.

2 Methodology

An active suspension system is a type of suspension system that uses actuators to adjust the stiffness and damping of the suspension. This allows the vehicle to maintain a smooth ride over rough terrain and improve handling and stability. However, designing a control system for an active suspension system is challenging due to the uncertainty and imprecision of the input variables, such as road conditions and vehicle speed.

To minimise the imposed vibrations from the pavement profile, a passive suspension system consisting of springs and dampers was used in the traditional technique. It is essential to be aware of the fact that the damping ratio of such a system cannot be modified in order to eliminate the imposed vibration because such a system does not have the capability to do so. As a result of recent developments in technology, active suspension systems that are fitted with efficient controllers that are capable of generating an active force are increasingly being put to use in order to mitigate the effect of the pavement condition, which, for the purposes of this investigation, has been interpreted as Gaussian white noise [26].

The fuzzy logic controller is then implemented in the active suspension system and continually monitors the input variables and adjusts the suspension settings as necessary. This allows the vehicle to maintain a smooth ride over rough terrain and improve handling and stability.

In order to accomplish this task, the purpose of this study has been outlined as the development of an efficient AI controller for the purpose of mitigating the unfavourable impact of the road profile. The steps of the adopted recommended approach have been outlined in the following sequence.

1. Develop the mathematical model of the quarter-car active suspension system. This includes the equations of motion for the vehicle and the suspension, as well as the parameters of the system such as the mass, spring constant, and damping coefficient.
2. Use the Simulink toolbox in MATLAB to create a simulation-based model of the suspension system. This involves creating blocks that represent the different components of the system, such as the vehicle body, the suspension, and the actuators, and connecting them together to create a dynamic model of the system.
3. Implement the fuzzy logic controller in the Simulink model using the fuzzy logic toolbox. This involves creating the fuzzy sets and rules that define how the input variables (e.g. road condition, vehicle speed) affect the output variables (e.g. suspension stiffness and damping).
4. Set up the simulation parameters and run the simulation. The simulation will use the mathematical model of the suspension system and the fuzzy logic controller to generate results for the vertical displacement of the vehicle.
5. Analyse the results of the simulation. Compare the results of the simulation with the results of the mathematical model and evaluate the performance of the fuzzy logic controller and compare it with the PID controller.

2.1 Active Suspension System

Figure 1 illustrates the quarter-car model. A quarter-car model is a simplified representation of a vehicle's suspension system that only includes one-fourth of the suspension system. It is utilised to model and study the dynamic behaviour of the suspension system, including vehicle body movement and tyre–road contact.

Typically, the quarter-car model consists of a mass representing the vehicle body, a spring representing the suspension spring, and a damper representing the suspension damper. In addition, the model contains a tyre that is modelled as a linear spring and damper.

The quarter-car model is a straightforward model that can be easily solved analytically, although it has several drawbacks. It is only capable of simulating the vertical motion of the vehicle and does not account for pitch and roll. In addition, the quarter-car model only represents one corner of the vehicle; hence, its results are only applicable to that particular corner. Despite these limitations, the quarter-car model is a

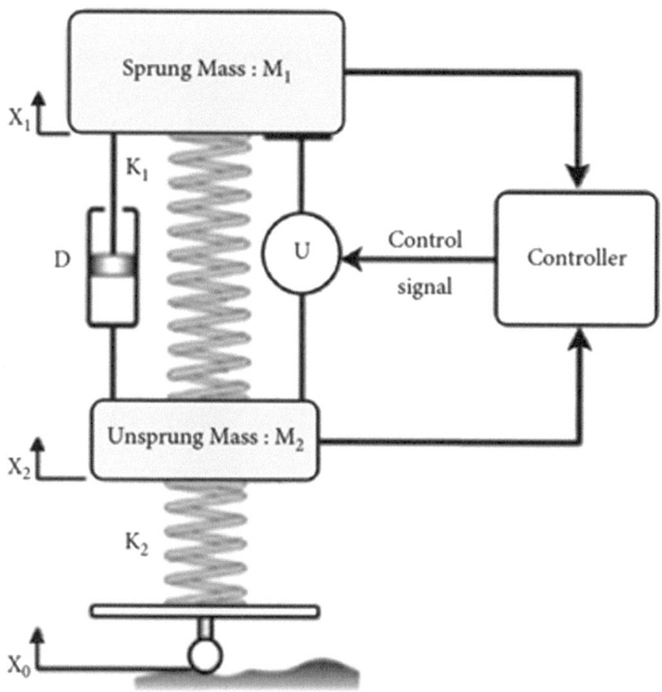

Fig. 1 Active suspension system

useful tool for comprehending the fundamental behaviour of the suspension system and for designing and optimising the suspension system.

In conclusion, the quarter-car model is a simplified mathematical representation of one corner of a vehicle's suspension system. It is utilised to model and study the dynamic behaviour of the suspension system, including vehicle body movement and tyre–road contact. The model is basic and can be easily solved analytically, although it has several limitations [8–10].

It is vital to note that some parameters, such as geometrical information, pitching and rolling angles, have not been accounted for, although the bulk of significant elements, such as load change, have.

In order to proceed with the design of the Simulink model, the equations of motion of the system need to be derived and then arranged in the two phases that immediately follow.

1st step: Deriving the equation of motion:

$$M_1\ddot{X}_1(t) + D\dot{X}_1(t) + K_1X_1(t) - D\dot{X}_2(t) - K_1X_2(t) = U(t) \quad (1)$$

$$M_2\ddot{X}_2(t) + D\dot{X}_2(t) + K_1X_2(t) + K_2X_2(t) - D\dot{X}_1(t) - K_1X_1(t) - K_2X_0(t) = -U(t) \quad (2)$$

2nd step: Rearranging the above equations based on $\ddot{x}_1$ *and* $\ddot{x}_2$, respectively.

$$\ddot{X}_1(t) = \frac{1}{M_1}(U(t) - D\dot{X}_1(t) - K_1X_1(t) + D\dot{X}_2(t) + K_1X_2(t) \quad (3)$$

$$\ddot{X}_2(t) = \frac{1}{M_2}[-U(t) - D\dot{X}_2(t) - K_1X_2(t) - K_2X_2(t) + D\dot{X}_1(t) + K_1X_1(t) + K_2X_0(t)] \quad (4)$$

$$\ddot{X}_2(t) = \frac{1}{M_2}\left[-U(t) - D\dot{X}_2(t) - K_1X_2(t) - K_2X_2(t) + D\dot{X}_1(t) + K_1X_1(t) + K_2X_0(t)\right]$$

$$\ddot{X}_2(t) = \frac{1}{M_2}\left[-U(t) - D\dot{X}_2(t) - K_1X_2(t) - K_2X_2(t) + D\dot{X}_1(t) + K_1X_1(t) + K_2X_0(t)\right]$$

As illustrated in Fig. 2, Eqs. 3 and 4 have been modelled in the Simulink environment. In this model, the U, which represents the force created by the controller, has been modelled as a step function, while the road profile has been modelled as a Sine wave. Because at this point in the process, we have not designed a controller that will generate the force U, we are able to think of the design as being that of a passive suspension system.

The next stage is to construct a feedback loop and a fuzzy logic controller to generate the needed force to minimise the impact of the imposed vibration on the vehicle caused by the pavement condition.

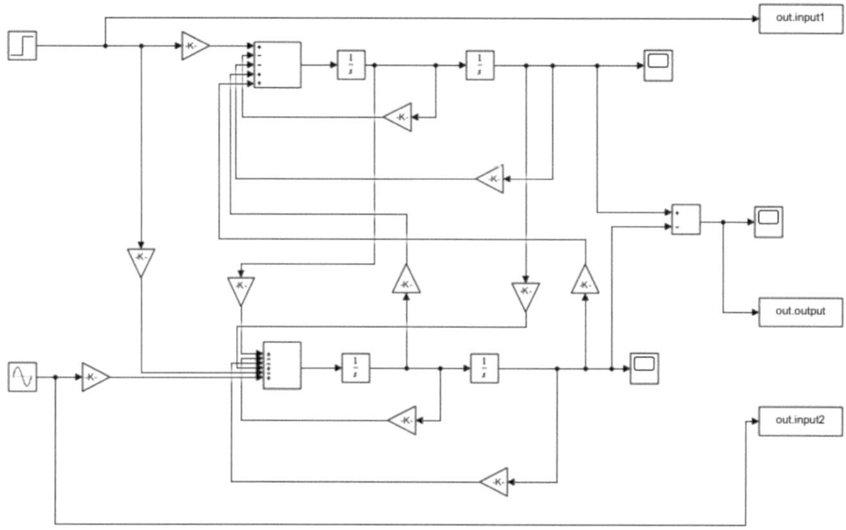

Fig. 2 Passive suspension system

2.2 *Fuzzy Logic Controller*

Fuzzy logic controllers are well suited for an active suspension system as they can handle imprecise or uncertain input variables. The design of a fuzzy logic controller for an active suspension system involves several steps:

1. Defining the input variables: The input variables for the controller include the current road condition (smooth, rough, bumpy), vehicle speed, and suspension deflection. These variables are used to determine the appropriate suspension settings.
2. Defining the output variables: The output variables for the controller include the suspension stiffness and damping. These variables are adjusted based on the input variables to maintain optimal suspension performance.
3. Creating fuzzy rules: Fuzzy rules are created to define how the input variables affect the output variables. For example, a rule might state that if the road condition is bumpy and the vehicle speed is high, then the suspension stiffness should be high and the damping should be low.
4 Using a fuzzy inference system: The fuzzy inference system processes the input variables and applies the fuzzy rules to determine the appropriate output variables for the active suspension.

5. Implementing a feedback control loop: A feedback control loop is implemented to continually monitor the input variables and adjust the output variables as necessary to maintain optimal suspension performance.

The process of creating the mapping from a given input to an output using fuzzy logic is referred to as a fuzzy inference system (FIS). A FIS is a method of reasoning that uses fuzzy logic to approximate human reasoning. It consists of a set of input variables (also known as antecedents), a set of output variables (also known as consequences), and a set of rules that relate the inputs to the outputs. The rules are usually in the form of 'if–then' statements, where the antecedents are the 'if' part and the consequences are the 'then' part. The fuzzy logic used in the FIS allows for partial truths and uncertainties, rather than strict true or false values. This makes FISs useful in a wide range of applications, such as control systems, decision-making, and pattern recognition. Once the mapping is set up, decisions and patterns can then be determined [27].

There are several types of fuzzy inference systems, including:

1. Mamdani FIS: This type of FIS uses a set of fuzzy if–then rules to infer an output based on the input. The fuzzy sets used in the antecedents and consequences of the rules are typically defined using membership functions.
2. Tsukamoto FIS: This type of FIS uses a set of linear equations to infer an output based on the input. The equations are defined by the slope and y-intercept of the membership functions.
3. Larsen FIS: This type of FIS uses a set of nonlinear equations to infer an output based on the input. The equations are defined by the coefficients of the membership functions.
4. Sugeno FIS: This type of FIS uses a set of linear or nonlinear equations to infer an output based on the input. The equations are defined by the coefficients of the membership functions and the input variables.
5. Hybrid FIS: This type of FIS combines different types of FISs to take advantage of their respective strengths.

It is important to note that the type of FIS used will depend on the specific problem being solved and the trade-offs between accuracy and complexity.

The FIS utilised in this chapter is the Mamdani technique that has been well-accepted as an effective way to capture expert knowledge.

The fuzzy Mamdani approach is a type of fuzzy logic control system that utilises fuzzy sets and fuzzy rules to make decisions. Developed by Lotfi Zadeh and named after the two researchers, Ebrahim Mamdani and S. Assilian, who first implemented it in 1974, the fuzzy Mamdani approach is widely used in control systems, decision-making, and artificial intelligence. In this article, we will discuss the basic principles and features of the fuzzy Mamdani approach and its applications.

The fuzzy Mamdani approach is based on the concept of fuzzy sets, which are sets that have a degree of membership rather than a binary membership. For example, a fuzzy set for temperature might include the values 'hot', 'warm', and 'cold', each with a degree of membership between 0 and 1. This allows for more flexibility and precision in the decision-making process.

Fuzzy rules are used in the fuzzy Mamdani approach to relate the input variables to the output variables. These rules are written in a natural language format, such as 'if temperature is hot and humidity is high, then the air conditioning should be turned on'. The rules are then processed by the fuzzy inference system, which uses the degree of membership of the input variables to determine the appropriate output.

The fuzzy Mamdani approach also utilises a defuzzification process, which converts the fuzzy output back into a crisp value that can be used by the control system. This is typically done using a method called centroid defuzzification, which calculates the centre of gravity of the fuzzy output.

The fuzzy Mamdani approach has many applications in control systems, decision-making, and artificial intelligence. It has been used in areas such as industrial process control, automotive systems, and robotics. For example, it has been used to control the temperature in HVAC systems, the speed of a vehicle, and the position of a robotic arm.

In conclusion, the fuzzy Mamdani approach is a powerful tool for making decisions in systems with imprecise or uncertain input variables. Its use of fuzzy sets and fuzzy rules allows for more flexibility and precision in the decision-making process, and its defuzzification process allows for the conversion of fuzzy output back into crisp values that can be used by control systems. With a wide range of applications, the fuzzy Mamdani approach is a valuable tool for engineers and researchers in control systems and artificial intelligence [28].

In this way, we are able to explain the level of knowledge in terms that are more relatable and easier to understand. However, Mamdani-type fuzzy inference is computationally intensive.

The fuzzy inference process in the Mamdani style is divided into five stages, as follows [28]:

1. Fuzzification: This is the process of converting crisp input values into fuzzy values. This allows the system to handle imprecise or uncertain input variables.
2. Rule evaluation: This is the process of applying fuzzy rules to the fuzzy input values. The rules are written in a natural language format, such as 'if temperature is hot and humidity is high, then the air conditioning should be turned on'.
3. Inference: This is the process of determining the appropriate output based on the fuzzy input values and the fuzzy rules. The fuzzy inference system uses the degree of membership of the input variables to determine the appropriate output.
4. Defuzzification: This is the process of converting the fuzzy output back into a crisp value that can be used by the control system. This is typically done using a method called centroid defuzzification, which calculates the centre of gravity of the fuzzy output.

5. Action: This is the final stage where the crisp value output from defuzzification is used to control the system.

The first step is to take the inputs and use membership functions to determine their degree of membership in each of the relevant fuzzy sets (fuzzification). Fuzzification is the process of converting crisp input values into fuzzy values. It is the first step in a fuzzy logic control system. The result is a fuzzy degree of linguistic membership (always the interval from 0 through 1). Fuzzification of the input is equal to searching a table or evaluating a function. As a result, each input undergoes fuzzification across all of the requisite membership functions. There are different algorithms for fuzzification; some of the most common are:

1. Triangular membership function: This algorithm uses a triangular shape to define the membership function for a fuzzy set. It is simple to implement and easy to understand, but it may not be suitable for all types of input variables.
2. Trapezoidal membership function: This algorithm uses a trapezoidal shape to define the membership function for a fuzzy set. It is more flexible than the triangular membership function and can be used for a wider range of input variables.
3. Gaussian membership function: This algorithm uses a Gaussian curve to define the membership function for a fuzzy set. It is useful for input variables that are continuous and have a normal distribution.
4. Generalised Bell membership function: This algorithm uses a bell-shaped curve to define the membership function for a fuzzy set. It is useful for input variables that have a wide range of values.
5. Sigmoid membership function: This algorithm uses a sigmoid-shaped curve to define the membership function for a fuzzy set. It is useful for input variables that have a range of values that are limited to a specific range.

The choice of fuzzification algorithm depends on the type of input variable, the range of its values, and the desired level of precision. Each algorithm has its own advantages and disadvantages, and the best algorithm for a specific application should be chosen based on the specific requirements of the system.

In this research, the triangular membership functions have been utilised.

After the inputs have been fuzzified, you will be able to determine the extent to which each component of the antecedent is fulfilled by each rule. A rule's antecedent may include more than one component, and when this happens, the fuzzy operator is used to get a single integer that reflects what happened in the rule antecedent. The output function is then applied with this value. As previously stated, any number of well-defined techniques may be used in place of the AND or OR operations.

When the disjunction of the rule antecedents is to be evaluated, the fuzzy OR operation should be used. In most cases, fuzzy expert systems make use of the classical fuzzy operation union, which is as follows:

$$\mu A \cup B(x) = \max [\mu A(x), \mu B(x)] \tag{5}$$

Similarly, we use the AND fuzzy operation intersection to assess the combination of the rule antecedents.

$$\mu A \cap B(x) = \min [\mu A(x), \mu B(x)] \tag{6}$$

The second stage involves putting the rule assessment technique into action. The AND method (minimum or α cut) and the product method are the two main techniques.

Typically, this is accomplished by removing the membership function at the antecedent truth level. This method is described by the phrase clipping (alpha-cut). Due to the slicing of the top of the membership function, the clipped fuzzy set is devoid of data. Clipping, on the other hand, is frequently favoured due to the fact that it involves less complex and faster mathematics and gives an aggregated output surface that is easier to defuzzify than other methods.

Instead of clipping, which is a common tactic, scaling is a better way to maintain the fuzzy collection's original form. By raising each membership degree by the truth value of the rule antecedent, the initial membership function of the rule consequence is changed. This method may be quite useful in fuzzy expert systems because it frequently leads to less information loss.

The fuzzy sets constituting a result are expressed using membership functions that are contextually weighted. The outcome is shaped by an antecedent-related function (a single number). As input for the implication process, the antecedent offers a single integer, and the result is a fuzzy set. Every rule is executed with consequence.

The outputs of the rules are combined in the third stage. All of the rules in a FIS must be tested in order for judgments to be made, so the rule outputs must be combined in some way. Aggregation is the process of combining the fuzzy sets representing each rule's outputs into a single fuzzy set.

Before the final defuzzification stage, there is only one instance of aggregation for each output variable. For each rule, the implication process returns a list of output functions that have been cut off. This list is what the aggregation process takes as input. Each output variable from the aggregation procedure is represented by a fuzzy set. It does not matter what order the rules are executed in as long as the aggregate method is commutative; this makes the order of execution irrelevant. The three most important methods are max, probor, and sum (sum of the rule output sets).

Defuzzification, the final step, is the process of converting the fuzzy output of a fuzzy logic control system back into a crisp value. It is the final step in a fuzzy logic control system and is used to determine the appropriate control action to take. There are different algorithms for defuzzification; some of the most common are:

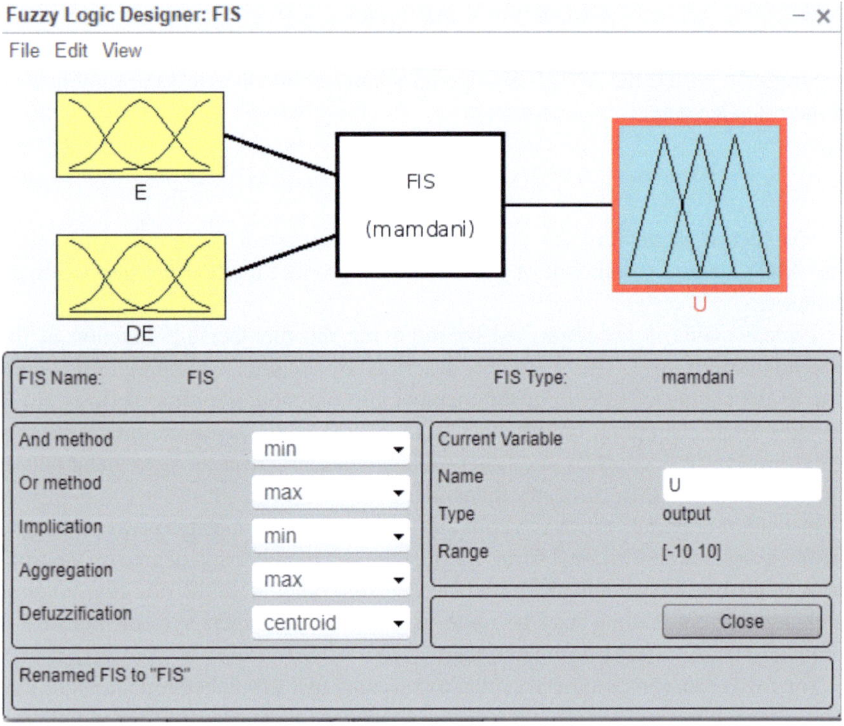

Fig. 3 Fuzzy logic controller

1. Centroid method: This algorithm calculates the centre of gravity of the fuzzy output and converts it into a crisp value. It is simple to implement and provides good results in most cases.
2. Maximum membership method: This algorithm selects the crisp value that corresponds to the maximum membership degree in the fuzzy output. It is simple to implement and easy to understand, but it may not always provide the best results.
3. Mean of maximum method: This algorithm calculates the mean of the maximum membership degree for each fuzzy set in the fuzzy output and converts it into a crisp value. It is a more complex algorithm than the centroid method, but it may provide more accurate results.
4. Weighted average method: This algorithm calculates the weighted average of the crisp values that correspond to the membership degrees in the fuzzy output. It is a more complex algorithm than the centroid method, but it may provide more accurate results.
5. Smallest of maximum (SOM) method: This algorithm selects the crisp value that corresponds to the smallest maximum membership degree in the fuzzy output.

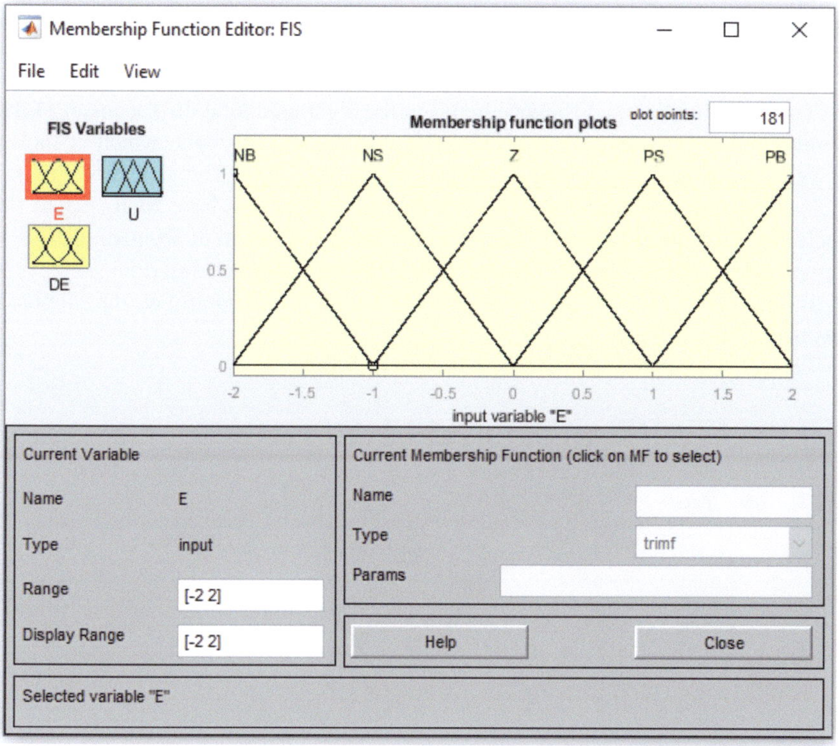

Fig. 4 Membership function plots for input E

It is simple to implement and easy to understand, but it may not always provide the best results.

The choice of defuzzification algorithm depends on the specific requirements of the system and the desired level of precision. Each algorithm has its own advantages and disadvantages, so the best algorithm for a specific application should be chosen based on the specific requirements of the system.

In this study, the centroid defuzzification approach has been utilised for the defuzzification (see Eq. 7).

$$COG = \frac{\int_a^b \mu_A(x)xdx}{\int_a^b \mu_A(x)dx} \quad (7)$$

In this study, the Mamdani technique was employed to build the fuzzy logic controller for the active suspension, as depicted in Fig. 3. The controller under examination has one output and two inputs.

The first input is the difference between the computed total displacement of the system and the required displacement (E), and the second input, which is shown as velocity, is the rate of change of this displacement (DE). The generated force is the output of the controller (U). Five membership functions have been created for each input and the output variable, as shown in Figs. 4, 5 and 6: negative big (NB), negative small (NS), zero (Z), positive small (PS), and positive big (PB).

Based on the defined two inputs and one output of the controller, the rule base has been constructed with 25 rules (see Fig. 7).

The rule base can be summarised as shown in Table 1.

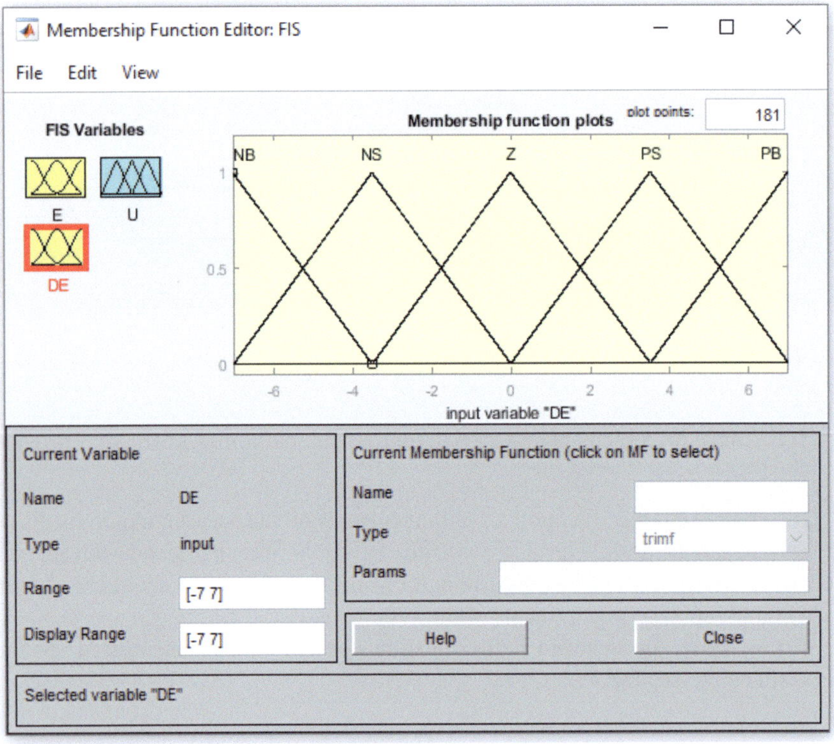

Fig. 5 Membership function plots for input DE

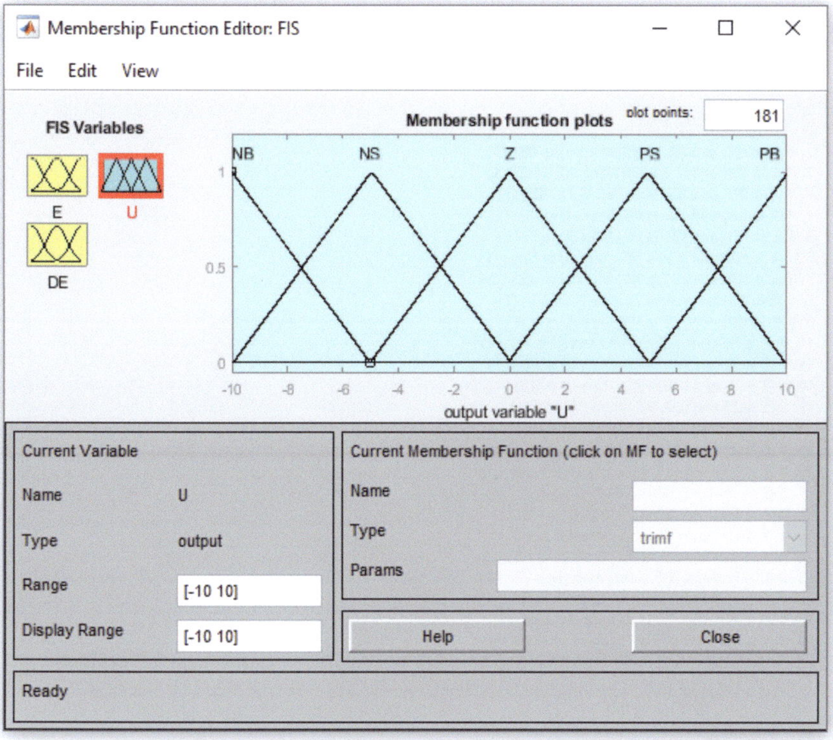

Fig. 6 Membership function plots for output U

Also, the rule base can be illustrated as the graphical representation of the rules, as shown in Fig. 8.

The surface based on the two inputs and the output has been illustrated in Fig. 9.

A proposed active suspension system based on the fuzzy logic controller has been modelled in the Simulink environment, as shown in Fig. 10.

In order to evaluate the performance of the proposed fuzzy logic controller compared to that of other well-known controllers, a PID controller, as depicted in Fig. 11, has been developed. The MATLAB PID control tuning toolbox was used to tune the proposed PID controller so that its parameters could be found.

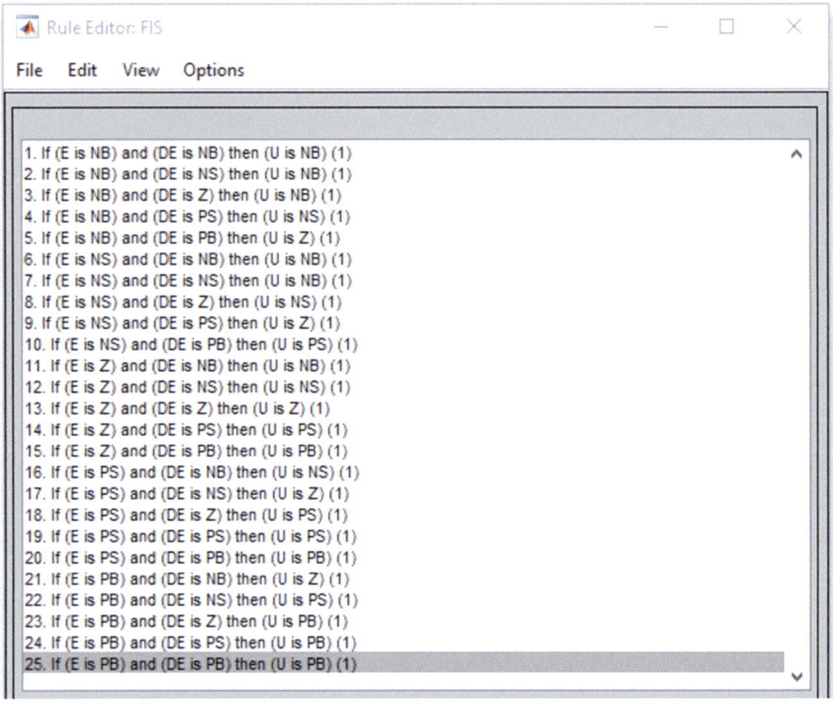

Fig. 7 Rule base

Table 1 Fuzzy rule base

EDE	NB	NS	Z	PS	PB
NB	NB	NB	NB	NS	Z
NS	NB	NB	NS	Z	PS
Z	NB	NS	Z	PS	PB
PS	NS	Z	PS	PS	PB
PB	Z	PS	PB	PB	PB

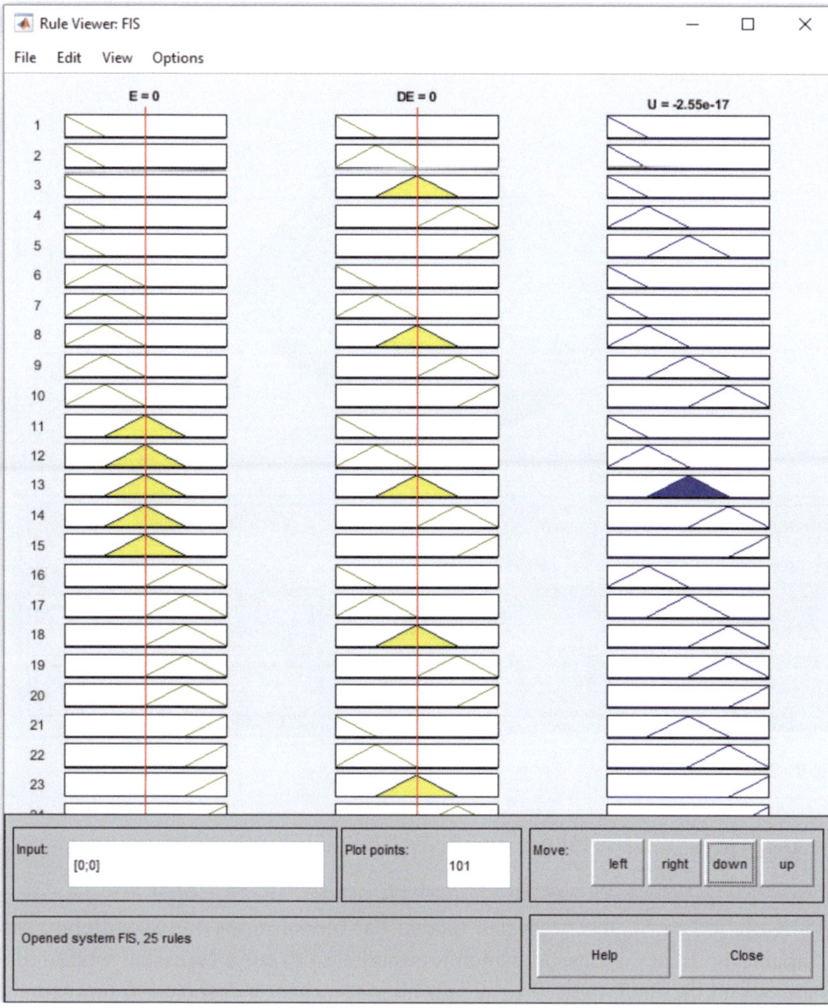

Fig. 8 Rule base illustration

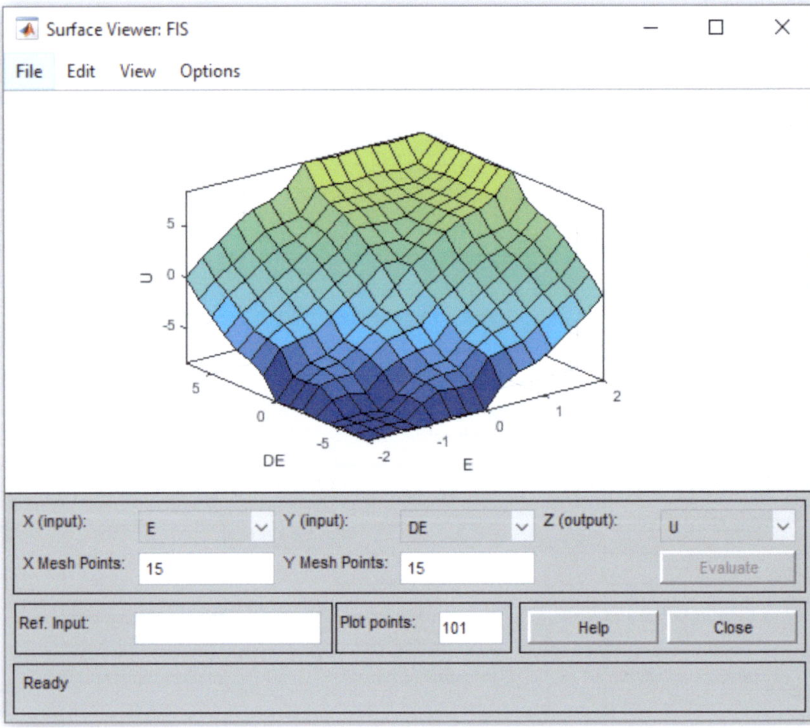

Fig. 9 Surface

3 Results

As already indicated, the suggested quarter-car model relies on two different sets of inputs: the road conditions, which are modelled using a Gaussian white noise function, and the force created by a controller specifically designed to mitigate the effects of forced vibrations on the automobile.

Figure 12 depicts the system's controlled behaviour based on the random pavement state. It is evident that the built fuzzy logic controller has an excellent performance to lessen the adverse influence of the road profile on the vehicle by generating suitable control signals (U) to control the system's output (displacement).

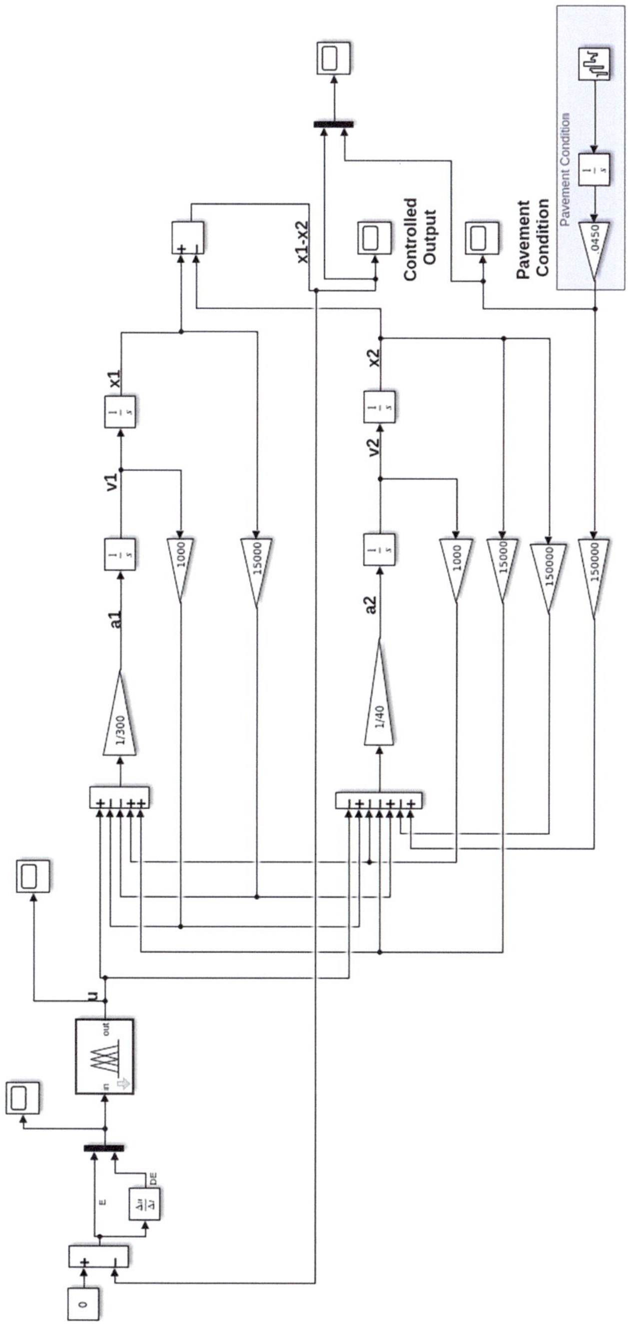

Fig. 10 Active suspension system with fuzzy logic controller

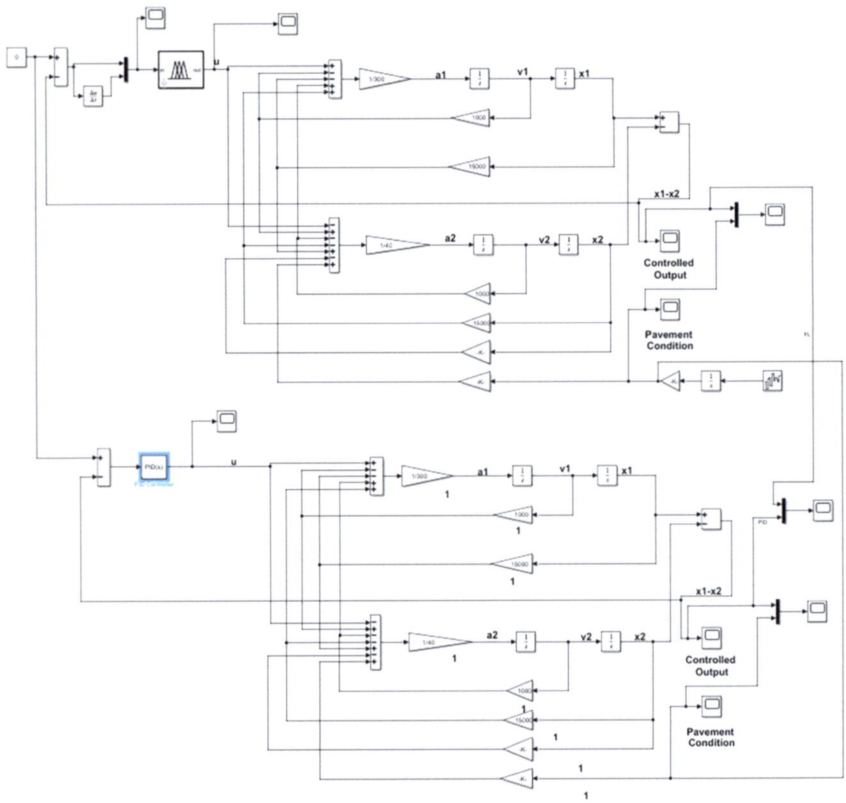

Fig. 11 Active suspension system with fuzzy logic and PID controllers

Figure 13 illustrates the output of the fuzzy logic controller, which is the force (U). As is evident from the road profile, the FLC has executed numerous control signals.

Figure 14 shows that, while both fuzzy logic and PID controllers are capable of reducing the undesirable vibrations imposed on the vehicle by the road profile, the designed FLC outperforms the PID controller.

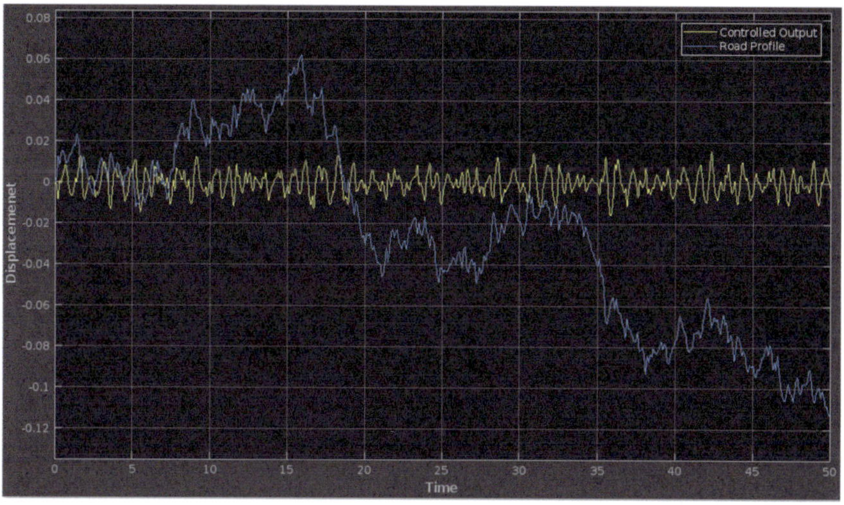

Fig. 12 Active suspension system performance equipped with a fuzzy logic controller based on the random pavement condition

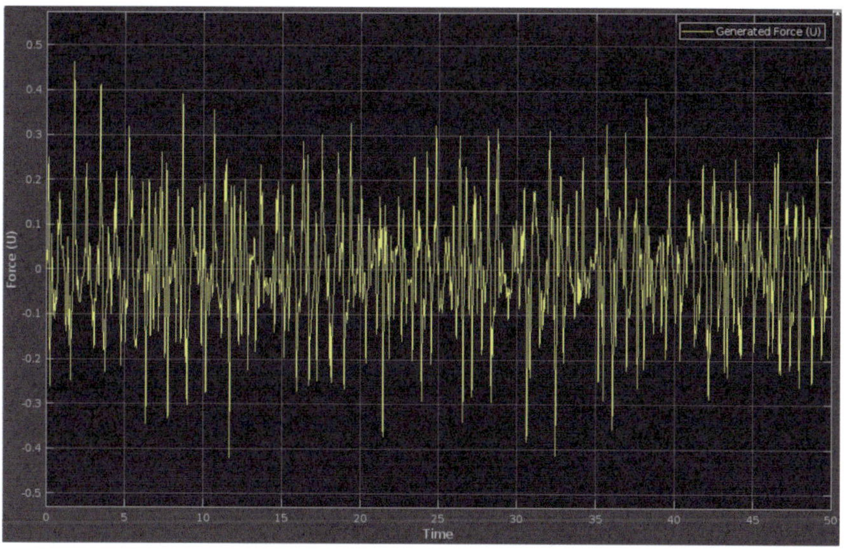

Fig. 13 Generated control signal (U) by the fuzzy logic controller

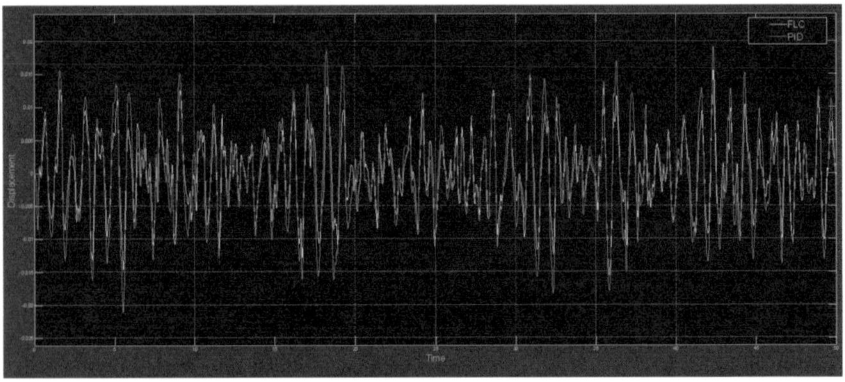

Fig. 14 Active suspension system with FL and PID controllers

4 Conclusion

An effective artificial intelligence controller for the active suspension system of a quarter-car model has been developed as part of the current research. To meet the challenge, the controller was designed using fuzzy logic, and the road input signal was simulated using Gaussian white noise. Modelling the recommended configuration with MATLAB and Simulink. The results demonstrate that the proposed fuzzy logic controller is capable of minimising a vehicle's vulnerability to adverse road conditions. The findings of a comparison between the performance of the fuzzy logic controller and the PID controller reveal that the FLC performs better. In conclusion, fuzzy logic controllers are an effective solution for controlling active suspension systems. They can handle imprecise or uncertain input variables and make decisions based on fuzzy rules. By following the steps outlined in this article, a fuzzy logic controller can be designed and implemented in an active suspension system to improve performance and provide a smooth ride for the occupants. Future work may employ other control systems, such as sliding mode controllers and Artificial Neural Network-based controllers, as controllers.

Conflicts of Interest The authors declare they have no conflicts of interest.
Funding The research has received no funding.

Data Availability The related data are presented within the manuscript.

References

1. Rao KD (2014) Modeling, simulation and control of semi active suspension system for automobiles under matlab simulink using pid controller. IFAC Proc Vol 47(1):827–831
2. Youn I, Wu L, Youn E, Tomizuka M (2015) Attitude motion control of the active suspension system with tracking controller. Int J Automot Technol 16(4):593–601

3. Rao KD, Kumar S (2015) Modeling and simulation of quarter car semi active suspension system using LQR controller. In: Proceedings of the 3rd international conference on frontiers of intelligent computing: theory and applications (FICTA) 2014. Springer, Cham, pp 441–448
4. Zhao F, Dong M, Qin Y, Gu L, Guan J (2015) Adaptive neural networks control for camera stabilization with active suspension system. Adv Mech Eng 7(8):1687814015599926
5. Bello MM, Akramin Shafie A, Khan R (2015) Active vehicle suspension control using full state-feedback controller. In: Advanced materials research, vol 1115. Trans Tech Publications Ltd., pp 440–445
6. Zhao F, Ge SS, Tu F, Qin Y, Dong M (2016) Adaptive neural network control for active suspension system with actuator saturation. IET Control Theory Appl 10(14):1696–1705
7. Deshpande VS, Shendge PD, Phadke SB (2016) Nonlinear control for dual objective active suspension systems. IEEE Trans Intell Transp Syst 18(3):656–665
8. Azizi A (2018) Computer-based analysis of the stochastic stability of mechanical structures driven by white and colored noise. Sustainability 10(10):3419
9. Azizi A (2020) A case study on computer-based analysis of the stochastic stability of mechanical structures driven by white and colored noise: utilizing artificial intelligence techniques to design an effective active suspension system. Complexity
10. Azizi A, Mobki H (2021) Applied mechatronics: designing a sliding mode controller for active suspension system. Complexity
11. Ting CS, Li THS, Kung FC (1995) Design of fuzzy controller for active suspension system. Mechatronics 5(4):365–383
12. Li H, Zhang Z, Yan H, Xie X (2018) Adaptive event-triggered fuzzy control for uncertain active suspension systems. IEEE Trans Cybern 49(12):4388–4397
13. Na J, Huang Y, Wu X, Su SF, Li G (2019) Adaptive finite-time fuzzy control of nonlinear active suspension systems with input delay. IEEE Trans Cybern 50(6):2639–2650
14. Yatak MÖ, Şahin F (2021) Ride comfort-road holding trade-off improvement of full vehicle active suspension system by interval type-2 fuzzy control. Eng Sci Technol Int J 24(1):259–270
15. Lin B, Su X, Li X (2019) Fuzzy sliding mode control for active suspension system with proportional differential sliding mode observer. Asian J Control 21(1):264–276
16. Li W, Xie Z, Zhao J, Wong PK, Li P (2019) Fuzzy finite-frequency output feedback control for nonlinear active suspension systems with time delay and output constraints. Mech Syst Signal Process 132:315–334
17. Golouje YN, Abtahi SM (2021) Chaotic dynamics of the vertical model in vehicles and chaos control of active suspension system via the fuzzy fast terminal sliding mode control. J Mech Sci Technol 35(1):31–43
18. Mrazgua J, Ouahi M (2019) Fuzzy fault-tolerant H∞ control approach for nonlinear active suspension systems with actuator failure. Procedia Comput Sci 148:465–474
19. Mohammadikia R, Aliasghary M (2019) Design of an interval type-2 fractional order fuzzy controller for a tractor active suspension system. Comput Electron Agric 167:105049
20. Zhang Z, Li H, Wu C, Zhou Q (2018) Finite frequency fuzzy H∞ control for uncertain active suspension systems with sensor failure. IEEE/CAA J Automatica Sinica 5(4):777–786
21. Xu Y, Xie Z, Zhao J, Li W, Li P, Wong PK (2021) Robust non-fragile finite frequency H∞ control for uncertain active suspension systems with time-delay using TS fuzzy approach. J Franklin Inst 358(8):4209–4238
22. Cao F, Sun H, Li Y, Tong S (2019) Fuzzy adaptive fault-tolerant control for a class of active suspension systems with time delay. Int J Fuzzy Syst 21(7):2054–2065
23. Wang W, Song Y, Chen J, Shi S (2018) A novel optimal fuzzy integrated control method of active suspension system. J Braz Soc Mech Sci Eng 40(1):1–10
24. Khodadadi H, Ghadiri H (2018) Self-tuning PID controller design using fuzzy logic for half car active suspension system. Int J Dyn Control 6(1):224–232
25. Abdul Zahra AK, Abdalla TY (2021) Design of fuzzy super twisting sliding mode control scheme for unknown full vehicle active suspension systems using an artificial bee colony optimization algorithm. Asian J Control 23(4):1966–1981

26. Azizi A, Ghafoorpoor Yazdi P (2019) White noise: applications and mathematical modeling. In: Computer-based analysis of the stochastic stability of mechanical structures driven by white and colored noise. Springer, Singapore, pp 25–36
27. Zadeh LA (1988) Fuzzy logic. Computer 21(4):83–93
28. Kickert WJM, Mamdani EH (1993) Analysis of a fuzzy logic controller. In: Readings in fuzzy sets for intelligent systems. Morgan Kaufmann, pp 290–297

The Project-Based Learning Approach: A Case Study on the Design of the Advanced Engineering Module to Meet Industry 5.0 Requirements

Aydin Azizi

Abstract To meet the needs of a knowledge-based society, the education system is in the midst of a rapid change. In tandem with the fifth industrial revolution, the emergence of inter-disciplinary engineering education has led to the convergence of numerous disciplines, necessitating a thorough discussion on the future of engineering education. To meet this demand, new interactive modules that are trans-disciplinary, interactive, and market-driven must be developed. It is widely acknowledged that project-based learning is a viable method for improving student learning in higher education institutions. This chapter discusses the use of project-based learning in the teaching of design skills to engineering students at Oxford Brookes University (OBU) in the UK. As a case study, a comprehensive explanation of how project-based learning (PBL), combined with constructive alignment and a blended learning approach, has been applied in ENGR6024: Advanced Control, a third-year module for the Mechanical Engineering Design degree programme, has been provided. In the case of the module assessment, the findings indicate that the module has been designed in an effective manner and offers an efficient mechanism for developing a wide range of abilities, including problem-solving, creativity, teamwork, and presenting skills.

Keywords Project-based learning · Industry 5.0 · Higher education · Engineering Module · Constructive alignment · Blended learning

UKPSF A1 · A2 · A3 · A4 · A5 · CK1 · CK2 · CK3 · CK4 · CK5 · CK6 · V1 · V2 · V3 · V4

OBU strategies EEA1AC4 · EEA2AC1

A. Azizi (✉)
School of Engineering, Computing and Mathematics, Oxford Brookes University, Oxford, UK
e-mail: aydin.azizi@brookes.ac.uk

A. Azizi (ed.), *Intelligent Mechatronic Control Systems*, Emerging Trends in Mechatronics, https://doi.org/10.1007/978-981-95-3925-3_5

1 Introduction

With higher expectations from industries, universities must provide their students with more advanced abilities, but at the same time, the revolution in information is affecting how students learn. In order to gain a competitive advantage in today's rapidly changing global market, Whitney et al. [1] believe that enterprises must create more value from their current resources and assets. According to Baillie et al. [2], companies need transformational workers who have the ability to analyse, criticise, and convey creative ideas in a team-based work environment. They also need employees who are self-motivated and have a positive attitude.

During recent years, colleges and universities have attempted to provide students with both the necessary cognitive information and professional competencies (such as problem-solving and collaboration) [3, 4]. These skill-related objectives, on the other hand, are difficult to accomplish since conventional learning has taken precedence, with instructors serving as 'the transmitters of knowledge' and students serving as 'the receivers of information' [5]. Because of this, students may struggle to fully participate in educational processes, which may result in just a shallow grasp of discipline knowledge on their part. Furthermore, institutions, particularly research universities, place a greater emphasis on developing students' research abilities rather than professional or transferable skills. This may result in a significant divide between what students learn at the university and what they will need to have for a successful career [6].

There are a number of professional accrediting and engineering assessment organisations that have set up task groups with a mission to enhance engineering curricula. One of the guiding principles of the United States Accreditation Board for Engineering and Technology (ABET) guidelines [7] was that instruction in design should incorporate at least some of the following: the use of open-ended questions, formulation of design problems, and consideration of alternatives.

Providing students with the chance to engage in actual problem-solving and knowledge creation in realistic professional settings has been proposed as a means of reversing the current situation. The use of project-based learning (PBL) is an appealing method of achieving this objective. The impacts of PBL and instructors' direct teaching on students' academic performance were compared by Chen et al. [8]. This research defines PBL as a learning process in which students are actively involved in the creation of genuine projects and the completion of real-world tasks. The findings revealed that project-based learning had a greater beneficial effect on students' academic performance than direct teaching did on the same measures.

Based on the definition proposed by researchers, PBL is referred to as project-based learning and is an inquiry-based instructional approach that involves learners in knowledge creation by requiring them to complete meaningful projects and create real-world outcomes [9, 10]. Six characteristics of PBL, as mentioned by Krajcik et al. [10], include questions that are challenging, emphasis on learning objectives, group involvement, the use of scaffolding technology, and the production of physical products. The production of artefacts that address genuine issues is the most important of these characteristics, which differentiates PBL from other student-centred pedagogies [11, 12]. This design process necessitates collaboration among learners in order to discover answers to real-world issues as part of the knowledge integration, application, and building processes. Students get feedback and assistance from instructors in order to aid them in their learning process.

Both PBL techniques and their effect on students' learning were addressed by Helle et al. [12]. Overall, it was discovered that the majority of the research used to determine practice details related to course description specifications, teacher criteria, and team size. Concerning the effect of PBL on student learning, the study revealed that few studies examined the impact on cognitive (e.g. knowledge) or emotional outcomes (e.g. motivation). In 2015, Ralph et al. [13] conducted another research in which they evaluated papers that used PBL in Science, Technology, Engineering, and Mathematics (STEM) teaching. It has been discovered that PBL aids in the development of both the knowledge and the abilities of the learners. Additionally, students believed that PBL facilitated group cooperation and discussion. Although some students indicated a lack of interest in working together, most expressed a willingness to do so. Reis et al. [14] used bibliometric and categorising techniques from the papers examined to conduct a literature review on PBL in engineering education. Examples of the top three terms utilised in bibliometrics were project-based learning, engineering education, and problem-based learning. The findings of the categorisation showed that more than 70% of the studies were conducted on undergraduates, and that the case study method was the most commonly utilised research technique in the studies. PBL has also been shown to enhance students' abilities, such as their academic knowledge, abilities, and motivation, but students have also said that the process of PBL has proven to be challenging for them (e.g. time consuming).

Chang et al. [15] investigated the PBL technique at Taylor's University School of Engineering, a project-based learning school where the syllabus for each undergraduate degree is structured to give every student the opportunity to apply theories learned in class. Belagra et al. [16] did research on PBL in higher education in 2018. This research aimed to determine how project-based learning and information and communication technologies affect student motivation. The study includes two second-year electrical engineering classrooms at the University Tahrir Mohamed in Béchar, Algeria. Integration of the tutorial with a project-based approach is likely to boost students' willingness to learn and master the subject of power electronics, according to the findings. Wu et al. [17] utilised Moodle as a learning platform and implemented project-based learning and SCAMPER teaching techniques in an information engineering project course in 2020. By engaging students in experimental activities for two semesters, this study evaluated cognitive, personal motivation, and personality differences between high- and low-creativity learners. In 2022, Rafael et al. [18] investigated the viability of a plan to teach this form of tool management using a project-based learning method in a distance learning environment. The methodology has been implemented cross-disciplinarily in two Master's degree courses linked to advanced design and manufacturing, thanks to the operation of the software for the product lifecycle management platform on virtual machines. The technique has yielded excellent pedagogical results in the teaching of industrial design and manufacturing-related skills. The virtual system has proven excellent efficiency, and students' professional development has progressed satisfactorily. The findings of all of these studies demonstrate the effectiveness of the PBL approach in designing successful modules.

Deep technical breakthroughs have occurred in the recent two decades, accompanied by significant disruptive advancements on both the software and hardware sides. Information, communication, and artificial intelligence (AI) are coming together, and many different ideas are crossing over with each other. These alterations are producing what is usually known as the digital transition, whose primary features are convergence and degradation. A convergence phenomenon is taking place, in which one level of convergence feeds another level of convergence, and eventually lays the groundwork for the birth of advanced and innovative technologies.

Industry 4.0 is the combination of the Internet and new technologies that drives a fundamental shift in industrial production. Industry 5.0 builds on this by putting research and innovation at the service of the shift to a sustainable, human-centred, and resilient industry. This paradigm shift occurs at the frontier of conventional fields [19].

The connections between different fields are becoming the heart of new technologies in a way that is neither multi-disciplinary nor inter-disciplinary, but rather trans-disciplinary. The boundaries between disciplines (and many other things) are collapsing; it is becoming harder to describe mechanical engineering, electrical engineering, and computer engineering. In actuality, engineering and the basic sciences (and even the social sciences) are difficult to define nowadays. In short, we are experiencing a phenomenon of confluence of multiple disciplines, prompting a comprehensive discussion on the future of engineering education. If we do not change the way we teach, we will be in trouble in 30 years [19].

Engineers of the twenty-first century are faced with the difficulty of managing change in a more competitive and global market, which is a must for businesses to exist. In this new era, engineering school graduates should be able to transition from technology to solutions and from solutions to operations. This demands a diverse set of skills [19].

To meet this need, it is essential and necessary to develop new trans-disciplinary, interactive, and market-driven modules. This chapter explains how the author of this chapter has combined various concepts from artificial intelligence, electromechanical systems, and decision-making techniques to design a new computer-based module.

This chapter explores project-based learning (PBL) at Oxford Brookes University and how I utilised it in teaching computer-based AI design skills to engineering students. A full explanation has been given as part of a case study on the use of PBL in ENGR6024: Advanced Control of the Engineering Programme, the module for the third-year students of the Mechanical Engineering Design degree programme. As for the module evaluation, the results show that the module provided an effective mechanism for developing a broad variety of different skills, including problem-solving abilities, creativity, computer-based design skills, utilising AI techniques, teamwork, and presentation skills.

After a short introduction about the project-based learning approach and the needs of higher education based on the fifth industrial revolution, the rest of the chapter has been organised in the following order. In the next section, the case study of the module and various parts of its body, such as the delivery method of the lectures and the assessment approach, have been discussed. The last two sections have been dedicated to the conclusion and bibliography.

2 Case Study (A1, A2, A4, V4)

In 2020, I started lecturing in the School of Engineering, Computing and Mathematics (ECM). I have had several teaching activities, such as designing new modules, lecturing, and running seminars for engineering students in Years 1–3. ENGR6024: Advanced Control is the most recent module for which I have been assigned to serve as the module leader and lecturer of the course to design, prepare, and deliver engaging lectures and seminars.

I developed ENGR6024 as a trans-disciplinary, interactive, and market-driven curriculum by mixing diverse concepts from artificial intelligence, electro-mechanical systems, and decision-making approaches.

This module expands students' understanding of control to include 'intelligent control' and 'decision-making theory'. The course content is separated into two sections: fuzzy logic and artificial neural network (ANN) approaches. Students examine the application of these artificial intelligence approaches to the modelling and control of dynamic systems, as well as the strategies required to create effective decision-making tools. Students investigate, model, and use these methodologies in a simulation environment using modern software such as MATLAB/Simulink and related packages, including control toolbox, fuzzy logic toolbox, PID controller toolbox, and neural networks toolbox, as needed to reach this aim.

For the first time, ENGR6024 has been offered to BEng Mechanical Engineering Design students in the second semester of the academic year 2021–2022. I designed the module in a manner that the lectures provide the background, set the scene, and develop the interest in the topic for each week. Typically, the core technology is also explained in the lectures. All the labs are computer-based experiments that involve the use of software such as MATLAB/Simulink.

In this module, I also make use of blended learning, with extensive learning material provided on the MOODLE site for the module. The module is assessed by 100% coursework, 40% assignment, and 60% mini-project. The assignment requires finding answers to questions using system models, training algorithms focusing on modelling a quarter car model utilising artificial neural networks. The mini-project contains a computer-based project about modelling and control of the quarter car model suspension system utilising MATLAB/Simulink software, focusing on fuzzy logic.

As regards Graduate Attributes, I directed most of this module to academic literacy, as it covers core knowledge and skills that are fundamental to the operation of utilising artificial intelligence techniques for the modelling and control of dynamic systems and for their use as decision-making tools; by far, the majority of the assessment is directed to this attribute.

Because a good deal of the module involves independent learning in the lab or at the computer, the attribute critical self-awareness and personal literacy is also developed. The assessment of this attribute is indirect, in that the learning outcome that relates most strongly to academic literacy cannot be mastered without this attribute.

Some important aspects of this module, particularly those relating to simulation and control, rely on sophisticated computer use, so a significant contribution towards digital and information literacy is also made. This is developed in the computer-based lab sessions and directly assessed in the final class test.

For the coursework, students work individually to design an artificial neural network model of a quarter car model. For the project Modelling and Control of an Active Suspension System, 18 students worked in eight groups. They developed a fuzzy logic controller to operate an active suspension system in the MATLAB-Simulink environment, where they exercised many modelling and control elements. All of the class sessions have been delivered face to face as synchronous lectures.

2.1 Teaching, Learning, and Assessment Strategy (A1, A2, A3, A4, CK1, CK2, CK3, CK4)

Lecturers and practical sessions on particular topic content, case studies, independent learning instruction, and assessment guidelines are all included in the teaching and learning activities. I was able to fulfil the Graduate Attributes in the following ways:

- **Academic Literacy**

Academic literacy has been addressed via the development of knowledge, skills, and an awareness of the engineering design process. This has been accomplished through the completion of a group report project as well as individual activities and assignments. The group project report and individual coursework were evaluated both orally and in writing.

- **Research Literacy**

The research literacy and digital information literacy skills have been acquired as a result of the individual and group design reports, which detail the design approach they used to create a computer-based solution. Additionally, there were particular activities in the field of simulation analysis that needed students to demonstrate research literacy in order to address issues analytically and to allow for cross-checking of computer simulations.

- **Critical Self-Awareness and Personal Literacy**

The critical self-awareness and personal literacy have been created via formative peer evaluation of module assignments, in which students evaluated their own work as well as the work of their peers. In addition, they organised themselves and their groups in order to complete the project successfully.

- **Digital Information Literacy**

Digital information literacy has been created as a result of the need to access computer-based academic research databases in order to get required information, such as access to the online MATHWORKS website.

- **Active Citizenship**

The active citizenship has been developed via learning outcomes, in which students acquired a knowledge of their position within a collaborative design project and how to communicate their work to a larger audience (i.e. to their designated class set).

Students have gotten the knowledge and skills that the learning objectives require by being taught in the form of lectures that are supplemented by seminars and group sessions. Students were assigned a study plan that guided them through the process of developing group and individual coursework submissions in the form of written reports and oral presentations. I graded the projects based on departmental standards in order to guarantee that their scores reflect the quality of work done. Students were made aware of the policies and procedures.

2.2 *Inclusive Learning Environment (A2, A4, V1, V2, CK3, CK4, EEA2AC1, EEA1AC4)*

In general, all sessions have been conducted based on the blended learning approach via utilising various tools, such as utilising computers to run software and interact with the lecturer, utilising the Zoom platform and its tools, etc. The following steps were made to create an inclusive learning environment for all students, taking into account their backgrounds and cohort composition.

I defined the project in accordance with the students' backgrounds; also, I delivered comprehensive and interactive lectures on each stage of the project during the semester. A PBL approach encourages students to become independent workers, critical thinkers, and lifelong learners by incorporating real-world context and technology into the curriculum. PBL is not merely a method of instruction; it is also a method of collaboration. If students learn to assume responsibility for their own education, they will have a solid foundation for working with others as adults.

In all sessions, I have provided the students with interactive lecture material in the form of the PowerPoint, lecture notes and an opportunity to work and practice in class with MATLAB software. It is important to mention that while students were following the PowerPoint slides at the same time the related step-by-step explanations were demonstrated in a software simulation environment. It is important to mention that all PowerPoint material is prepared in non-white background slides to help students with special needs.

I delivered all sessions face to face, with synchronous sessions split into two parts (lecture and seminar). In each lecture, the first hour began with reviewing previous sessions, followed by teaching and showing how to simulate a dynamic system and design a controller step by step in the Simulink environment. And for the remaining 2 h, the activities were performed in a computer-based environment.

Students were given the opportunity to ask questions, and I was able to ask them about each of the key parts of the lecture content to ensure that they understood it. Additionally, I explained the idea and processes to each of the students.

Students were required to implement all lecture-demonstrated procedures in the Simulink environment throughout the 2-h session. This method enables students to maximise their takeaways from the lecture and ensures they can complete all stages independently.

At Week 1, students made groups of 2–3 people, and they were working on their design and assigned tasks in groups during the practical sessions. During lectures and practical sessions, students had access to computers, which gave me the opportunity to design the session in a more interactive way. I prepared a step-by-step simulation manual for each session, so students, by following the steps, were able to design their model in the MATLAB environment. Additionally, I provided them with the MATLAB file of the design task for each week. This approach enabled students to discuss each step which was unclear for them with me and gave me the opportunity to provide more information and explanations for students who needed more support. Additionally, they could compare their final design with the provided correct simulation file and discuss their design during and after the sessions with me. It enabled me to monitor students and ensured that everyone engaged and left the session with maximum takeover. I utilised this approach as a tool to determine whether or not the learning objectives were accomplished and that all students completed the tasks. Additionally, students were permitted to ask questions during and in the middle of lectures at any point which was not clear to them. This gave me the opportunity to gauge their grasp of the lecture topic. Additionally, I explained the concept and actions individually to each student.

At the conclusion of the session, students were asked to review the entire session by completing some questions on the course's Moodle website, in order to document and determine whether the learning objectives had been fulfilled and that all students had finished the steps. This enabled me to determine if students have grasped all of the information and are capable of completing the session's defined tasks. Additionally, it assisted me in determining the students' perceptions of the session's quality and overall usefulness.

It is important to mention that all sessions have been recorded via the Zoom platform and have been available after the session for students to watch. Additionally, I allowed students who had issues to attend a particular session and be on campus to take part in the session by joining the lecture on the Zoom platform.

It is important to mention that students were given the opportunity to have individual weekly meetings with me as their instructor by booking a time slot via Google Calendar during my drop-in hours. In addition, the Blue Marking Card adjustment was available to students who had an inclusive support plan (ISP) specifying this adjustment.

2.3 *Role of Moodle (A2, A3, A4, CK3)*

Moodle was used to assist the module in the following ways:

Prior to each session, at the start of each week, I sent students an announcement outlining the subjects that would be addressed. Additionally, 48 h before the session, all course material (lecture notes, PowerPoint presentations and step-by-step practical approach) was uploaded on the Moodle page of the course under the tab of the related week. This allowed students to have clean notes during the lecture, in addition to having the answer to the Simulink modelling for the practical session.

Recordings of the sessions were posted after each session. Before posting the recordings, I had a quick review of all closed captions in Panopto to ensure the recorded captions convey the intended meanings.

Besides being a means of communication and posting lecture materials, Moodle has been used to get the feedback of the students about the course. As a final step, students were invited to reflect on the entire session by answering a few questions on the course's Moodle website, which was used to document and verify whether or not all of the learning objectives had been met. This allowed me to judge whether or not students had mastered the material and were capable of fulfilling the session's assigned tasks. In addition, it helped me determine how the students evaluated the session's quality and overall value.

2.4 *Opportunities for Formative Feedback (A1, A2, A3, A4, V3)*

There were numerous opportunities for students to complete work, have it evaluated, and receive feedback, all while knowing that this work did not count towards their grade. It was just to assist them in improving! This formative assessment was accomplished through the following mechanisms:

Computer-based experiments: In this module, students conducted a series of experiments. They all entailed answering questions, doing actions, simulating and developing virtual models, and documenting findings. They conferred with the instructor about the results and the responses they provided. The information given to students in the lectures and the practical sessions was tested throughout the module during the practical sessions. Furthermore, in Weeks 5 and 10, special group meetings were held in which each group received verbal feedback on their idea designs as well as comments on their development.

Constructive alignment: students tackled a set of problems in class, during lectures, as group projects, and shared their findings with other groups. They discussed the results with me at the session's conclusion. Throughout the practical sessions, as they worked on their group project, I offered specific advice on areas that required improvement, tasks that had fallen behind, and areas of excellent work. Finally, in Week 12, students got formative and summative comments on their group presentation.

In addition, it is important to mention that posting and asking questions via Moodle to exchange thoughts, observations, and issues with faculty and other students and to share the responses in order to increase the knowledge base of all participants and to foster the development of a self-sustaining community of autonomous learners was utilised as an effective tool for this module.

2.5 *Feedback (A5, CK5, CK6)*

To enhance the quality of the designed module, I solicited feedback at three distinct stages: before, during, and after its launch.

The first stage occurred during the summer of 2021, while I was designing course materials. During that period, every 2 weeks, I was drafting lecture notes and seminar materials for the second semester of the academic year 2021–2022 and soliciting comments from OBU experts in the field, which helped me ensure the quality and quantity of the materials to be presented.

Before beginning the lectures in February 2022, I uploaded the course content, including the assessments and coursework, to the defined module box on Google Drive to be reviewed by the internal examiner, and I amended the assessment description based on the input.

During the module delivery period, to enhance the teaching approach and know about the points that need attention, feedback was sought in two ways. First, the students were asked to participate in giving feedback via Moodle, and the second way one of the senior staff members attended the session and provided detailed feedback on the provided material and teaching approach as a peer assessment.

The second feedback approach is related to the feedback received from students. It is important to mention that the feedback of the students has been collected in Weeks 7 and 11 by asking students to answer various questions in Moodle. The questionnaires for these weeks were completely different; while the target of the first one was the quality of the particular session, the second one was about evaluating the whole module.

After the final evaluation, I presented the internal examiner with three samples of submissions from each of the three categories of attained marks: high, medium, and low. The grades were approved and given to the exam board committee, which comprised OBU faculty and external examiners, who then approved the final grades.

2.6 *Assessment Information (A3)*

Coursework in this subject was assessed through a combination of 40% an assignment and 60% a mini-project. The project entailed finding answers to queries using system models and artificial neural network-focused techniques. Focusing on fuzzy logic, the mini-project was a computer-based modelling and control of a dynamic system using MATLAB/Simulink. To assess the oral presentations, I designed the rubric shown in Fig. 1.

In the same manner, I designed two Rubrics for written reports of the artificial neural networks and fuzzy logic projects (see Figs. 2 and 3).

To gauge the collective effect of each team member, a method of peer evaluation will be executed for the next academic year (2022–2023) across 3 weeks in Weeks 5, 8, and 11. As part of this process, the group will assign a percentage contribution to each individual member. All group members will be required to sign the paper confirming their agreement with the assigned percentages. The cumulative contribution of all group members will be equal to 100%. To encourage the team to collaborate well, half of the group's report grade will be carried over, and the other half will be graded based on individual participation.

Take, as an example, a group of four members for a report mark of 62%. Therefore, for an equal share of work, all the members should have contributed 25% each (see Table 1).

For example, if the group mark is 62%, Student 1 would get half of the report mark carried over (i.e. 31%), even if his/her individual contribution was only 20%. Given that there were four people in the group, a reasonable contribution would have been 25%. Therefore, the other half of the report mark is scaled accordingly (i.e. he/she receives (20/25) * half the report mark). After adding the marks together, a final grade of 55.8% is calculated. In a similar vein, Student 3 gets a score of 31 + [(15/25) * 31] = 49.6%, and so on.

If a student has an individual support plan, every reasonable attempt will be made to offer an alternate assessment that meets the requirements. This will be determined on an individual basis in consultation with Wellbeing, the Programme Lead, and the Subject Co-ordinator. If the module is constrained by PSRB standards, every effort will be taken to assist the student during the assessment process.

Students Names		total mark	0.0
Group Number			

Weightage	Criterion	Mark distribution						Marks obtaine
		0	1 to <39	40 to <50	50 to <60	60 to <70	70 to 100	
0.3	Oral Presentation Quality • Were the speakers prepared, enthusiastic, and persuasive? • Was the overall purpose of the project explained clearly, including a clear statement of the aim and objectives? • Was the information delivered clearly? • Did the presenters show confidence in answering Questions?	Poor in every aspect	Inadequate quality of delivery	Satisfactory: Overall, can make out what is being presented. Covers slides at an inappropriate pace, may be monotonic. Doesn't highlight appropriate sections. Shows absolutely no interest in topic presented. Presenter mumbles, talks very fast, and speaks too quietly for a majority of students to hear & understand.	Overall, can make out what is being presented. Covers slides at a reasonable pace, may be monotonic. Doesn't highlight appropriate sections. Shows some negativity toward topic presented.Presenter's voice is low. The pace is much too rapid/slow. Audience members have difficulty hearing presentation.	Presented at a right pace, highlighted appropriate sections. Not monotonic. Occasionally shows positive feelings about topic. presented at a good pace so audience members can hear presentation. Does not read off slides. Presenter's voice is clear. The pace is a little slow or fast at times. Most audience members can hear presentation	Excellent quality presentation, very confident, excellent delivery. Rightly highlights appropriate sections. Demonstrates a strong, positive feeling about topic during entire presentation.Uses a clear voice and speaks at a good pace so audience members can hear presentation. Does not read off slides	0
0.3	Quality: • Were the slides and other visual aids clear? • Were they readable and highlighted the main points? • Were any artifacts used that added significantly to the presentation?	Poor in every aspect	Inadequate slides	just about satisfies requirements of presentation. There are some concerns with the qulity of slides, figures, tables and other visual aspects.	Good: appropriate quantity of text, figures and tables correctly presented and readable.	Very good: additionally appropriate use of animations, artefacts etc. The colour combination is visually appealing that helps readability.	Excellent quality slides and visual aides	0
0.4	Value of Content • Was the material presented of significant Quantity? • Did the content include sufficient critical analysis? • Was the project a success?	Not presented	Inadequately addressed	below average results: can't explain significant number of aspects. Some objectives are not addressed adequately but some meaningful outcomes be achieved. Not all parameters given.	satisfactory results: can't explain some aspects. Some significant outcomes can still be achieved. All objectives are addressed. Most of the parameters given.	good results: minor difficulty in interpretation and analysis of the results. The approaches used are appropriate and nothing is missing. All objectives addressed. All parameters reuired to conduct the study are given.	All options explored, evidenced by excellent literature review. Makes a very well informed choice of methods. Excellent description with details addressing all objectives.	0

Comments:

Fig. 1 Rubric: oral presentation

3 Conclusion and Future Work

In this chapter, the effectiveness of the project-based learning approach based on the Industry 5.0 needs has been explored by introducing the procedure of designing a new module, ENGR6024, offered in OBU for engineering students.

ENGR6024 has been delivered to BEng Mechanical Design Students for the first time in the second semester of the academic year 2021–2022. The aim of the course was to enhance the design skills of students via working individually and in groups for two design projects. These students were working on two individual and group projects. They practised several modelling and control aspects via utilising AI-based techniques. They have learned to design a plant simulator via ANN for the quarter car model, individual project, and design an effective controller utilising fuzzy logic theory to control an active suspension system in MATLAB-Simulink environment, group project.

Although teaching and delivering such a complex module for the first time had some challenges, students' performance was satisfactory, and all of them passed the

Student Name

Student number

Total Mark 0.0

Weightage	Criterion	Mark distribution 0	1 to <40	40 to <50	50 to <60	60 to <70	70 to 100	Marks obtai
0.1	Introduction: • Explanation of the topic • Demonstrates familiarity with the background literature on this topic • Clearly states the application of the project within the relevant field of study • Design Specification / Objectives	Not presented	Inadequatley addressed	Introduction is unclear with lots of low level data, some of it irrelevant. The target is vaguely defined. the objectives are presented but all of them may not be appropriate or some may be missising. There may be shortcomings in achieving the project outcomes.	Acceptable overview of the problem and the subject area that would benefit from industrial and research relevance being identified. the objectives are presented that address all aspects of the project. Some of the targets and measurables are not well defined.	Good overview of the problem and the subject area with industrial and research relevance well developed. wide range of literature sources which can benefit further from being put in the context of the project. The objectives are presented that address all aspects of the project. The targets and measurables are well defined.	Excellent overview of the problem and the subject area with industrial and research relevance well developed. wide range of literature sources used well, put in the context of the project. Excellent formation of objectives and their presentation.	0
0.3	Generating MATLAB Data Sets • Various concept ideas explored • Appropriate selection method/s applied • Described advantages / disadvantages of the final utilized approach • Various datasets based on behaviour of the compound pendulum in response to the different inputs.	Not presented	Inadequately addressed	The approach can produce results to an average level but there are significant errors in the theory and/or experimental approach. Inadequate use of literature. May not have a model validation process in place. The presented results can't explain significant number of aspects. With significant additional information results can be faithfully reproduced. Some objectives are not addressed adequately but some meaningful outcomes can still be achieved. Not all parameters given.	There are some errors in the theory and/or experimental approach and/or appropriateness of applied methods. Adequatley supported by published literature. Have a model validation process but may not address everything expected.The presented results can't explain some aspects. With some additional information results can be faithfully reproduced. Some significant outcomes can still be achieved. All objectives are addressed. Most of the parameters given.	Most of it is sound. There are very minor concerns in the theory and/or experimental approach but that are not expected to alter the outcome. Very well supported by published literature. The implications of assumptions are analysed to a very good level but may have some shortcomings. A detailed model validation pocess is in place. the presented results are good but there is minor difficulty in interpretation and analysis of them. With minor additional information results can be faithfully reproduced. The approaches used are appropriate and nothing is missing. All objectives addressed. All parameters reuired to conduct the study are given.	The methods used are theoretically, experimentaly very sound and are appropriate. Very well supported by the published literature. The implications of any assumptions are analysed in depth. Model validation is correctly incorporated.Interim results that are excellently described and analysed. Excellent description with details addressing all objectives so that the study outcomes can be faithfully reproduced.	0
0.4	Generating MATLAB ANN compound pendulum model • Appropriate theoretical framework • Design an ANN plant approximator • Train the proposed ANN • Analyse the performance and reliability of the ANN model • Explore and analyse the performance of the designed ANN by changing the number of	Not presented	Inadequately addressed	The approach can produce results to an average level but there are significant errors in the theory and/or experimental approach. Inadequate use of literature. May not have a model validation process in place. The presented results can't explain significant number of aspects. With significant additional information results can be faithfully reproduced. Some objectives are not addressed adequately but some meaningful outcomes can still be achieved. Not all parameters given.	There are some errors in the theory and/or experimental approach and/or appropriateness of applied methods. Adequatley supported by published literature. Have a model validation process but may not address everything expected.The presented results can't explain some aspects. With some additional information results can be faithfully reproduced. Some significant outcomes can still be achieved. All objectives are addressed. Most of the parameters given.	Most of it is sound. There are very minor concerns in the theory and/or experimental approach but that are not expected to alter the outcome. Very well supported by published literature. The implications of assumptions are analysed to a very good level but may have some shortcomings. A detailed model validation pocess is in place. the presented results are good but there is minor difficulty in interpretation and analysis of them. With minor additional information results can be faithfully reproduced. The approaches used are appropriate and nothing is missing. All objectives addressed. All parameters reuired to conduct the study are given.	The methods used are theoretically, experimentaly very sound and are appropriate. Very well supported by the published literature. The implications of any assumptions are analysed in depth. Model validation is correctly incorporated.Interim results that are excellently described and analysed. Excellent description with details addressing all objectives so that the study outcomes can be faithfully reproduced.	0
0.2	Overall Structure of the Report • Structure of the report and literary accuracy • Quality of the general presentations • Figures, Graphs, manufacturing drawings with self-sufficient details • Use of references consistent with the reference Citation Guidelines	Poor in every aspect	Inadequately addressed	just about satisfies requirements of over all structure and personal learning and development evalouation at this level . There are some concerns with the qulity of figures, tables. There is a significant lack of referring to scientific articles.	appropriate quantity of text, figures and tables correctly presented and readable. The references should be improved. Appropriate personal learning and development evalouation.	good use of figures and tables ,good set of scientific refrences which can be improved by citing the papers publishied during past 5 years. Good personal learning and development evalouation.	Excellent personal learning and development evalouation. Excellent quality of figures, tables and ecellent use of set of scientific references.	0

Comments

Fig. 2 Rubric: ANN written report

course. We conducted a student survey in Week 11, and the results indicate that the majority of students were satisfied with the module. To enhance the student learning experience and accommodate a wider cohort, the module has been integrated with the Advanced Electronics module and is now offered as ENGR6034 Electronics & Control II, which commenced in the 2025–2026 academic year.

The findings indicate that the module offered an efficient mechanism for developing a wide range of abilities, including problem-solving, creativity, teamwork, and presenting skills.

Students Names		Total Mark	0.0
Group number			

Weightage	Criterion	Mark distribution						Marks obtained
		0	1 to <40	40 to <50	50 to <60	60 to <70	70 to 100	
0.1	Introduction: • Explanation of the topic • Demonstrates familiarity with the background literature on this topic • Clearly states the application of the project within the relevant field of study • Design Specification / Objectives	Not presented	Inadequatley addressed	Introduction is unclear with lots of low level data, some of it irrelevant. The target is vaguely defined. the objectives are presented but all of them may not be appropriate or some may be missising. There may be shortcomings in achieving the project outcomes.	Acceptable overview of the problem and the subject area that would benefit from industrial and research relevance being identified. the objectives are presented that address all aspects of the project. Some of the targets and measurables are not well defined.	Good overview of the problem and the subject area with industrial and research relevance well developed. wide range of literature sources which can benefit further from being put in the context of the project. The objectives are presented that address all aspects of the project. The targets and measurables are well defined.	Excellent overview of the problem and the subject area with industrial and research relevance well developed. wide range of literature sources used well, put in the context of the project. Excellent formation of objectives and their presentation.	0
0.3	Generating MATLAB-Simulink active suspension model • Present the mathematical model of an active suspension system and its components. • Design The Simulink model of the proposed suspension system. • Simulate and analyse the behaviour of the active suspension system based on the various road disturbances.	Not presented	Inadequately addressed	The approach can produce results to an average level but there are significant errors in the theory and/or experimental approach. Inadequate use of literature. May not have a model validation process in place. The presented results can't explain significant number of aspects. With significant additional information results can be faithfully reproduced. Some objectives are not addressed adequately but some meaningful outcomes can still be achieved. Not all parameters given.	There are some errors in the theory and/or experimental approach and/or appropriateness of applied methods. Adequatley supported by published literature. Have a model validation process but may not address everything expected.The presented results can't explain some aspects. With some additional information results can be faithfully reproduced. Some significant outcomes can still be achieved. All objectives are addressed. Most of the parameters given.	Most of it is sound. There are very minor concerns in the theory and/or experimental approach but that are not expected to alter the outcome. Very well supported by published literature. The implications of assumptions are analysed to a very good level but may have some shortcomings. A detailed model validation pocess is in place. the presented results are good but there is minor difficulty in interpretation and analysis of them. With minor additional information results can be faithfully reproduced. The approaches used are appropriate and nothing is missing. All objectives addressed. All parameters reuired to conduct the study are given.	The methods used are theoretically, experimentaly very sound and are appropriate. Very well supported by the published literature. The implications of any assumptions are analysed in depth. Model validation is correctly incorporated.Interim results that are excellently described and analysed. Excellent description with details addressing all objectives so that the study outcomes can be faithfully reproduced	0
0.4	Generating Fuzzy Logic Controller • Appropriate theoretical framework. • Define inputs and outputs. • Design membership functions of the inputs and outputs. • Design Rule Base. • Design the FIS based on the designed inputs/outputs and rule base. • Analyse the performance and reliability of the FLC. • Correct interpretation of the results. • Logical conclusions drawn from the analysis	Not presented	Inadequately addressed	The approach can produce results to an average level but there are significant errors in the theory and/or experimental approach. Inadequate use of literature. May not have a model validation process in place. The presented results can't explain significant number of aspects. With significant additional information results can be faithfully reproduced. Some objectives are not addressed adequately but some meaningful outcomes can still be achieved. Not all parameters given.	There are some errors in the theory and/or experimental approach and/or appropriateness of applied methods. Adequatley supported by published literature. Have a model validation process but may not address everything expected.The presented results can't explain some aspects. With some additional information results can be faithfully reproduced. Some significant outcomes can still be achieved. All objectives are addressed. Most of the parameters given.	Most of it is sound. There are very minor concerns in the theory and/or experimental approach but that are not expected to alter the outcome. Very well supported by published literature. The implications of assumptions are analysed to a very good level but may have some shortcomings. A detailed model validation pocess is in place. the presented results are good but there is minor difficulty in interpretation and analysis of them. With minor additional information results can be faithfully reproduced. The approaches used are appropriate and nothing is missing. All objectives addressed. All parameters reuired to conduct the study are given.	The methods used are theoretically, experimentaly very sound and are appropriate. Very well supported by the published literature. The implications of any assumptions are analysed in depth. Model validation is correctly incorporated.Interim results that are excellently described and analysed. Excellent description with details addressing all objectives so that the study outcomes can be faithfully reproduced	0
0.2	Overall Structure of the Report • Structure of the report and literary accuracy • Quality of the general presentations • Figures, Graphs,	Poor in every aspect	Inadequately addressed	just about satisfies requirements of over all structure at this level. There are some concerns with the qulity of figures, tables. There is a significant lack of referring to scientific articles.	appropriate quantity of text, figures and tables correctly presented and readable. The references should be improved.	good use of figures and tables ,good set of scientific refrences which can be improved by citing the papers publishied during past 5 years.	Excellent quality of figures, tables and ecellent use of set of scientific references.	0

Comments:

Fig. 3 Rubric: fuzzy logic written report

Table 1 Example of a group self-assessment

Group member	Contribution (%)	Report mark	50% Report mark fwd	50% Report mark scl	Total
Student 1	20	62	31	31*(20/25) = 24.8	55.8
Student 2	30	62	31	31*(30/25) = 37.2	68.2
Student 3	15	62	31	31*(15/25) = 18.6	49.6
Student 4	35	62	31	31*(35/25) = 43.4	74.4
Total	100%	–	–	–	–

Conflicts of Interest The authors declare they have no conflicts of interest.
Funding The project has received no funding.

Data Availability The related data are presented within the manuscript.

References

1. Whitney J (1992) Serious creativity Using the power of lateral thinking to create new ideas: By Edward de Bono (Harper Business, 1992). $23.00, hardcover. Pergamon, p 338
2. Baillie C, Fitzgerald G (2000) Motivation and attrition in engineering students. Eur J Eng Educ 25(2):145–155
3. Vogler JS et al (2018) The hard work of soft skills: augmenting the project-based learning experience with interdisciplinary teamwork. Instr Sci 46(3):457–488
4. Talmi I, Hazzan O, Katz R (2018) Intrinsic motivation and 21st-century skills in an undergraduate engineering project: the formula student project. High Educ Stud 8(4):46–58
5. Alorda B, Suenaga K, Pons P (2011) Design and evaluation of a microprocessor course combining three cooperative methods: SDLA, PjBL and CnBL. Comput Educ 57(3):1876–1884
6. Holmes MH (2012) Introduction to perturbation methods, vol 20. Springer Science & Business Media
7. Prados JW (1998) Engineering education in the united states: past, present, and future
8. Chen C-H, Yang Y-C (2019) Revisiting the effects of project-based learning on students' academic achievement: a meta-analysis investigating moderators. Educ Res Rev 26:71–81
9. Brundiers K, Wiek A (2013) Do we teach what we preach? an international comparison of problem-and project-based learning courses in sustainability. Sustainability 5(4):1725–1746
10. Krajcik, J.S. and P.C. Blumenfeld, *Project-based learning*. 2006: na.
11. Blumenfeld PC et al (1991) Motivating project-based learning: sustaining the doing, supporting the learning. Educ Psychol 26(3–4):369–398
12. Helle L, Tynjälä P, Olkinuora E (2006) Project-based learning in post-secondary education–theory, practice and rubber sling shots. High Educ 51(2):287–314
13. Ralph RA (2016) Post secondary project-based learning in science, technology, engineering and mathematics. J Technol Sci Educ 6(1):26–35
14. Reis ACB, Barbalho SCM, Zanette ACD (2017) A bibliometric and classification study of project-based learning in engineering education. Production 27
15. Chung EC (2019) The design spine: the core design modules behind Taylor's School of Engineering project-based learning. In: Engineering grand challenges in scholar programs. Springer, Singapore, pp 17–31
16. Belagra M, Draoui B (2018) Project-based learning and information and communication technology's integration: impacts on motivation. Int J Electr Eng Educ 55(4):293–312
17. Wu TT, Wu YT (2020) Applying project-based learning and SCAMPER teaching strategies in engineering education to explore the influence of creativity on cognition, personal motivation, and personality traits. Thinking Skills Creativity 35:100631
18. Sola-Guirado RR, Guerrero-Vacas G, Rodríguez-Alabanda Ó (2022) Teaching CAD/CAM/CAE tools with project-based learning in virtual distance education. Educ Inf Technol 27(4):5051–5073
19. Broo DG, Kaynak O, Sait SM (2022) Rethinking engineering education at the age of industry 5.0. J Ind Inf Integr 25:100311

Developing a Data Envelopment Analysis Model for Performance Evaluation of Green Supply Chain Management

Farzad Zaare Tajabadi and Sahand Daneshvar

1 Introduction

With the increasing concerns about the environment in the recent decade, environmental pollution should be paid attention to in the processes besides industry development. The environmental and economic benefits of product remanufacturing and consumer environmental consciousness have pushed numerous manufacturing and retail enterprises to produce and sell green products [43]. All the solutions to this problem should be combined and reviewed in a comprehensive supply chain procedure framework. Supply chain management (SCM) is an important factor that is directly related to a company's efficiency and competitive position. SCM entails planning product manufacture from raw materials to customer delivery [1]. Supply chain management is essential to every company organization, and proper planning will assure economic, environmental, and social sustainability [6]. Greening the supply chain is a new concept. Green supply chain management (GSCM), in particular, has seen considerable expansion in recent years [48]. Green supply chains rely

The original version of the chapter has been revised. A correction to this chapter can be found at https://doi.org/10.1007/978-981-95-3925-3_8

F. Zaare Tajabadi · S. Daneshvar
Department of Industrial Engineering, Engineering Faculty, Eastern Mediterranean University, Famagusta, 99628 North Cyprus via Mersin 10, Türkiye
e-mail: sahand.daneshvar@emu.edu.tr

F. Zaare Tajabadi (✉)
Department of Industrial Engineering, Faculty of Engineering, Near East University, Nicosia, 99138 North Cyprus via Mersin 10, Türkiye
e-mail: farzad.zaare@neu.edu.tr

Research Center for Science, Technology and Engineering (BILTEM), Near East University, Nicosia, 99138 North Cyprus via Mersin 10, Türkiye

A. Azizi (ed.), *Intelligent Mechatronic Control Systems*, Emerging Trends in Mechatronics, https://doi.org/10.1007/978-981-95-3925-3_6

primarily on cutting costs, enhancing efficiency, and generating "green" or environmentally friendly products [63]. According to this concept, the purchaser uses his/her purchasing power to demand better environmental performance from the upstream supplier in the supply chain. This means that, in most cases, the purchaser is a big company that has a facilitating role for its suppliers. These suppliers are usually companies of small or medium size, and they help them to become environmentally friendly organizations. In recent years, following the rapid industrialization of most of the developed economies, environmental losses have been given a great deal of attention, and all over the world, governments have begun to apply environmental protection rules. Although from a merely economic point of view and regardless of some side factors, industry and industrial activities, constituting a major share of national income and a large percentage of the present human force of society, are considered the main basis of development and growth of countries. Moreover, as a major indicator of development, from a sustainability view, the benefits of such a development which is resulted from not regarding environmental aspects of exploitation of national resources and environmental protection are not justifiable in the long term, because from an environmental point of view, destructive consequences of industrial activities will finally lead to health damage, lack of working motivation, and, above all, gradual or sudden decrease in life quality in societies and residential areas. This imposes further costs on society in addition to the apparent and tangible costs.

In recent years, nearly 70% of the world's leading firms have prioritized sustainability in their work plans [31]. According to the reports of leading firms, the success of their sustainability activities is also dependent on collaboration with supply chain (SC) participants [22, 36]. Green supply chain management (GSCM) entails incorporating environmental and economic goals into the supply chain's operating plan. This type of integration reduces carbon footprint while enhancing financial return and profitability [54]. Bowen et al. [5] define GSCM as the "integration of the company's purchase plan with the environmental activities in SCM, to improve the environmental performance of suppliers and customers." Concerns of product design, usage, reuse, disassembly, and final disposal are also included in GSCM [51], in addition to storage, transportation, supplier development to satisfy green purchasing criteria, and encouragement for the adoption of environmental certifications such as ISO 14000 [51, 53]. Zhu and Sarkis [74] consider GSCM to be the integration of environmental thinking with SC operations management, beginning with product design and progressing through raw material selection, manufacturing processes, transportation and delivery, and the end consumer arriving at the final destination after usage. According to Large and Gimenez Thomsen [41], GSCM comprises the design process, raw material selection, green procurement, production, distribution, and reverse logistics (Fig. 1).

Therefore, regarding the emphasis on organizational efficiency to effectively and properly use the resources for achieving organizational goals, and also due to the requirements of national and international rules about environmental issues, proper compromise and compatibility between the two goals of economic growth and environmental protection, and integrating the two important issues of efficiency and

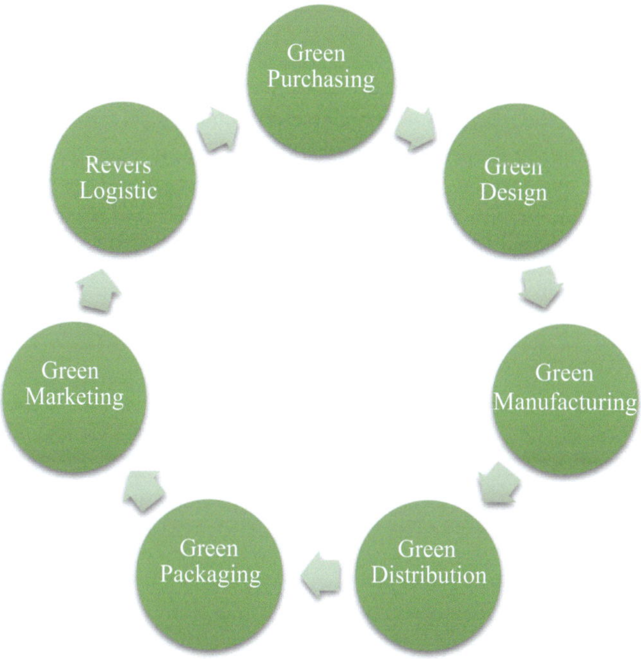

Fig. 1 Functions of GSCM

environmental protection under the title "Green Efficiency Program," can be of great importance.

This study poses the following research questions: How can we measure efficiency in green supply chains? How can we distinguish the potential decision-making units (DMUs) among inefficient ones that can be efficient with less cost and effort according to benchmark DMUs? And how can we verify the effect of green factors on the decision-making unit's efficiency values?

The following can be considered as the contributions of the current study: efficiency evaluation of the SCs as DMUs with different products, production costs per unit, and chain sizes to introduce the efficient DMUs and consider them as a benchmark; finding the potential DMUs (those that can become efficient) among the inefficient DMUs that can become efficient after modification according to the benchmarks; and, finally, verifying the effectiveness of three green factors, such as service level, emissions (CO_2), and the size of the chains (arcs) on the efficiency of the whole system before and after modifications of the potential DMUs, and comparing the results. The expectations include the positive effect of service level (customer satisfaction) and arcs (as the size of the transportation factor) and the negative effect of CO_2 emissions (environmental effect) on the efficiency value of DMUs. Moreover, the changes in the mentioned effects after improvement adjustments will help the decision-makers identify the true performance of the green supply chains.

1.1 Research Structure

The research design is organized as follows: Sect. 2 contains a literature review of previous relevant research in efficiency improvement and performance evaluation in supply chain and green supply chain management. Section 3 introduces the methodology procedure and explains it in detail. In Sect. 4, a case study is considered, and some analytical discussions are provided. The conclusion and results are presented in Sect. 5 with guidelines for future studies (Fig. 2).

2 Literature Review

This section provides an overview of the previous research that has direct consequences on efficiency improvement and performance evaluation in supply chain and green supply chain management. It also aims to study the literature in the backgrounds on supply chain management, methods of solving multiobjective problems of

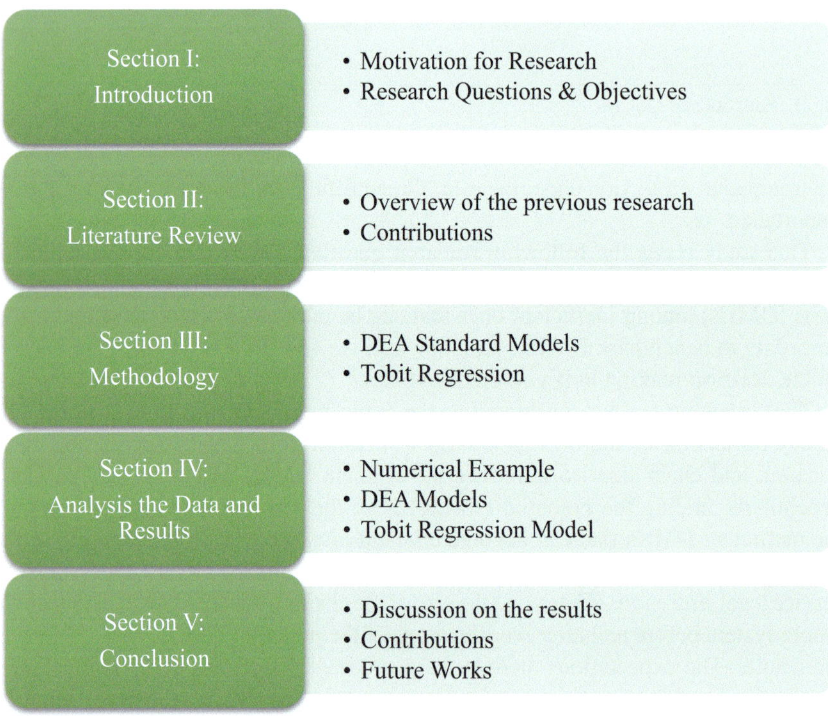

Fig. 2 Research framework

supply chain network designing, green supply chain management, developed benchmark approaches, and standard data envelopment analysis (DEA) models to find out the research gap and distinguish the recent approaches, methods, and models that are discussed for evaluating the efficiency of supply chain and green supply chain management.

2.1 Background of Supply Chain Management

In the 1990s, along with the improvement of production capacities, the managers of industries found that the raw materials received from different suppliers had a significant role in increasing organizational capabilities to meet the customers' needs. This also significantly affected the organization's concentration, supply bases, and resourcing strategies. Also, the managers found that only manufacturing a qualitative product is not enough, and in fact, supplying the products with the criteria considered by the customers (when, where, and how) and their desired quality and cost led to new challenges. Under such conditions, as a conclusion of the mentioned changes, they found that they would not be enough to manage their organization for a long time. In managing the network, they needed to be involved with the customers in managing the network of all the factories and companies that supply their organization inputs directly or indirectly, as well as the network of companies engaged in delivery and after-sales services. With such an attitude, the approaches of "supply chain" and "supply chain management" appeared.

Supply chain management (SCM) is considered one of the important management activities in most organizations. In general, all of the activities of an organization that involve the suppliers, factories, warehouses, retailers, and final consumers and require managing the goods, finance, and information are referred to as the supply chain (SC) in the literature [44]. Due to the complexity of mutual relationships between different components of SC, the supply chain is exposed to a wide range of risks, making decision-making a challenging problem for managers. Undoubtedly, one of the most important and difficult aspects of the supply chain structure of goods and services is making decisions about facility location for each layer of the chain. This is considered one of the strategic decisions, usually, its change is not justifiable in the short term or even the midterm period [29]. There are examples that there is uncertainty and risk in the reliability of chain facilities due to phenomena such as natural disasters, changes in capital holders, mistakes of the labor force, climatic conditions, etc. In October 2001, the prevalence of anthrax virus in the Washington branch of the US Postal Service led to the closure of the 633,000-square-foot office (facility) and, as a result, the loss of a major part of the capacity of the chain. Hurricane Katrina in 2005 resulted in the fact that, after many years, parts of Central Louisiana are still faced with a shortage of materials due to the breakdown of some factories and warehouses [57]. Quarrels among labor forces in September 2002 led to the closure of western ports of America, while the retailers intended to store a great deal of goods before the beginning of seasonal holidays. In this situation, production was stopped for a while in some factories. Therefore, when making strategic decisions, uncertainties of the real world should be considered so that, at the time of their occurrence, the system can continue functioning with the minimum loss.

2.2 Definition of Supply Chain

Different sources have presented very comprehensive definitions of supply chains. For example, in the American Production and Inventory Control Society (APICS) dictionary defines the term "supply chain" as follows:

A supply chain includes all the processes linking the suppliers and consumers to each other. Such processes begin with the initial raw materials and end with the final consumer receiving the completed product. In addition, the term supply chain was mentioned in the American Production and Inventory Control Society (APICS) dictionary as follows: The functions inside and outside a company enable the value chain to produce products and deliver customer service.

In this definition, value chain means the functions inside a company that add value to the products or services sold by the organization to receive money.

In another definition, the supply chain was defined as follows:

A network of interconnected organizations through the upstream and downstream relations in different processes and activities produces value as a product or service for the final customer [19]. This definition indicates that a supply chain is a group of three or more organizations that are directly present in the flow of products or services to final customers. The integration of supply chain management seems necessary to ensure coordination in the supply chain.

However, some other definitions were presented from the supply chain dimensions of the supply chain. Here is a review of some of these definitions:

Supply chain management is a set of methods used for effectively integrating the suppliers, manufacturers, warehouses, and stores to produce the required products at a certain amount, place, and time to minimize the total chain costs and satisfy customer needs at a high service level.

A supply chain refers to a set of activities coordinating the following goals [37]:

- Having access to new materials
- Transferring and transforming the raw materials into certain products
- Creating the value added in these products
- Distributing these products to retailers and customers
- Facilitating the exchange of information between different institutions in the business environment (including suppliers, manufacturers, distributors, etc.)

In this definition, the supply chain includes two main processes:

- Materials management (internal logistics)
- Physical distribution and support (external logistics).

Materials management deals with the discharge and storage of raw materials, parts, and procurement. Specifically, materials management includes all steps of purchase, internal control, materials planning for production, circulating inventory control, warehouse, transportation, and final product distribution. On the other hand, physical distribution and support involve all external logistics activities in customer service delivery. These activities include taking orders, inventory settlement, warehousing, locating, external transportation, pricing, advertising, product locating, and life-cycle support. As a result of combining materials management with physical distribution and support, a supply chain cannot be a linear chain of one-by-one business relations but a texture of multiple relations and networks.

Harland [34] explained supply chain management as the business process management and internal relations of an organization with direct and indirect suppliers (first-rank, second-rank, etc.) and customers over the supply chain with supply chain institutions. He defined supply chain management as a tool for linking the chain of each supply and manufacturing process until the final consumer, including the organizational boundaries. Based on this extensive definition, supply chain management includes all activities and institutions in the supply chain from raw materials extraction until the end of their useful life.

According to Beamon [10], supply chain management is defined as the set of common integrated processes between a variety of firms (e.g., suppliers, manufacturers, distributors, and sellers) working together to achieve the following goals:

- Achieving the raw materials
- Converting the raw materials into certain final products
- Delivering these final products to sellers and finally to customers

Supply chain management was defined by Simchi-Levi et al. [61] as follows:

SCM is a set of approaches used for the efficient integration of suppliers, manufacturers, warehouses, and stores to minimize system costs and satisfy the needs of customers so that the products are produced and distributed at the right place and time. The abovementioned definitions indicate that supply chain management depends on all partners of the chain and considers the whole supply chain. The purpose of supply chain management is to increase clarity among the members and ensure that their goals are aligned with each other. For this purpose, all chain elements should cooperate; otherwise, the chain will be disrupted.

2.3 *Background of Research on Methods of Solving Multiobjective Problems of Supply Chain Network Design*

Since supply chain management is based on integrating suppliers, manufacturers, warehouses, and stores, supply chain management encompasses all activities of the company at various levels, from the strategic to the tactical and operational. All of these activities should be taken into account when designing the supply chain. To design a supply chain, decision-making should be conducted in five areas: production inventory, facilities, location, transportation, and information. Chopra and Meindl [20] referred to these five areas as functional levers and stated that the required potentials of a supply chain can be created by managing these levers. Efficient supply chain management requires an understanding of each lever and its function. Each lever has a direct effect on the supply chain and activates certain potentials in the chain. An appropriate combination of accountability and efficiency of each lever increases the efficiency and reduces the operational and inventory costs in the supply chain.

There are a few types of research in the literature on supply chain design with a multiobjective optimization approach. Sabri and Beamon [50] used the ε-limitation method for solving the model after designing an integrated three-purpose model for minimizing the cost, maximizing the service rate, and increasing the flexibility regarding the uncertainty of delivery time and demands for the product. Chan et al. [13] proposed a multiobjective genetic algorithm for demand-oriented distribution problems in the supply chain network. The goals of the mentioned problem included optimization of system costs, total delivery time, and efficiency rate of the manufacturers' capacity. Chen and Lee [17] designed a multi-period, multistage, and multi-product scheduling model in a supply chain network about the uncertainty of demand and product price with the goal of fair distribution of profits among all the stakeholders, keeping the inventory and service rate at the optimum level, and stable decision making regarding the instability of demand. For solving the problem, they used a two-stage fuzzy decision-making method. By using ε-limitation and branch-and-bound techniques, [30] solved their two-purpose model (to maximize the profit in the determined period and increase the level of customer satisfaction), which belonged to the category of linear mixed-integer random planning.

Considering operational costs, service level, and resource efficiency as the goals of the study of production and distribution problems, [13] proposed a hybrid approach based on the combination of genetic algorithms by the hierarchical method (AHP).

2.4 Background of Green Supply Chain Management

In 1996, Michigan State University's Industrial Research Association introduced green supply chain management. Environmental protection is achieved through the use of this new management model. From the perspective of the product life cycle, green supply chain management includes all the stages of raw materials extraction, product design, manufacturing, sales, transportation, usage, and recycling. Using green supply chain management and technology, the company can decrease the negative environmental effects and achieve the optimum use of resources and energy. Greening the supply chain is the process of considering the environmental criteria or observations throughout the supply chain. Green supply chain management integrates supply chain management with environmental requirements in all the stages of designing the product, choosing and supplying the raw materials, manufacturing and building, distributing and transporting, delivering to the customer, and managing recycling after consumption for maximizing the efficiency of energy and resource consumption besides improving the function of the whole supply chain [52]. During the review of the environmental impacts of supply chain activities, the effects of products on the environment are examined from a holistic perspective, including the analysis of the life cycle of a product from its beginning to its end of life. The ecological effects of every activity throughout the product life cycle (the science of the organisms' habits and their interactions with the environment) are measured and considered in the design process, including product conception, design, raw material preparation, manufacturing and construction, assembly, storage, packaging, transportation, and subsequent use of the product [71].

Kuo et al. [39] aimed to create a green supplier selection model that combines an artificial neural network (ANN) and two multi-attribute decision analysis (MADA) methods: data envelopment analysis (DEA) and analytic network process (ANP). [32] introduced 19 environmental criteria in their article and classified them into five groups: purchase management, research and development management, process management, quality control of input materials, and system management. They then chose the suppliers using the network analysis process technique.

In their study, [16] chose 18 criteria, the most important of which include environmental criteria, environmental management systems, supplier's profitability, and close relationships with the supplier, and then, using fuzzy theory, the criteria were converted to definite numbers.

In a case study on the printed circuit board in Taiwan [16] sought to implement green supply chain management for selecting the supply. They developed two classes of criteria—environmental and nonenvironmental—in the studied company, determined the weights of the criteria based on qualitative and quantitative factors, and finally used the Gray analysis method for rating the suppliers. Shaik and Abdul-Kader [56] used a framework of environmental, green, and organizational criteria for selecting the green supplier. They created a hierarchy for evaluating the criteria and subcriteria of suppliers, which led to the compilation of an appropriate liability strategy by the managers.

For a two-echelon supply chain, [62] presented an NL-IP model aiming to minimize total costs, maximize demand served, and minimize transportation pollution. Two metaheuristic algorithms, non-dominated sorting genetic algorithm II (NSGA-II) and non-dominated ranking genetic algorithm (NRGA), were developed to solve the problem, and the Taguchi method was used to set the parameters. Forghani et al. [28] modified classic stochastic data envelopment analysis (SDEA) model by manipulating weak efficient hyperplanes. The suggested model was applied to the environmental efficiency of sustainable development goals in Latin America and the Caribbean (LAC) countries, resulting in better discrimination. A summary of the research on green supply chain management is presented in Table 1.

According to Table 1, this study emphasizes three criteria in green supply chain management such as green transportation, green production, and green production cost to reduce the emissions and cost of production. This study also considers the

Table 1 Criteria of the green supply chain in the previous studies

Criterion	Component	Authors (year)
Green supply and purchase	Choosing the supplier regarding the environmental criteria, providing the materials in environmentally friendly packs, having environmental certificates such as ISO 14000, holding seminars for informing the suppliers about environmental issues, supporting the suppliers in improving their environmental performance, requiring the suppliers to observe environmental rules, and buying recyclable materials	[25, 39, 46, 58, 66, 75, 77]
Green designing	Designing the products regarding the reduction of material or energy consumption and recyclability of the products, designing the product for reducing or avoiding the consumption of dangerous materials or inappropriate production processes, and designing the products to reduce their environmental effects	[24, 25, 39, 45, 46, 67, 76]
Green production	A commitment of the senior and junior managers to observing the environment-related rules, having qualified environmental management, having environmental certificates such as EUP, ROHS, and ODC, using materials with less harm to the environment, using devices with less pollution to the environment, controlling the release of dangerous gases such as ammonia and CO_2, using appropriate methods for removing wastewater, having an appropriate environmental position to other manufacturers, having a low occurrence of environmental incidents, decreasing noise pollution, holding environmental educational programs for the staff and the managers, focusing on reducing the wastes and optimizing the use of materials, and using environmentally friendly equipment and technology	[24, 32, 39, 45, 46, 64, 67, 76, 77]

(continued)

Table 1 (continued)

Criterion	Component	Authors (year)
Green packing	Using recyclable packs and containers, using environmentally friendly materials in packing the products, using labels to show the level of accordance of the product with environmental standards, and using labels to show the recyclability of the product	[24, 25, 39, 45, 46, 67, 76, 77]
Green transportation and distribution	Marketing the products relies on environmental issues such as emphasizing environmental certificates, increasing the consumers' environmental awareness, choosing clean transportation methods, returning the products to the company for recycling, having a better competitive situation than other competitors, and choosing the distribution networks and customers with an emphasis on environmental criteria	[24, 25, 39, 45, 46, 67, 76, 77]
Green production costs	The costs of eliminating dangerous and harmful materials, costs of producing environmentally friendly products, costs of offering environmentally friendly packs, and costs of informing the staff about environmental products	[25, 39, 45, 46, 67, 76]

efficacy of some green factors on the whole system's efficiency like the service level to meet customer satisfaction regarding ordering, green product cost and delivery, and emissions (CO_2) caused by vehicles through the chain and arcs, which indicate the size of the chain for transportation.

2.5 *Background of Benchmarking in GSCM*

Benchmarking likely had its origins in the textile mills of the 1800s [7]. Over time, it has experienced significant advancements, particularly concerning the emergence of quality management principles. The Xerox[1] Corporation began using benchmarking as an effective and practical management tool during the 1980s, when the company was losing market share and experiencing significant pressure from competitors, particularly Japanese companies. It was Xerox's success that encouraged many other

[1] Xerox Corporation is a globally operating American multinational company that offers a wide range of paper-based solutions, services, and information technology products on a global scale.

corporations to adopt this revolutionary approach to increase performance levels and production efficiency to gain a competitive advantage [11]. Several studies have found that it improves performance [70], eliminates trial and error, increases new product development efficiency [33], and results in improved customer satisfaction [9]. In this regard, there are various definitions of benchmarking in the literature. Benchmarking is the process of searching for best practices and trying to emulate them [55]. Benchmarking has quickly become a standard practice among top companies. GSCM benchmarking is the practice of comparing a company's green goods, services, and procedures along the supply chain to the relevant indicators of successful enterprises or chains. As a result, GSCM benchmarking encompasses a wide variety of factors like processes, products, performances, and strategies.

Data envelopment analysis (DEA) has been a widespread and accepted tool for measuring efficiency and performance for many years. According to [59], one of the standard outputs of DEA can be a benchmark for inefficient DMUs, which, by implication, can reach the desired point. In standard DEA models, mainly inefficient DMUs are revealed based on previous data. This means that if we have selected efficient DMUs as a benchmark, they are suitable for now but may not remain efficient in the future.

da Costa et al. [21] recognized sustainability indexes for benchmarking the performance and decreasing the environmental effects of the product life cycle with a benchmarking method. Radovanov et al. [47] operated a two-stage DEA model for benchmarking and improving the sustainability performance of tourism-driving services. A majority of sustainability research has examined the relationship between energy, environmental, and economic factors while focusing less on how social factors impact safety performance [2, 72, 73].

In essence, benchmarking is a strategy that aims to improve performance by studying and adopting the best practices of other organizations to address the needs of the company. According to [4], benchmarking helps companies focus and get closer to their marketplaces and consumers. According to Boxwell [8], the benchmarking process can be categorized into three approaches: training, management, and comprehensive. The training approach aims to enhance employees' understanding of competition, whereas the management approach focuses on addressing weaknesses and enhancing processes at the level of factory laborers. The overall approach involves creating an all-inclusive benchmarking process within the organization. Regardless of the approach used, all organizations share two goals: They are not satisfied with the current situation and want to boost their competitiveness. By way of explanation, benchmarking helps organizations think differently and look for the best attitude to achieve their objectives [3].

Rostamzadeh et al. [49] identified eight major applications such as product planning, service sector, hotel industry, transportation, education, maintenance, distribution, and environmental factors. The highest recent development among all the applications has been applied in both the service sector and transportation. In addition, the results illustrate that DEA, as a suitable efficiency evaluation and performance measurement, has the highest potential for future benchmarking research, while the production specification between inputs and outputs has been difficult to gain.

2.6 *Background on the Application of DEA in SCM and GSCM*

The concepts of DEA were proposed by [15, 27] in their paper "*Measuring the Efficiency of Decision-Making Units*," which foremost employed linear programming (LP) for the experimental bound of the production technology estimation. Thenceforth, many papers and articles have been published about DEA or the different areas in which DEA can be used to address the issues. Data envelopment analysis (DEA) is an implementation for measuring the efficiency of decision-making units (DMUs) with multiple inputs and outputs. The DEA is a linear technique and a free parameter programming method. This method has been used in such a situation to compare the inputs and outputs of DMUs that represent production units with each other. Also, DEA is a suitable tool for measuring the performance and evaluating the efficiency of manufacturers or units. The old and most popular statistical methods could be evaluated by comparing the specifications of manufacturers due to the average characteristics of the manufacturing unit. However, DEA, as an extreme-point method, compares the specifications of each DMU and evaluates them with only the best one. Moreover, the development of the DEA method began with the measurement of productivity with input and output in the usual way and then introduced the method of estimating comparative efficiency with different inputs as a ratio of the weighted output to input. So, in the past decade, it has been used for many goals like in the economy, environmental issues, selecting the best supplier, and many more. According to [42], DMUs impart on several bases like transportation, business, universities, hospitals, etc. Classic DEA considers the DMUs as a black box that their structures denied and performance evaluation of DMUs just related to the inputs and outputs. However, in most scenarios, DMUs have a network design, where the output of the first stage becomes the input for the next level.

Färe and Grosskopf [26] introduced the overall efficiency measurement of new methods with DEA models. Overall efficiency can be considered as an average weighted value regardless of the state of efficiency for each stage. In recent years, many studies have been done on DMUs that are considered efficient for each stage. DEA is used to assist in common benchmarking activities and to give management guidance [23]. According to several experiments, this method is an effective way of evaluating performance, benchmarking, and improving the company's performance. In consequence, since DEA is first suggested by [15], it is widely used in benchmarking studies. DEA also positively impacts on defining functions and operating efficiency of different firms [53].

Evaluation of the efficiency of green supply chain management should consider the effect of environmental factors besides the profitability of the firms. These factors may lead to the failure of green supply chains in meeting the green principles and the expectation of customers who care about the environment. Hence, regular evaluation of the efficiency of green supply chains could impact the performance regarding environmental issues. When these factors are considered simultaneously, green supply chain management can be planned more efficiently to minimize the environmental effects and production costs. Thus, this work aims to present a new approach to evaluating the efficiency of green supply chains when some of them are inefficient but have the potential to become efficient through their benchmarks, while environmental factors are considered. An important novelty in this approach is that by benchmarking the efficient DMUs, we can distinguish the potential DMUs and modify them to become efficient while following the green principles' factors at the same time. In addition, the comparison of results before and after modification of potential DMUs through green factors and their effect on the whole system efficiency has been applied.

3 Methodology

This section introduces the standard data envelopment analysis (DEA) and Tobit regression models, with brief historical origins and how they were developed and used in the research. Also, the relationship between the two methods and the way of approach is shown in a methodology diagram at the end of this section, with an explanation.

3.1 *DEA Efficiency Analysis*

3.1.1 Standard CCR Model

DEA is a method for evaluating efficiency that employs a nonparametric approach and has been widely applied in operations research and general management [15]. Productivity is a function of economic factors that determine output and input [27]. It is important to identify both desirable and undesirable outcomes in environmental evaluation [60]. The following five efficiency measurements were identified by Zhou et al. [72, 73] after analyzing more than 100 energy and environment research papers: (1) radial efficiency, (2) nonradial efficiency, (3) slacks-based efficiency, (4) hyperbolic efficiency, and (5) directional distance function efficiency. We assume that each DMU_j has multiple inputs $x_{i,j}$ and multiple outputs $y_{k,j}$.

To institutionalize these assumptions, we have used the following notations:

DMU_j, by $j = 1, 2, 3, \ldots, n$, defines the j-th DMU.

x_{ij}, by $i = 1, 2, 3, \ldots, m$ and $j = 1, 2, 3, \ldots, n$, defines the ith input of the jth DMU.

y_{kj}, by $k = 1, 2, 3, \ldots, s$ and $j = 1, 2, 3, \ldots, n$, defines the kth output of the jth DMU.

v_i, by $i = 1, 2, 3, \ldots, m$, defines the weight of the ith input.

u_k, by $k = 1, 2, 3, \ldots, s$, defines the weight of the kth output (Fig. 3).

The relative efficiency measure is defined as follows:

$$\text{Efficiency} = \frac{\sum_k u_k y_{k,j}}{\sum_i v_i x_{i,j}} \tag{1}$$

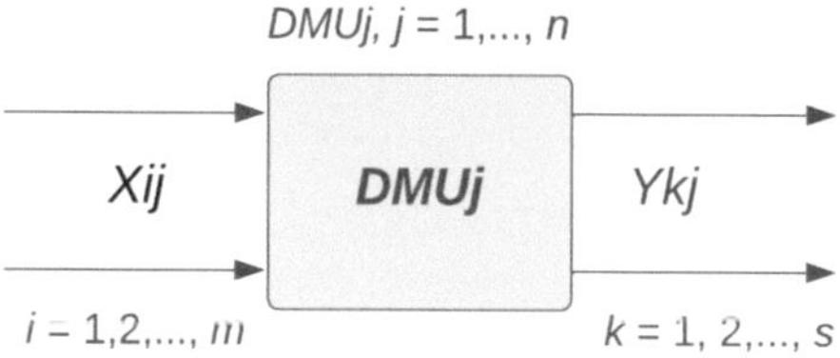

Fig. 3 Inputs and outputs of *j*th DMU

where *u* and *v* are weights. The efficiency is often scaled to range between [0, 1].

The weights have a flaw: Assigning a standard value to them across all DMUs is quite arbitrary. The key notion underlying DEA is that each DMU_{j0} is given the ability to select its weights. It is achieved by solving the following optimization issue. According to [15], the efficiency score in the CCR model is 1, and inefficiency is 1/efficiency. Whenever the DMU's score exceeds one, it is considered inefficient. For output-oriented DMUs, most packages provide a score ranging from 0 to 1. The objective is to maximize the efficiency of DMU_{j0} while keeping all other DMUs' efficiencies less than or equal to 1 [38].

$$\begin{aligned} &\text{Maximize}\,\theta_0 = \frac{\sum_k u_k y_{k,j0}}{\sum_i v_i x_{i,j0}} \\ &\text{Subject to}\ \frac{\sum_k u_k y_{k,j}}{\sum_i v_i x_{i,j}} \le 1 \forall j \\ &u_k, v_i \ge 0 \end{aligned} \tag{2}$$

It is not an LP model. Therefore, a simple change is made to fix the denominator to a constant value of 1, which can be set as a constraint on the v_j weights, as shown below:

$$\begin{aligned} &\text{Maximize} \sum_k u_k y_{k,j0} \\ &\text{Subject to} \sum_i v_i x_{i,j0} = 1 \\ &\sum_k u_k y_{k,j} \le \sum_i v_i x_{i,j}\ \forall j \\ &u_k, v_i \ge 0 \end{aligned} \tag{3}$$

It can be noted that the decision variables are *u* and *v* as weights. In some cases, the dual model is preferable for primal models with many rows and columns. The dual DEA model can be represented as:

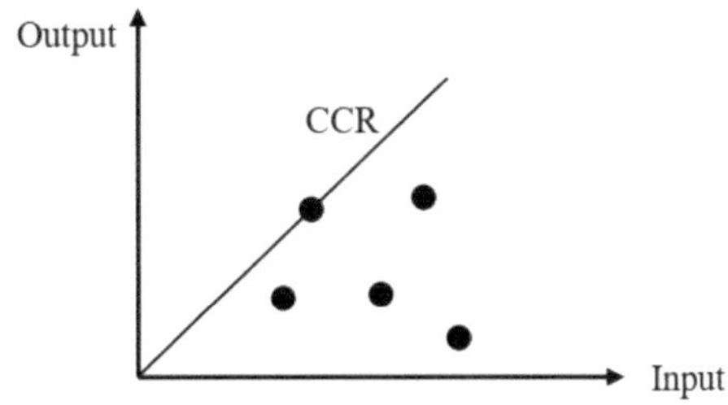

Fig. 4 CCR frontier

$$\begin{aligned}
&\text{Minimize } z_0 = \theta_{j0} \\
&\sum_j \lambda_j y_{k,j} \geq y_{k,j0} \\
&\theta_{j0} x_{i,j0} \geq \sum_j \lambda_j x_{i,j} \\
&\lambda_j \geq 0
\end{aligned} \tag{4}$$

Model (4) will be used to compare the inefficient DMUs and efficient DMUs as benchmark units. Cross-efficiency will determine the potential DMUs that can become efficient after some adjustments in their input or output values.

The DEA model has been proposed in other forms. We discussed the CCR model above [15] as one of the standard DEA models (Fig. 4).

3.1.2 Two-Stage DEA Model

Two-stage in a DMU means the procedure is divided into two stages with intermediate values in between. Initially, inputs are used to make intermediate values, which are then used to make outputs. The intermediate values are used in the second stage to produce the final outputs. First-stage outputs are the only inputs to the second stage, which is a critical assumption. The following notations have been used to institutionalize these assumptions:

DMU_j, by $j = 1, 2, 3, \ldots, n$, defines the jth two-stage DMU.

x_{ij}, by $i = 1, 2, 3, \ldots, m$, and $j = 1, 2, 3, \ldots, n$, defines the ith input of the jth DMU in stage 1.

z_{dj}, by $d = 1, 2, 3, \ldots, D$, and $j = 1, 2, 3, \ldots, n$, defines the dth output of the j-th DMU in stage 1 and the d-th input of the jth DMU in stage 2.

y_{rj}, by $r = 1, 2, 3, \ldots, s$, and $j = 1, 2, 3, \ldots, n$, defines the rth output of the jth DMU in stage 2.

v_i, by $i = 1, 2, 3, \ldots, m$, defines the ith input factor in stage 1.

η_d^1, by $d = 1, 2, 3, \ldots, D$, defines the dth output factor in stage 1.

η_d^2, by $d = 1, 2, 3, \ldots, D$, defines the dth input factor in stage 2.

u_r, by $r = 1, 2, 3, \ldots, s$, defines the rth output factor in stage 2.

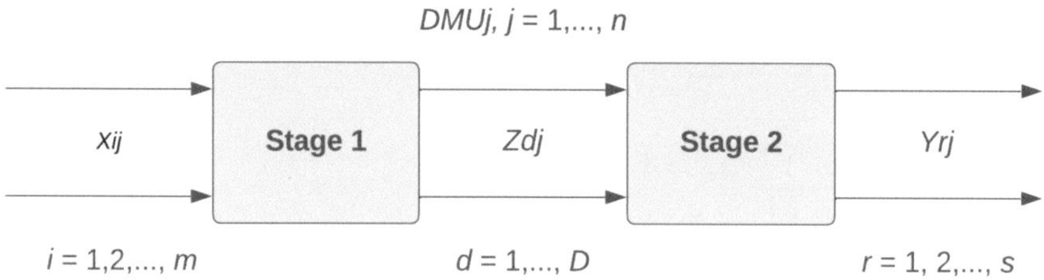

Fig. 5 Two-stage DEA model

Figure 5 illustrates the first stage of evaluating each DMU, starting with *m* inputs and producing *D* output units. Intermediary units are created from *D* output units and become inputs to the next level. Hence, the DMUs must be evaluated in the second stage to determine their ability to convert *D* intermediate units into *r* final outputs.

In the first and second stages, the constant returns to scale (CRS) efficiency of the *j*-th DMU, in each case $j = 1, 2\ldots, n$, is as follows:

$$\theta_j^{1*} = \sum_{d=1}^{D} \eta_d^1 z_{dj} / \sum_{i=1}^{m} v_i x_{ij} \text{ and } \theta_j^{2*} = \sum_{r=1}^{s} u_r y_{rj} / \sum_{d=1}^{D} \eta_d^2 z_{dj}$$

The overall CRS efficiency of DMU_j is defined by [40] as the result of the two individual efficiencies, namely $\theta_j^* = \theta_j^{1*} \times \theta_j^{2*}$. Since the two stages have a series relationship, they assumed that $\eta_d^1 = \eta_d^2 = \eta_d$ for $d = 1\ldots, D$. As a result of their research, they developed a two-stage DEA model to measure the total performance of each DMU over some time:

$$\begin{aligned}
&\theta_0^* = \text{Maximize}\left[\frac{\sum_{d=1}^{D} \eta_d z_{d0}}{\sum_{i=1}^{m} v_i x_{i0}} \times \frac{\sum_{r=1}^{s} u_r y_{r0}}{\sum_{d=1}^{D} \eta_d z_{d0}} = \frac{\sum_{r=1}^{s} u_r y_{r0}}{\sum_{i=1}^{m} v_i x_{i0}}\right] \\
&\text{Subject to } \frac{\sum_{d=1}^{D} \eta_d z_{dj}}{\sum_{i=1}^{m} v_i x_{ij}} \le 1, \quad j = 1, 2, \ldots, n, \\
&\frac{\sum_{r=1}^{s} u_r y_{rj}}{\sum_{d=1}^{D} \eta_d z_{dj}} \le 1, \quad j = 1, 2, \ldots, n, \\
&\eta_d, v_i, u_r \ge 0, \quad d = 1, 2, \ldots, D; i = 1, 2, \ldots, m; \text{ r } = 1, 2\ldots, s,
\end{aligned} \tag{5}$$

while DMU_0 indicates the DMU undergoing inspection. A linear program (LP) can be generated for the result of the model (5) using [14] transformation:

$$
\begin{aligned}
&\theta_0^* = \text{Maximize} \sum_{r=1}^{s} u_r y_{r0} \\
&\text{Subject to} \sum_{i=1}^{m} v_i x_{i0} = 1 \\
&\sum_{d=1}^{D} \eta_d z_{dj} - \sum_{i=1}^{m} v_i x_{ij} \le 0, \qquad j = 1 \ldots, n, \\
&\sum_{r=1}^{s} u_r y_{rj} - \sum_{d=1}^{D} \eta_d z_{dj} \le 0, \qquad j = 1 \ldots, n, \\
&\eta_d, v_{i,} u_r \ge 0, \qquad d = 1 \ldots, D; i = 1 \ldots, m; r = 1 \ldots, s.
\end{aligned} \tag{6}
$$

After obtaining the total efficiency θ_0^*, the two individual efficiencies θ_0^{1*} and θ_0^{2*} values can be determined by solving one of the following LP models, and the other by using $\theta_0^2 = \theta_0^* / \theta_0^{1*}$ or $\theta_0^1 = \theta_0^* / \theta_0^{2*}$.

Using DEA in two stages, Kao and Hwang presented the following model for measuring DMUs' overall CRS efficiency:

$$
\begin{aligned}
&\theta_0^{1*} = \text{Maximize} \sum_{d=1}^{D} \eta_d z_{dj} \\
&\text{Subject to}\, \theta_0^* = \sum_{r=1}^{s} u_r y_{r0} \\
&\sum_{i=1}^{m} v_i x_{i0} = 1, \\
&\sum_{d=1}^{D} \eta_d z_{dj} - \sum_{i=1}^{m} v_i x_{ij} \le 0, \qquad j = 1 \ldots, n, \\
&\sum_{r=1}^{s} u_r y_{rj} - \sum_{d=1}^{D} \eta_d z_{dj} \le 0, \qquad j = 1 \ldots, n, \\
&\eta_d, v_i, u_r \ge 0, \qquad d = 1 \ldots, D; i = 1 \ldots, m; r = 1 \ldots, s,
\end{aligned} \tag{7}
$$

or

$$\theta_0^{2*} = \text{Maximize} \sum_{r=1}^{s} u_r y_{r0}$$

$$\begin{aligned}
\text{Subject to } & \sum_{d=1}^{D} \eta_d z_{dj} = 1, \\
& \sum_{r=1}^{s} u_r y_{r0} - \theta_0^* \sum_{i=1}^{m} v_i x_{i0} = 0, \\
& \sum_{d=1}^{D} \eta_d z_{dj} - \sum_{i=1}^{m} v_i x_{ij} \leq 0, \quad j = 1 \ldots, n, \\
& \sum_{r=1}^{s} u_r y_{rj} - \sum_{d=1}^{D} \eta_d z_{dj} \leq 0, \quad j = 1 \ldots, n, \\
& \eta_d, v_i, u_r \geq 0, \quad d = 1 \ldots, D; i = 1 \ldots, m; r = 1 \ldots, s.
\end{aligned} \tag{8}$$

In [18], the overall efficiency is not defined as the product of the two individual efficiencies. The overall efficiency of a two-stage process is estimated as the weighted sum of its two individual efficiencies. The multiplicative and additive models can both be applied equally well to aggregate their components. Regarding the two-stage DEA model, the following recall was issued:

$$\begin{aligned}
\theta_0^* = \text{Max} & \left[w_1 \cdot \frac{\sum_{d=1}^{D} \eta_d z_{d0}}{\sum_{i=1}^{m} v_i x_{i0}} + w_2 \cdot \frac{\sum_{r=1}^{s} u_r y_{r0}}{\sum_{d=1}^{D} \eta_d z_{d0}} \right] \\
\text{Subject to } & \frac{\sum_{d=1}^{D} \eta_d z_{dj}}{\sum_{i=1}^{m} v_i x_{ij}} \leq 1, \quad j = 1 \ldots, n, \\
& \frac{\sum_{r=1}^{s} u_r y_{rj}}{\sum_{d=1}^{D} \eta_d z_{dj}} \leq 1, \quad j = 1 \ldots, n, \\
& \eta_d, v_i, u_r \geq 0, \quad d = 1 \ldots, D; i = 1 \ldots, m; r = 1 \ldots, s,
\end{aligned} \tag{9}$$

where w_1 and w_2 are user-specified weights satisfying $w_1 + w_2 = 1$. Rather than being decision variables, these weights represent functions of decisions, reflecting the relative contributions of the performances of DMU_0 to its overall outcome in two stages. By setting:

$$\begin{aligned}
w_1 &= \frac{\sum_{i=1}^{m} v_i x_{i0}}{\sum_{i=1}^{m} v_i x_{i0} + \sum_{d=1}^{D} \eta_d z_{d0}} \text{ and} \\
w_2 &= \frac{\sum_{d=1}^{D} \eta_d z_{d0}}{\sum_{i=1}^{m} v_i x_{i0} + \sum_{d=1}^{D} \eta_d z_{d0}}
\end{aligned} \tag{10}$$

The objective function of (9), which is presumed to reverberate the relative sizes of the two stages, can then be rewritten as follows:

$$\theta_0 = \left(\sum_{d=1}^{D} \eta_d z_{d0} + \sum_{r=1}^{s} u_r y_{r0}\right) \Big/ \left(\sum_{i=1}^{m} v_i x_{i0} + \sum_{d=1}^{D} \eta_d z_{d0}\right).$$

Thus, the aliquot programming model (9) can be interpreted as the linear programming (LP) model, as shown below:

$$\begin{aligned}
\theta_0^* = \text{Maximize} & \sum_{d=1}^{D} \eta_d z_{d0} + \sum_{r=1}^{s} u_r y_{r0} \\
\text{Subject to} & \sum_{i=1}^{m} v_i x_{i0} + \sum_{d=1}^{D} \eta_d z_{d0} = 1, \\
& \sum_{d=1}^{D} \eta_d z_{dj} - \sum_{i=1}^{m} v_i x_{ij} \leq 0, \quad j = 1 \ldots, n, \\
& \sum_{r=1}^{s} u_r y_{rj} - \sum_{d=1}^{D} \eta_d z_{dj} \leq 0, \quad j = 1 \ldots, n, \\
& \eta_d, v_i, u_r \geq 0, \quad d = 1 \ldots, D; i = 1 \ldots, m; r = 1 \ldots, s.
\end{aligned} \tag{11}$$

Following the determination of the optimal objective value of (11) at the end of the process, the efficiency of the two levels can be analyzed in a manner identical to [40] study. For instance, θ_0^{1*} can be defined by solving the following linear programming (LP) model:

$$\begin{aligned}
\theta_0^{1*} = \text{Maximize} & \sum_{d=1}^{D} \eta_d z_{d0} \\
\text{Subject to} & \sum_{i=1}^{m} v_i x_{i0} = 1, \\
& \left(1 - \theta_0^*\right) \sum_{d=1}^{D} \eta_d z_{d0} + \sum_{r=1}^{s} u_r y_{r0} = \theta_0^*, \\
& \sum_{d=1}^{D} \eta_d z_{dj} - \sum_{i=1}^{m} v_i x_{ij} \leq 0, \quad j = 1 \ldots, n, \\
& \sum_{r=1}^{s} u_r y_{rj} - \sum_{d=1}^{D} \eta_d z_{dj} \leq 0, \quad j = 1 \ldots, n, \\
& \eta_d, v_i, u_r \geq 0, \quad d = 1 \ldots, D; i = 1 \ldots, m; r = 1 \ldots, s.
\end{aligned} \tag{12}$$

θ_0^2 can be figured out by $\theta_0^2 = \left(\theta_0^* - w_1^* \theta_0^{1*}\right) / w_2^*$, where w_1^* and w_2^* are the weights gained from the model (10) by way of (9). θ_0^{2*} can also be defined by solving the following linear programming (LP) model:

$$\theta_0^{2*} = \text{Maximize} \sum_{r=1}^{s} u_r y_{r0}$$

$$\begin{aligned} &\text{Subject to} \sum_{d=1}^{D} \eta_d z_{d0} = 1, \\ &\theta_0^* \sum_{i=1}^{m} v_i x_{ij} - \sum_{r=1}^{s} u_r y_{r0} = 1 - \theta_0^*, \\ &\sum_{d=1}^{D} \eta_d z_{dj} - \sum_{i=1}^{m} v_i x_{ij} \le 0, \quad j = 1\ldots, n, \\ &\sum_{r=1}^{s} u_r y_{rj} - \sum_{d=1}^{D} \eta_d z_{dj} \le 0, \quad j = 1\ldots, n, \\ &\eta_d, v_i, u_r \ge 0, \quad d = 1\ldots, D; i = 1\ldots, m; r = 1\ldots, s. \end{aligned} \tag{13}$$

θ_0^1 can be obtained by $\theta_0^1 = \left(\theta_0^* - w_2^* \theta_0^{2*}\right)/w_1^*$.

Wang and Chin [68] generalized the two-stage DEA model of [18], which can be used to compute both constant returns to scale (CRS) and variable returns to scale (VRS) efficiency scores. To reflect the relative importance of each stage in the process, they assigned a weight to each one by letting $\lambda_1 > 0$ and $\lambda_2 > 0$ to set of comparative significance weights of both stages like $\lambda_1 + \lambda_2 = 1$. Henceforth, the total output and input of DMU_j can be gained as $\lambda_1 \sum_{d=1}^{D} \eta_d z_{dj} + \lambda_2 \sum_{r=1}^{s} u_r y_{rj}$ and $\lambda_1 \sum_{i=1}^{m} v_i x_{ij} + \lambda_2 \sum_{d=1}^{D} \eta_d y_{dj}$. It is necessary to calculate both the first-stage efficiency as well as the second-stage efficiency to calculate the CRS efficiency of the entire two-stage process (Fig. 6).

The total efficiency of DMU_0 can be estimated by either of the following models:

$$\begin{aligned} &\theta_0^* = \text{Max}\lambda_1 \sum_{d=1}^{D} \eta_d z_{d0} + \lambda_2 \sum_{r=1}^{s} u_r y_{r0} \\ &\text{Subject to}\, \lambda_1 \sum_{i=1}^{m} v_i x_{i0} + \lambda_2 \sum_{d=1}^{D} \eta_d z_{d0} = 1, \\ &\sum_{d=1}^{D} \eta_d z_{dj} - \sum_{i=1}^{m} v_i x_{ij} \le 0, \quad j = 1\ldots, n, \\ &\sum_{r=1}^{s} u_r y_{rj} - \sum_{d=1}^{D} \eta_d z_{dj} \le 0, \quad j = 1\ldots, n, \\ &\eta_d, v_i, u_r \ge 0, \quad d = 1\ldots, D; i = 1\ldots, m; r = 1\ldots, s \end{aligned} \tag{14}$$

and

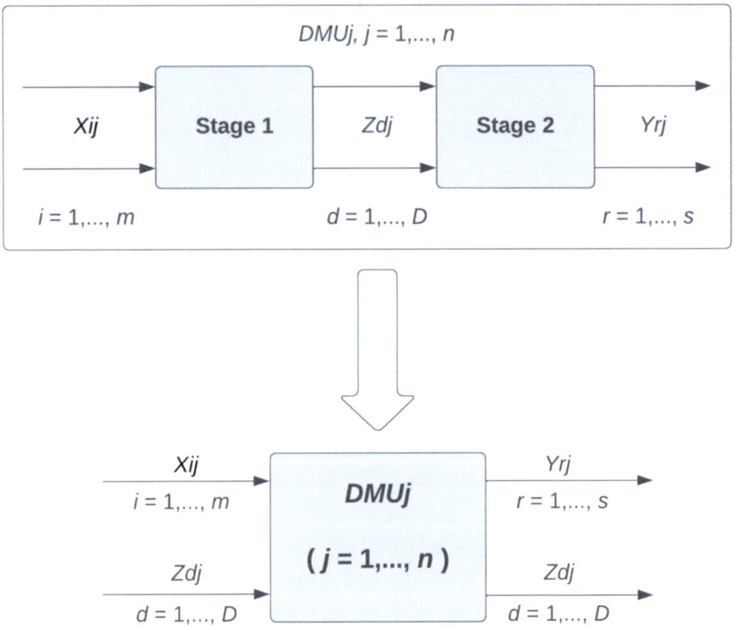

Fig. 6 Conversion of two-stage procedure to single process

$$\begin{aligned}
\theta_0^* &= \text{Maximize}\, \lambda_1\left(\sum_{d=1}^{D}\eta_d z_{d0} + \sigma^1\right) + \lambda_2\left(\sum_{r=1}^{s} u_r y_{r0} + \sigma^2\right) \\
&\text{Subject to}\, \lambda_1 \sum_{i=1}^{m} v_i x_{i0} + \lambda_2 \sum_{d=1}^{D}\eta_d z_{d0} = 1, \\
&\sum_{d=1}^{D}\eta_d z_{dj} - \sum_{i=1}^{m} v_i x_{ij} + \sigma^1 \le 0, \qquad j = 1\ldots, n, \\
&\sum_{r=1}^{s} u_r y_{rj} - \sum_{d=1}^{D}\eta_d z_{dj} + \sigma^2 \le 0, \qquad j = 1\ldots, n, \\
&\eta_d, v_i, u_r \ge 0, \qquad d = 1\ldots, D; i = 1\ldots, m; r = 1\ldots, s,
\end{aligned} \tag{15}$$

where σ^1 and σ^2 are free variables.

While the overall efficiency θ_0^* is obtained, θ_0^{1*} can be determined by solving the following linear programming (LP) models for different assumptions regarding returns to scale:

$$
\begin{aligned}
&\theta_0^{1*} = \text{Maximize} \sum_{d=1}^{D} \eta_d z_{d0} \\
&\text{Subject to} \sum_{i=1}^{m} v_i x_{i0} = 1, \\
&\quad \left(\lambda_1 - \lambda_2 \theta_0^*\right) \sum_{d=1}^{D} \eta_d z_{d0} + \lambda_2 \sum_{r=1}^{s} u_r y_{r0} = \lambda_1 \theta_0^*, \\
&\quad \sum_{d=1}^{D} \eta_d z_{dj} - \sum_{i=1}^{m} v_i x_{ij} \le 0, \qquad j = 1\ldots, n, \\
&\quad \sum_{r=1}^{s} u_r y_{rj} - \sum_{d=1}^{D} \eta_d z_{dj} \le 0, \qquad j = 1\ldots, n, \\
&\quad \eta_d, v_i, u_r \ge 0, \qquad d = 1\ldots, D; i = 1\ldots, m; r = 1\ldots, s
\end{aligned} \tag{16}
$$

or

$$
\begin{aligned}
&\theta_0^{1*} = \text{Maximize} \sum_{d=1}^{D} \eta_d z_{d0} + \sigma^1 \\
&\text{Subject to} \sum_{i=1}^{m} v_i x_{i0} = 1, \\
&\quad \left(\lambda_1 - \lambda_2 \theta_0^*\right) \sum_{d=1}^{D} \eta_d z_{d0} + \lambda_1 \sigma^1 + \lambda_2 \sum_{r=1}^{s} u_r y_{r0} + \lambda_2 \sigma^2 = \lambda_1 \theta_0^*, \\
&\quad \sum_{d=1}^{D} \eta_d z_{dj} - \sum_{i=1}^{m} v_i x_{ij} + \sigma^1 \le 0, \qquad j = 1\ldots, n, \\
&\quad \sum_{r=1}^{s} u_r y_{rj} - \sum_{d=1}^{D} \eta_d z_{dj} + \sigma^2 \le 0, \qquad j = 1\ldots, n, \\
&\quad \eta_d, v_i, u_r \ge 0, \qquad d = 1\ldots, D; i = 1\ldots, m; r = 1\ldots, s,
\end{aligned} \tag{17}
$$

where σ^1 and σ^2 are free variables.

Likewise, θ_0^{2*} can be defined by solving one of the following linear programming (LP) models for different return-to-scale assumptions:

$$
\begin{aligned}
\theta_0^{2*} = \text{Maximize} \sum_{r=1}^{s} u_r y_{r0} \\
\text{Subject to} \sum_{d=1}^{D} \eta_d z_{d0} = 1, \\
\lambda_2 \sum_{r=1}^{s} u_r y_{r0} - \lambda_1 \theta_0^* \sum_{i=1}^{m} v_i x_{i0} = \lambda_2 \theta_0^* - \lambda_1, \\
\sum_{d=1}^{D} \eta_d z_{dj} - \sum_{i=1}^{m} v_i x_{ij} \leq 0, \quad j = 1 \ldots, n, \\
\sum_{r=1}^{s} u_r y_{rj} - \sum_{d=1}^{D} \eta_d z_{dj} \leq 0, \quad j = 1 \ldots, n, \\
\eta_d, v_i, u_r \geq 0, \quad d = 1 \ldots, D; i = 1 \ldots, m; r = 1 \ldots, s
\end{aligned}
\tag{18}
$$

or

$$
\begin{aligned}
\theta_0^{2*} = \text{Maximize} \sum_{r=1}^{s} u_r y_{r0} + \sigma^2 \\
\text{Subject to} \sum_{d=1}^{D} \eta_d z_{d0} = 1, \\
\lambda_2 \sum_{r=1}^{s} u_r y_{r0} + \lambda_2 \sigma^2 - \lambda_1 \theta_0^* \sum_{i=1}^{m} v_i x_{i0} + \lambda_1 \sigma^1 = \lambda_2 \theta_0^* - \lambda_1, \\
\sum_{d=1}^{D} \eta_d z_{dj} - \sum_{i=1}^{m} v_i x_{ij} + \sigma^1 \leq 0, \quad j = 1 \ldots, n, \\
\sum_{r=1}^{s} u_r y_{rj} - \sum_{d=1}^{D} \eta_d z_{dj} + \sigma^2 \leq 0, \quad j = 1 \ldots, n, \\
\eta_d, v_i, u_r \geq 0, \quad d = 1 \ldots, D; i = 1 \ldots, m; r = 1 \ldots, s,
\end{aligned}
\tag{19}
$$

where σ^1 and σ^2 are free variables.

In [40], the assumption that $\eta_d^1 = \eta_d^2 = \eta_d$ is vital to transform the two-stage DEA model into a linear programming (LP) problem and compute the efficiency scores of DMUs. Veritably, with a lack of this assumption, it is not possible to cancel the factors $\sum_{d=1}^{D} \eta_d^1 z_{d0}$ and $\sum_{d=1}^{D} \eta_d^2 z_{d0}$ in the objective function $\theta_0^* = \theta_0^{1*} \times \theta_0^{2*}$ of model (6). Aside from that, assuming that $\eta_d^1 = \eta_d^2 = \eta_d$ also connects the two levels of the chain to transform it into a particular process rather than two autonomous one-stage processes evaluated by two nonaligned CCR models. Accordingly, [18, 68] also assigned the same weight to mediator outputs when they are considered intermediate inputs.

3.2 Tobit Regression Analysis

Tobin [65] established the Tobit regression model, which is a statistical model based on linear assumptions that are employed when information on the dependent variable is unavailable for every perception due to censoring. Holden [35] has tested the normality assumptions in the Tobit model. A significant portion of the information has been altered due to the skewness of the continuous dependent variable to one side. As a result, by altering, the regression is enabled. The population's conventional Tobit regression model is specified as:

$$\begin{aligned} y^* &= x\text{ß} + u, \qquad u \sim N(0, \sigma^2) \\ y &= \max(0, y^*) \end{aligned} \tag{20}$$

y^* is a vector of the dependent variable
x is a vector of the independent variable
ß is a vector coefficient measured by Tobit regression analysis
u is a vector of fault with regard to normal distribution.

The relationship between the dependent and independent variables has been investigated using the Tobin regression model.

As part of this study, we analyzed how some of the green principles' factors can affect the whole supply chain efficiency. The two factors that we are focusing on are environmental and economic. The three selected subfactors are:

1. Service level: the percentage of the orders from customers that need to be satisfied and the time they need to wait to receive the service or product
2. Emissions (CO_2): the average of each truck emissions produced per day in kilograms
3. Arcs: the total number of arcs in each chain (size of the chain for transportation factor).

In this test, the efficiency of each supply chain was considered as a dependent variable, and the three mentioned subfactors like service level, emissions, and arcs were considered independent variables to examine their effects on the supply chain efficiency. Therefore, the following is used for Tobit regression analysis:

$$\mathit{Efficiency} = \beta_1(\mathit{Service\ Level}) + \beta_2(\mathit{Emission}) + \beta_3(\mathit{Arcs}) + u \tag{21}$$

The Tobit test provides a better understanding of the changes in efficiency based on some factors of green principles.

3.3 Methodology Description

To present the novelty of this research and better describe the process of introducing the approach, a sequence of steps is illustrated in Fig. 7.

As Fig. 7 illustrates, in the first step, data are collected from an official source, which is mentioned in the references at the end of the research, and the inputs and outputs are then specified. The second step is efficiency determination; this determines both efficient and inefficient DMUs by applying standard DEA models. The third step is benchmarking. Inefficient DMUs are analyzed using the efficient DMUs as benchmarks. This produces potential DMUs (those that compare favorably well with the efficient DMUs) and strictly inefficient DMUs (those that fail to compare with those on the frontier). In the last step, after modification, the inefficient DMUs become efficient, and then the effects of environmental factors on the whole system are measured and the results before and after the modification of inefficient DMUs are compared using Tobit regression analysis.

The mentioned methodology and approach have been applied to the data in the numerical examples and are explained in detail in the next section.

4 Analysis of Data and Results

In this section, the main theoretical contributions of this research are presented by a numerical example with real-world data, and the results are discussed. This section has two main parts; the first part represents the application of standard DEA models to find the initial efficient DMUs and, using a developed benchmark approach, attempts

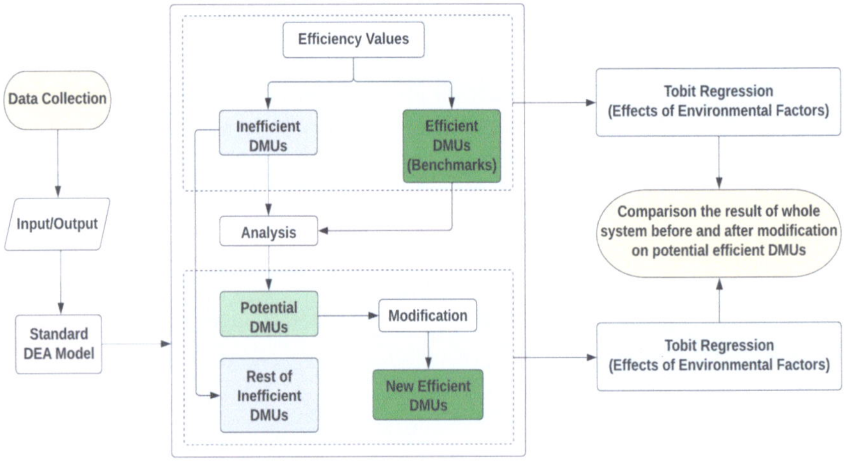

Fig. 7 Methodology summary diagram

Table 2 Summary of collected data for 38 DMUs under evaluation

	Dist	Manuf.	Part	Retail	Transp.	Av. CGS
Min	0.0001%	0.1%	23.3%	0.0001%	0.1%	$3.12
Max	64.2%	100.0%	99.1%	6.0%	62.8%	$150,816.00
Av	8.9%	29.5%	70.6%	1.8%	6.7%	$8,706.01
Var	0.025675	0.082809	0.043919	0.0003146	0.021063	–

Willems [69]

to increase the number of efficient DMUs. In the second part, Tobit regression model is applied to verify the effect of environmental factors on the whole system before and after the modification of inefficient DMUs.

4.1 Numerical Example

There are 38 supply chains from different companies that differ in size, material, and final product, which are analyzed using the CCR and two-stage models to find which of them can be considered efficient DMUs [69]. Then, for those selected as efficient, the cross-efficiency analysis on the benchmarks was applied to find out which inefficient DMUs can become efficient. First, the data are categorized, and the chains are specified with their structure according to their final product profit. For this purpose, in the CCR model, inputs including distribution, manufacturing, supplier, retailer, and transportation and output as the average cost of goods sold (ACGS) for each chain are considered. For the two-stage model, ACGS is used as an input and sales profit as an output. Table 2 shows the minimum, maximum, average, and variance of the cost metrics for each stage and their percentage of the total ACGS.

4.2 DEA Models

The two-stage and CCR input-oriented models, as standard DEA models, are applied to the extracted data to identify the efficient DMUs. Table 3 shows the results of the two-stage DEA under the CRS assumption by model (6), while Tables 4 and 5 show the result from model (14) with different weights for each stage. Table 6 illustrates the results of the CCR model (3) to find out the efficient DMUs. Model selection is made based on constant returns to scale assumption for the DMUs under evaluation. This means that any radial increase in the input vector results in a proportional radial increase in the output vector. The model (6) was applied to each ACGS chain because even the financial part is an essential part of each chain and management besides greening the supply chain. The results for the efficiency of 38 firms under the CRS assumption using model (6) are shown below:

Table 3 Efficiency of 38 firms under CRS assumption

DMU	Overall efficiency θ_0^*	First-stage efficiency θ_0^{1*}	Second-stage efficiency θ_0^{2*}
1	1.0000	0.0228	0.9000
2	1.0000	1.0000	1.0000
3	0.3875	0.3227	0.1010
4	0.2587	0.1176	0.0028
5	0.0578	0.0036	0.0008
6	0.0287	0.0287	0.0000
7	1.0000	1.0000	1.0000
8	0.0025	0.0012	0.0016
9	0.0194	0.0100	0.0002
10	1.0000	1.0000	0.0352
11	1.0000	1.0000	1.0000
12	0.0007	0.0007	0.0002
13	1.0000	1.0000	0.0022
14	0.2972	0.2972	0.0001
15	0.0106	0.0026	0.0001
16	0.0813	0.0812	0.0022
17	0.0003	0.0003	0.0002
18	0.0012	0.0012	0.0006
19	0.0321	0.0298	0.0009
20	0.0046	0.0046	0.0031
21	0.0065	0.0057	0.0001
22	0.0297	0.0244	0.0000
23	0.0425	0.0210	0.0007
24	1.0000	1.0000	0.0065
25	1.0000	1.0000	0.0155
26	1.0000	1.0000	0.0775
27	0.0213	0.0196	0.0005
28	0.0052	0.0052	0.0010
29	0.0001	0.0001	0.0000
30	0.0002	0.0002	0.0000
31	1.0000	1.0000	0.2314
32	0.0008	0.0008	0.0000
33	0.0001	0.0001	0.0000
34	0.0630	0.0630	0.0009
35	0.0000	0.0000	0.0000
36	0.2542	0.0614	0.0036
37	0.2475	0.0211	0.0035
38	0.0115	0.0086	0.0048

Table 4 Efficiency of 38 firms under CRS assumption for a certain set of stage weights ($\lambda_1 = 2/5$, $\lambda_2 = 3/5$)

DMU	Overall efficiency θ_0^*	First-stage efficiency θ_0^{1*}	Second-stage efficiency θ_0^{2*}
1	0.9409	1.0000	0.9409
2	0.9786	0.9621	0.9417
3	1.0000	0.9926	0.9926
4	0.9304	0.9543	0.8871
5	0.9452	0.9441	0.8922
6	0.9083	0.9975	0.9058
7	0.9559	0.9868	0.9433
8	0.9697	0.9869	0.9569
9	0.9058	1.0000	0.9058
10	0.9814	0.9981	0.9796
11	0.9426	0.9982	0.9407
12	0.9927	0.9980	0.9907
13	0.9378	0.9853	0.9241
14	0.9647	0.9717	0.9362
15	0.9739	0.9738	0.9477
16	0.9533	0.9853	0.9393
17	0.9758	0.9923	0.9683
18	0.9979	0.9842	0.9823
19	0.9599	0.9927	0.9528
20	0.9691	0.9869	0.9552
21	0.9766	0.9867	0.9636
22	0.9356	0.9785	0.9146
23	0.9836	0.9991	0.9830
24	0.9906	0.9911	0.9821
25	0.9732	0.9879	0.9604
26	0.9814	0.9845	0.9665
27	0.9565	0.9835	0.9405
28	0.9675	0.9926	0.9609
29	0.9734	0.9924	0.9655
30	0.9628	0.9835	0.9475
31	0.9785	0.9856	0.9648
32	0.9615	0.9916	0.9531
33	0.9763	0.9836	0.9602
34	0.9538	0.9953	0.9491
35	0.9757	0.9852	0.9611
36	0.9680	0.9899	0.9587

(continued)

Table 4 (continued)

DMU	Overall efficiency θ_0^*	First-stage efficiency θ_0^{1*}	Second-stage efficiency θ_0^{2*}
37	0.9717	0.9791	0.9499
38	0.9865	0.9989	0.9789

Table 3 above illustrates the two-stage results. DMUs 1, 2, 7, 10, 11, 13, 24, 25, 26, and 31 are overall efficient, while DMUs 2, 7, and 11 are efficient in each stage and the rest of the DMUs are not efficient in one or both stages. To check alternative results, model (14) was applied two times by imposing different weights for the stages, and the results are provided in Tables 4 and 5. Testing the model with different weights produces slightly different results. Some DMUs are always efficient overall or at least in one stage, while the others cannot be considered as efficient DMUs generally. By applying model (14), the results with different weights for stages are discussed. The following shows the results for the efficiency of 38 firms under the CRS assumption for a certain set of stage weights with $\lambda_1 = 2/5$ and $\lambda_2 = 3/5$.

As shown in Table 4 with the mentioned stage weights, the results get closer to having more efficient DMUs but still need improvements. The only overall efficient DMU is 3 and, the rest are not efficient in their overall stage. Some DMUs like 1 and 9 are efficient in their first stage and not efficient in their second and overall stages. It seems that with the abovementioned weights for each stage, there are less amount of efficient DMUs, although they are close to becoming efficient by slight improvements.

After that, the efficiency of the 38 firms under the CRS assumption was tested for a certain collection of equal stage weights using model (14) with $\lambda_1 = 1/2$ and $\lambda_2 = 1/2$. The results of the test are shown below. More efficient DMUs are expected at this time because of equal weights for each stage.

According to the results in Table 5, DMUs 3, 6, 8, 9, 13, 18, 27, 34, and 38 are overall efficient, while DMUs 6, 13, and 38 are efficient in their all stages, and the rest of the DMUs are efficient in one or two stages. With the mentioned weights for each stage in this test, there are more efficient DMUs than in previous results, as expected. Also, some of the DMUs have the potential to become efficient with a slight improvement in one or two of their stages. Moreover, the CCR model (3) was applied to 38 firms, and the results were compared with previous results.

As obvious from the results of the CCR model (3) in Table 6, there are nine efficient DMUs. The efficient DMUs are 13, 14, 24, 28, 31, 33, 34, 36, and 38. They are selected as benchmarks for cross-efficiency verification. The optimal weights of each efficient DMU are used to compute the relative efficiency value (Eq. 1) for each inefficient DMU.

Table 7 presents the cross-efficiency values for each inefficient DMU concerning the benchmark DMUs. The eleventh column of this table shows that each one of the inefficient DMUs can be efficient, concerning the optimal weights of the benchmark DMUs (the number of 100 in each row). A DMU is considered potentially efficient

Table 5 Efficiency of 38 firms under CRS assumption for a certain set of stage weights ($\lambda_1 = 1/2$, $\lambda_2 = 1/2$)

DMU	Overall efficiency θ_0^*	First-stage efficiency θ_0^{1*}	Second-stage efficiency θ_0^{2*}
1	0.9409	1.0000	0.9409
2	0.9786	0.9621	0.9417
3	1.0000	0.9926	0.9926
4	0.9304	0.9543	0.8871
5	0.9452	0.9441	0.8922
6	1.0000	1.0000	1.0000
7	0.9486	0.9811	0.9310
8	1.0000	0.9866	0.9866
9	1.0000	0.9572	0.9572
10	0.9523	0.9848	0.9375
11	0.9744	1.0000	0.9744
12	0.9425	1.0000	0.9425
13	1.0000	1.0000	1.0000
14	0.9652	1.0000	0.9652
15	0.9838	1.0000	0.9838
16	0.9562	1.0000	0.9562
17	0.9877	1.0000	0.9877
18	1.0000	0.9982	0.9982
19	0.9725	1.0000	0.9725
20	0.9796	0.9667	0.9468
21	0.9779	1.0000	0.9779
22	0.9915	1.0000	0.9915
23	0.9437	0.9875	0.9323
24	0.9757	1.0000	0.9757
25	0.9714	1.0000	0.9714
26	0.9849	1.0000	0.9849
27	1.0000	0.9867	0.9867
28	0.9649	0.9696	0.9341
29	0.9892	0.9946	0.9839
30	0.9516	1.0000	0.9516
31	0.9687	1.0000	0.9687
32	0.9667	0.9655	0.9336
33	0.9791	1.0000	0.9791
34	1.0000	0.9817	0.9817
35	0.9694	1.0000	0.9694
36	0.9989	1.0000	0.9989

(continued)

Table 5 (continued)

DMU	Overall efficiency θ_0^*	First-stage efficiency θ_0^{1*}	Second-stage efficiency θ_0^{2*}
37	0.9846	1.0000	0.9846
38	1.0000	1.0000	1.0000

Table 6 CCR model results

DMU	Efficiency	DMU	Efficiency
01	54.49	20	35.4
02	62.12	21	54.87
03	41.97	22	48.3
04	79.14	23	60.95
05	38.94	24	100
06	66.9	25	65.58
07	37.26	26	37.66
08	39.22	27	57.56
09	52.48	28	100
10	86.21	29	36.84
11	39.05	30	61.84
12	68.1	31	100
13	100	32	69.19
14	100	33	100
15	57.1	34	100
16	85.57	35	36.43
17	78.84	36	100
18	27.32	37	34.04
19	59.05	38	100

when its efficiency value is near 100 and/or its relevant efficiency value reaches 100 more than four times in cross-efficiency verification. Moreover, as a supportive factor for the above criteria, the summation of cross-efficiency values in each line compute for inefficient DMUs as the last column of Table 7, shows which one of the inefficient DMUs where more efficient according to the benchmark DMUs weights. Therefore, the eleventh and twelfth columns are investigated to find the potential DMUs.

Based on the mentioned criteria, six DMUs can be considered as potentially efficient DMUs. The six DMUs are 4, 5, 10, 16, 17, and 27.

The potentially efficient DMUs can become efficient with small adjustments in the value of their inputs. The target results of the CCR model provide the necessary changes in the inputs of the potential DMUs to make them efficient. Table 8 presents the target results of the input-oriented CCR model.

Table 7 Result of cross-efficiency verification

DMU	13	14	24	28	31	33	34	36	38		Eff	Total
01	100	2.23	18.67	100	34.57	1.27	100	100	20.35	4	54.49	477.09
02	100	100	22.65	27.97	43.58	9.7	18.1	74.4	100	3	62.12	496.4
03	100	8.21	100	55.3	7.17	100	70.66	29.75	100	4	41.97	571.09
04	100	13.95	100	86.45	56.48	15.95	67.97	100	100	4	79.14	640.8
05	74.24	3.31	100	100	100	3.26	100	100	28.17	5	38.94	608.98
06	100	10.73	64.52	44.83	9.07	100	41.72	34.93	100	3	66.9	505.8
07	100	7.39	100	61.35	6.53	33.95	100	27.87	100	4	37.26	537.09
08	100	8.21	100	55.3	7.17	100	70.66	29.75	100	4	39.22	571.09
09	78.25	14.27	100	89.8	100	14.79	70.8	100	100	4	52.48	667.91
10	100	11.06	100	100	24.2	12.35	100	70.68	100	5	86.21	618.29
11	100	100	22.65	27.97	43.58	9.7	18.1	74.4	100	3	39.05	496.4
12	100	8.5	100	55.54	7.58	100	67.97	31.05	100	4	68.1	570.64
15	78.25	14.27	100	89.8	100	14.79	70.8	100	100	4	57.1	667.91
16	100	13.95	100	86.45	56.48	15.95	67.97	100	100	4	85.57	640.8
17	100	13.95	100	86.45	56.48	15.95	67.97	100	100	4	78.84	640.8
18	100	13.95	100	86.45	56.48	15.95	67.97	100	100	4	27.32	640.8
19	100	7.39	100	61.35	6.53	33.95	100	27.87	100	4	59.05	537.09
20	100	7.39	100	61.35	6.53	33.95	100	27.87	100	4	35.4	537.09
21	100	3.48	100	100	50.07	3.48	100	100	29.91	3	54.87	586.94
22	100	100	31.75	30.23	61.41	41.34	19.65	81.67	100	3	48.3	566.05
23	100	2.23	18.67	100	34.57	1.27	100	100	20.35	4	60.95	477.09
25	78.25	14.27	100	89.8	100	14.79	70.8	100	100	3	65.58	667.91
26	100	2.23	18.67	100	34.57	1.27	100	100	20.35	4	37.66	477.09
27	74.24	3.31	100	100	100	3.26	100	100	28.17	5	57.56	608.98
29	100	10.4	20.42	100	37.86	2.48	75.34	100	100	4	36.84	546.5
30	100	100	30.61	28.79	40.64	54.91	18.91	71.43	100	3	61.84	545.29
32	100	8.21	100	55.3	7.17	100	70.66	29.75	100	3	69.19	571.09
35	100	7.39	100	61.35	6.53	33.95	100	27.87	100	4	36.43	537.09
37	100	100	30.61	28.79	40.64	54.91	18.91	71.43	100	3	34.04	545.29
Total	23	4	18	8	4	5	11	14	24			

If an inefficient potential DMU keeps its output at the same level and decreases its inputs as given in Table 8, it can perform efficiently. For example, DMU 10, with a 13.79% decrease in the value of inputs 2 and 3 (manufacturer and supplier), can become efficient.

Table 8 Target results of CCR model for potential DMUs

DMU	Input 1 (%)	Input 2 (%)	Input 3 (%)	Input 4 (%)	Input 5 (%)
4	–89.4	–20.86	–20.86	0	0
5	–61.06	–610.6	–61.06	–61.06	0
10	0	–13.79	–13.79	0	0
16	–97	–14.43	–14.43	0	0
17	–94.16	–21.16	–21.16	–21.16	–21.16
27	–42.44	–42.44	–42.44	–42.44	0

4.3 Tobit Regression Model

In this part, after recognizing the potential DMUs and finding out each of them should improve in which inputs and how much to become efficient. Then Tobit Regression was used to verify the effects of the green factors that we considered before as service level, emission (CO_2), and arcs which express the size of the chain, on the efficiency before and after the modification of potential DMUs to see which one and how much can affect the total efficiency of the whole supply chain. The results of Tobit regression before recognizing the potential DMUs are shown in Table 9.

As shown in the efficiency column of Table 9, the service level has the most positive effect on improving the efficiency of the DMUs, while emissions have the most negative effect on the DMUs' efficiency. On the other hand, the number of arcs in the supply chain network has a slightly positive effect to make a DMU more efficient. The numbers in the coefficient column show the changes in the efficiency value of DMUs with respect to one-unit change in each green factor. For example, if CO_2 emissions increase by a unit, the total efficiency value will decrease by 0.1193%; conversely, if service level increases by one unit, the total efficiency value increases by approximately 3%. Moreover, the effect of one-unit increases in under-evaluation green factors on total efficiency value is 0.0477. There is an inverse reaction to green factors when the total efficiency value of the supply chain system is increased. Therefore, after making necessary adjustments on input values of potential DMUs and by nature increasing the total efficiency value, there may some changes in the effect of the green factors on the efficiency of the whole system. The first Tobit test

Table 9 Results of Tobit regression (for the whole DMUs before modification of potential DMUs)

Efficiency	Coefficient	Std. err.	t	p >|t|	[95% conf. interval]	
Service level	3.058122	2.432441	1.26	0.217	–1.879996	7.99624
Emission (CO_2)	–0.1192653	0.2147632	–0.56	0.582	–0.5552578	0.3167272
Arc	0.4362361	0.2238707	1.95	0.059	–0.0182455	0.8907178
_cons	–2.297785	2.335146	–0.98	0.332	–7.038383	2.442814
Var (e.Efficiency)	0.0477093	0.0109453			0.0299458	0.07601

was performed for six selected potential DMUs to see the changes and effects of green factors on their efficiency. The results are shown in Table 10.

This table illustrates the significant positive effect of service level, CO_2 emissions, and arcs on the selected DMUs' efficiency values. It shows that efficiency improvement on these DMUs will be caused an improvement in their network and relations between the stages and customer satisfaction of their service level dramatically. But from the other side, this will have a positive undesirable effect on CO_2 emission.

The next Tobit test was done for all of the DMUs after adjusting the inputs of the potential DMUs and improving their efficiency. The results are summarized in Table 11. For the new set of supply chains as in Table 9, service level and CO_2 emissions have more positive and negative effects on the total efficiency level of the system, respectively. On the other hand, the arcs have a slightly less positive effect. Moreover, the variety of the total efficiency values of the DMUs regarding the green factors increased from 0.047 to 0.058. This means that the efficiency scores of DMUs are now more sensitive to green factors.

To verify which one of the potential DMUs has the best situation to invest in for improvement and become efficient besides ranking them based on their efficiency values, priority verification is done again by the Tobit test. The efficiency verity of each one of the potential DMUs is measured regarding the benchmark DMUs separately and the results are shown in Table 12.

The third column of the above table shows the effect of the green factors on the efficiency value of each one of the potential DMUs. Based on these values and their

Table 10 Results of Tobit regression (for six potential DMUs)

Efficiency	Coefficient	Std. err.	t	p > |t|	[95% conf. [interval]	
Service level	9.254766	6.93133	1.34	0.274	–12.80382	31.31335
Emissions (CO_2)	2.07091	0.7734005	2.68	0.075	–0.3903959	4.532215
Arcs	3.377883	3.981558	0.85	0.459	–9.29321	16.04898
_cons	–8.676349	6.86242	–1.26	0.295	–30.51563	13.16294
Var (e.Efficiency)	0.0125101	0.0072227			0.001992	0.0785645

Table 11 Results of Tobit regression (for the whole DMUs after modification of potential DMUs)

Efficiency	Coefficient	Std. err.	t	p > |t|	[95% conf. [interval]	
Service level	5.13511	2.683682	1.91	0.064	–0.3130544	10.58327
Emissions (CO_2)	0.1338235	0.2369456	–0.56	0.576	–0.6148486	0.3472016
Arcs	0.2947164	0.2469937	1.19	0.241	–0.2067075	0.7361404
_cons	–4.233967	2.576338	–1.64	1.109	–9.464211	0.996276
Var (e.Efficiency)	0.0580738	0.0133231			0.0364513	0.0925227

Table 12 Results of Tobit regression for each potential DMU with efficient DMUs

Rank	DMU	Var (e.Efficiency)	Std. err.	Efficiency
1	10	0.0008452	0.000378	86.21
2	16	0.0011205	0.0005011	85.57
3	4	0.0016289	0.0007284	79.14
4	17	0.003404	0.0015223	78.84
5	27	0.0145833	0.0065218	57.56
6	5	0.0277673	0.0124179	38.94

standard deviation, DMU 10 has the best potential for efficiency improvement. As seen in Table 12, the mentioned ranking method matches the priority indicated by the efficiency values of the DMUs.

Thus, DMU 10 is the best candidate among the six potential DMUs for investment aimed at improvement and adjustment to become efficient. In this case, because of the fewer effects of the green principles' factors, it can become efficient and more robust compared to the other potential DMUs.

5 Conclusion

This section outlines the theoretical contributions of this research. The numerical example results with comparisons and the conclusions reached in this study are fully explained in detail for each part. Also, suggestions for potential future studies based on the research are discussed.

This study aimed to develop general standard DEA models with the help of a benchmark approach that can distinguish the efficient (supply chains) DMUs from the under-evaluated DMUs, according to green supply chain management principles.

In the first step, the two-stage and CCR models were applied to 38 DMUs, which represent the companies' chains, and the results from model (6) are shown in Table 3 under the CRS assumption. It showed that ten DMUs were overall efficient, while three DMUs were efficient in all stages, and the rest of them were inefficient in one or two stages. In Tables 4 and 5, the results from model (14) are illustrated by imposing different weights for each stage under the CRS assumption for 38 firms. With weights $\lambda_1 = 2/5$ and $\lambda_2 = 3/5$, there is only one overall efficient DMU, while two others were efficient in their first stage; the rest were not efficient. This indicates that with the mentioned weights for each stage, the number of efficient DMUs decrease. Then, by changing the weights for each stage to $\lambda_1 = 1/2$ and $\lambda_2 = 1/2$, it was expected that more efficient DMUs would be obtained due to equal weighting for the stages.

The results showed that nine DMUs were efficient in the overall stage, and three DMUs were efficient in their first and second stages; the rest of them were inefficient. Based on this expectation, the number of efficient DMUs increased by implying equal weights for each chain. Also, according to Table 6, results from the CCR model (3) application show that nine DMUs were efficient initially. The outcomes from previous results revealed that some DMUs are robust and efficient under CRS assumption even with different weights while others can become efficient or inefficient with different weights for each stage, and some DMUs cannot be considered efficient in any case.

Moreover, the initial efficient DMUs were considered as benchmarks for the second step by applying cross-efficiency on the selected ones to find out which DMUs from non-selected ones have the potential to become efficient. According to the results, six DMUs can become efficient with a slight improvement in one or some inputs, according to the target results of the CCR model (Table 7). After that, the Tobit regression model was applied to investigate the effects of some green principles' factors like service level, emissions (CO_2), and arcs to measure their effects on supply chain efficiency before the modification of potential DMUs (Table 9).

In general, the service level has the most positive effect on efficiency. In contrast, the emissions have the most negative impact on the efficiency of the supply chains. Also, the number of arcs for each chain, which is the network size, has a slight positive effect on the supply chain efficiency. However, when the Tobit regression model was applied to the six potential DMUs, all three green factors had a positive effect on their efficiency improvement, including emissions as an undesirable effect on the efficiency. After the modification of six potential DMUs to become efficient, there are now 15 efficient DMUs in whole supply chains among the 38 firms.

In the end, the Tobit regression model was used to find out the effect of total efficient DMUs on the whole supply chain's efficiency after the modification of potential DMUs. Table 11 illustrates that the general idea about the green factors' effect on the total efficiency of the supply chains was correct, but we should be aware that improving the efficiency of supply chains is necessary to maintain the robustness of green principles' factor efficiency. Also, the mentioned six potential DMUs were compared with each other to recognize which of them has more value and a better situation to invest for improving its efficiency. It was found that DMU 10, according to efficiency and standard deviation, is the best candidate among the potential DMUs to invest for its efficiency improvement (Table 12).

As expected, service level (customer satisfaction) and arcs (as the size of the transportation factor) had a positive effect, and CO_2 emissions (environmental effect) had a negative effect on the efficiency value of whole DMUs. But for potentially efficient DMUs, the effect of the mentioned green factors was positive. This means that there is a benefit in the green aspect to improve the potential DMUs and make them efficient. Moreover, after necessary adjustment on potential DMUs, the Tobit regression results show that the sensitivity of the whole system is increased regarding the under-consideration of green factors. This indicates on importance of priority to modify the potential DMUs according to their ranks which are provided by the efficiency values and Tobit test results.

For future studies, researchers can use stochastic data in time for each stage can be used in the supply chain to find an efficient supply chain in the first phase, and in the second phase, more green principles can be considered to see their effect on the whole supply chain efficiency and verify the difference between with or without applying the green factors on calculation. Also, other models like stochastic DEA (SDEA) can be applied depending on the nature of the data.

References

1. Andiappan V, Foo DCY, Tan RR (2022) Automated targeting for green supply chain planning considering inventory storage losses, production and set-up time. J Ind Prod Eng 39(5):341–352. https://doi.org/10.1080/21681015.2021.1991015
2. Ang BW, Zhang FQ (2000) A survey of index decomposition analysis in energy and environmental studies. Energy 25:1149–1176. https://doi.org/10.1016/s0360-5442(00)00039-6
3. Alosani MS, Al-Dhaafri HS, Yusoff RZB (2016) Mechanism of benchmarking and its impact on organizational performance. Int J Bus Manag 11(10):172. https://doi.org/10.5539/ijbm.v11n10p172
4. Asher M, Kanji GK (1996) 100 methods for total quality management. Sage Publication
5. Bowen FE, Cousins PD, Lamming RC, Farukt AC (2001) The role of supply management capabilities in green supply. Prod Oper Manag 10:174–189. https://doi.org/10.1111/j.1937-5956.2001.tb00077.x
6. Bui TD, Tsai FM, Tseng ML, Tan RR, Yu KDS, Lim MK (2021) Sustainable supply chain management towards disruption and organizational ambidexterity: a data driven analysis. In Sustainable production and consumption, vol 26. Elsevier B.V., pp 373–410. https://doi.org/10.1016/j.spc.2020.09.017
7. Bogan CE, English MJ (1994) Benchmarking for best practice: winning through innovative adaption. McGraw-Hill
8. Boxwell RJ Jr (1994) Benchmarking for competitive advantage. McGraw-Hill
9. Brah SA, Lin Ong A, Madhu Rao B (2000) Understanding the benchmarking process in Singapore. Int J Qual Reliab Manag 17(3):259–275. https://doi.org/10.1108/02656710010306157
10. Beamon BM (1998) Supply chain design and analysis: models and methods. Int J Prod Econ 55(3):281–294. https://doi.org/10.1016/S0925-5273(98)00079-6
11. Camp RC (1989) Benchmarking: the search for industry best practice that lead to superior performance. ASQ Quality Press
12. Chan FT, Chung S (2004) A multi-criterion genetic algorithm for order distribution in a demand driven supply chain. Int J Comput Integr Manuf 17:339–351. https://doi.org/10.1080/09511920310001617022
13. Chan FTS, Chung SH, Wadhwa S (2004) A heuristic methodology for order distribution in a demand driven collaborative supply chain. Int J Prod Res 42:1–19. https://doi.org/10.1080/0020754031000103293
14. Charnes A, Cooper WW (1962) Programming with linear fractional functionals. Naval Res Logistics Q 9(3–4):181–186. https://doi.org/10.1002/NAV.3800090303
15. Charnes A, Cooper W, Rhodes E (1978). Measuring the efficiency of decision making units. Company Eur J Oper Res 2
16. Chen CC, Tseng ML, Lin YH, Lin ZS (2010) Implementation of green supply chain management in uncertainty. In: IEEM2010—IEEE international conference on industrial engineering and engineering management, pp 260–264. https://doi.org/10.1109/IEEM.2010.5674461
17. Chen CL, Lee WC (2004) Multi-objective optimization of multi-echelon supply chain networks with uncertain product demands and prices. Comput Chem Eng 28(6–7):1131–1144. https://doi.org/10.1016/j.compchemeng.2003.09.014

18. Chen Y, Cook WD, Li N, Zhu J (2009) Additive efficiency decomposition in two-stage DEA. Eur J Oper Res 196(3):1170–1176. https://doi.org/10.1016/j.ejor.2008.05.011
19. Christopher M (1998) Logistics and supply chain management strategies for reducing cost and improving service. Financial Times, London Prentice-Hall
20. Chopra S, Meindl P (2001) Supply chain management strategy, planning, and operation. Prentice-Hall, Upper Saddle River, NJ
21. da Costa LG, Ferreira JCE, Kumar V, Garza-Reyes JA (2022) Benchmarking of sustainability to assess practices and performances of the management of the end of life cycle of electronic products: a study of Brazilian manufacturing companies. Clean Technol Environ Policy 24(4):1173–1189. https://doi.org/10.1007/s10098-020-01947-3
22. de Sousa Jabbour ABL (2015) Understanding the genesis of green supply chain management: lessons from leading Brazilian companies. J Clean Prod 87(1):385–390. https://doi.org/10.1016/j.jclepro.2014.09.034
23. Donthu N, Hershberger EK, Osmonbekov T (2005) Benchmarking marketing productivity using data envelopment analysis. J Bus Res 58(11):1474–1482. https://doi.org/10.1016/J.JBUSRES.2004.05.007
24. Diabat A, Govindan K (2011) An analysis of the drivers affecting the implementation of green supply chain management. Resour Conserv Recycl 55(6):659–667. https://doi.org/10.1016/j.resconrec.2010.12.002
25. ElTayeb TK, Zailani S, Jayaraman K (2010) The examination on the drivers for green purchasing adoption among EMS 14001 certified companies in Malaysia. J Manuf Technol Manag 21(2):206–225. https://doi.org/10.1108/17410381011014378
26. Färe R, Grosskopf S (2000) Network DEA. Socioecon Plann Sci 34(1):35–49. https://doi.org/10.1016/S0038-0121(99)00012-9
27. Farrell MJ (1957) The measurement of productive efficiency. Source: J R Stat Soc Ser A (General) 120(3). https://www.jstor.org/stable/2343100
28. Forghani D, Ibrahim MD, Daneshvar S (2022) Improving weak efficiency frontier in a variable returns to scale stochastic data envelopment analysis model. RAIRO—Oper Res 56(4):2159–2179. https://doi.org/10.1051/ro/2022100
29. Ghosh S, Murthy N, Park BJ, Park BJ (2004) Integrating product platform development with supply chain configuration in a global environment. https://ink.library.smu.edu.sg/lkcsb_research/2087
30. Guillén G, Mele FD, Bagajewicz MJ, Espuña A, Puigjaner L (2005) Multiobjective supply chain design under uncertainty. Chem Eng Sci 60(6):1535–1553. https://doi.org/10.1016/j.ces.2004.10.023
31. Herrmann FF, Barbosa-Povoa AP, Butturi MA, Marinelli S, Sellitto MA (2021) Green supply chain management: Conceptual framework and models for analysis. Sustainability (Switzerland) 13(15). https://doi.org/10.3390/su13158127
32. Hsu CW, Hu AH (2009) Applying hazardous substance management to supplier selection using analytic network process. J Clean Prod 17(2):255–264. https://doi.org/10.1016/j.jclepro.2008.05.004
33. Hong S-M, Paterson G, Mumovic D, Steadman P (2014) Improved benchmarking comparability for energy consumption in schools. Build Res Inf 42(1):47–61. https://doi.org/10.1080/09613218.2013.814746
34. Harland CM (1996) Supply chain management: relationships, chains and networks. Br J Manag 7(SPEC. ISS.):S63–S80. https://doi.org/10.1111/J.1467-8551.1996.TB00148.X
35. Holden D (2004) Testing the normality assumption in the Tobit model. J Appl Stat 31:521–532
36. de Sousa Jabbour AB, Jabbour CJ, Latan H, Teixeira AA, de Oliveira JH (2015) Reprint of Quality management, environmental management maturity, green supply chain practices and green performance of Brazilian companies with ISO 14001 certification: Direct and indirect effects. Transp Res Part E: Logistics Transp Rev 74:139–151. https://doi.org/10.1016/j.tre.2014.12.011
37. Johnson GM, Malucci L (1999) Shift to supply chain reflects more strategic approach, APICS-The performance advantage, pp 28–31

38. Kalvelagen E (2004) Efficiency solving DEA model with GAMS. http://amsterdamoptimization.com/models/dea/dea.gms
39. Kuo RJ, Wang YC, Tien FC (2010) Integration of artificial neural network and MADA methods for green supplier selection. J Clean Prod 18(12):1161–1170. https://doi.org/10.1016/j.jclepro.2010.03.020
40. Kao C, Hwang SN (2008) Efficiency decomposition in two-stage data envelopment analysis: an application to non-life insurance companies in Taiwan. Eur J Oper Res 185(1):418–429. https://doi.org/10.1016/j.ejor.2006.11.041
41. Large RO, Gimenez Thomsen C (2011) Drivers of green supply management performance: evidence from Germany. J Purch Supply Manag 17(3):176–184. https://doi.org/10.1016/j.pursup.2011.04.006
42. Lambert DM, Cooper MC (2000) Issues in supply chain management. Ind Mark Manag 29(1):65–83. https://doi.org/10.1016/S0019-8501(99)00113-3
43. Mondal C, Giri BC (2022) Investigating strategies of a green closed-loop supply chain for substitutable products under government subsidy. J Ind Prod Eng 39(4):253–276. https://doi.org/10.1080/21681015.2021.1974962
44. Nahmias S (1993) Production and operations analysis. Richard D. Irwin. Inc., Homewood, Illinois
45. Ninlawan C, Seksan P., Tassapol K, Pilada W (2010) The implementation of green supply chain management practices in electronics industry. In: World congress on engineering 2012, vol 2182. London, UK, pp 1563–1568. International Association of Engineers
46. Olugu EU, Wong KY, Shaharoun AM (2010) A comprehensive approach in assessing the performance of an automobile closed-loop supply chain. Sustainability 2(4):871–889. https://doi.org/10.3390/su2040871
47. Radovanov B, Dudic B, Gregus M, Horvat AM, Karovic V (2020) Using a two-stage DEA model to measure tourism potentials of EU countries and Western Balkan countries: an approach to sustainable development. Sustainability (Switzerland) 12(12). https://doi.org/10.3390/SU12124903
48. Rajeev A, Pati RK, Padhi SS, Govindan K (2017) Evolution of sustainability in supply chain management: a literature review. J Clean Prod 162:299–314. Elsevier Ltd. https://doi.org/10.1016/j.jclepro.2017.05.026
49. Rostamzadeh R, Akbarian O, Banaitis A, Soltani Z (2021) Application of dea in benchmarking: a systematic literature review from 2003–2020. Technol Econ Dev Econ 27(1):175–222. https://doi.org/10.3846/tede.2021.13406
50. Sabri EH, Beamon BM (2000) A multi-objective approach to simultaneous strategic and operational planning in supply chain design. Omega 28(5):581–598. https://doi.org/10.1016/S0305-0483(99)00080-8
51. Sarkis J (2003) A strategic decision framework for green supply chain management. J Clean Prod 11(4):397–409. https://doi.org/10.1016/S0959-6526(02)00062-8
52. Sarkis J (2006) The adoption of environmental and risk management practices: relationships to environmental performance. Ann Oper Res 145(1):367–381. https://doi.org/10.1007/s10479-006-0040-9
53. Sarkis J, Zhu Q, Lai KH (2011) An organizational theoretic review of green supply chain management literature. Int J Prod Econ 130(1):1–15. https://doi.org/10.1016/J.IJPE.2010.11.010
54. Sellitto MA, Bittencourt SA, Reckziegel BI (2015) Evaluating the implementation of GSCM in industrial supply chains: two cases in the automotive industry. Chem Eng Trans 43:1315–1320. https://doi.org/10.3303/CET1543220
55. Shabani A, Saen RF, Torabipour SMR (2012) A new benchmarking approach in Cold Chain. Appl Math Model 36(1):212–224. https://doi.org/10.1016/j.apm.2011.05.051
56. Shaik M, Abdul-Kader W (2011) Green supplier selection generic framework: a multi-attribute utility theory approach. Int J Sustain Eng 4(1):37–56. https://doi.org/10.1080/19397038.2010.542836

57. Sheffi Y (2005) Fixing government after Katrina. https://web.mit.edu/sheffi/www/documents/genmedia.FixingGovAfterKatrina.pdf
58. Shen L, Olfat L, Govindan K, Khodaverdi R, Diabat A (2013) A fuzzy multi criteria approach for evaluating green supplier's performance in green supply chain with linguistic preferences. Resour Conserv Recycl 74:170–179. https://doi.org/10.1016/j.resconrec.2012.09.006
59. Stewart TJ (2010) Goal directed benchmarking for organizational efficiency. Omega 38(6):534–539. https://doi.org/10.1016/j.omega.2010.01.004
60. Sueyoshi T, Goto M (2011) Methodological comparison between two unified (operational and environmental) efficiency measurements for environmental assessment. Eur J Oper Res 210(3):684–693. https://doi.org/10.1016/j.ejor.2010.10.030
61. Simchi-Levi D, Kaminsky S, Simchi-Levi E (2003) Designing and managing the supply chain concepts, strategies and cases
62. Tajabadi FZ, Kazemi A (2016) To develop an integrated model for green supply chain. J Fund Appl Sci 8(3):1340. https://doi.org/10.4314/jfas.v8i3s.158
63. Tseng ML, Islam MS, Karia N, Fauzi FA, Afrin S (2019) A literature review on green supply chain management: Trends and future challenges. In: Resources, conservation and recycling, vol 141. Elsevier B.V., pp 145–162. https://doi.org/10.1016/j.resconrec.2018.10.009
64. Tseng ML, Chiu ASF (2013) Evaluating firm's green supply chain management in linguistic preferences. J Clean Prod 40:22–31. https://doi.org/10.1016/J.JCLEPRO.2010.08.007
65. Tobin J (1958) Estimation of relationships for limited dependent variables. Econometrica 26(1):24. https://doi.org/10.2307/1907382
66. Walker H, Jones N (2012) Sustainable supply chain management across the UK private sector. Supply Chain Manag 17(1):15–28. https://doi.org/10.1108/13598541211212177
67. Walker H, di Sisto L, McBain D (2008) Drivers and barriers to environmental supply chain management practices: lessons from the public and private sectors. J Purch Supply Manag 14(1):69–85. https://doi.org/10.1016/j.pursup.2008.01.007
68. Wang YM, Chin KS (2010) Some alternative DEA models for two-stage process. Expert Syst Appl 37(12):8799–8808. https://doi.org/10.1016/j.eswa.2010.06.024
69. Willems SP (2008) Real-world multi-echelon supply chains used for inventory optimization. Manuf Serv Oper Manag 10(1):19–23. https://doi.org/10.1287/MSOM.1070.0176
70. Yasin MM (2002) The theory and practice of benchmarking: then and now. Benchmarking: Int J 9(3):217–243. https://doi.org/10.1108/14635770210428992
71. Zanjirani Farahani R, Asgari N, Davarzani H (2009) Supply chain and logistics in national, international and governmental environment: concepts and models. Springer, Berlin/Heidelberg, Germany
72. Zhou P, Ang B, Poh K (2008) Measuring environmental performance under different environmental DEA technologies. Energy Econ 30:1–14. https://doi.org/10.1016/j.eneco.2006.05.001
73. Zhou P, Ang B, Poh K (2008) A survey of data envelopment analysis in energy and environmental studies. Eur J Oper Res 189:1–18. https://doi.org/10.1016/j.ejor.2007.04.042
74. Zhu Q, Sarkis J (2006) An inter-sectoral comparison of green supply chain management in China: drivers and practices. J Clean Prod 14(5):472–486. https://doi.org/10.1016/j.jclepro.2005.01.003
75. Zhu Q, Sarkis J (2007) The moderating effects of institutional pressures on emergent green supply chain practices and performance. Int J Prod Res 45(18–19):4333–4355. https://doi.org/10.1080/00207540701440345
76. Zhu Q, Sarkis J, Lai KH (2007) Green supply chain management: pressures, practices and performance within the Chinese automobile industry. J Clean Prod 15(11–12):1041–1052. https://doi.org/10.1016/J.JCLEPRO.2006.05.021
77. Zhu Q, Sarkis J, Lai KH (2008) Confirmation of a measurement model for green supply chain management practices implementation. Int J Prod Econ 111(2):261–273. https://doi.org/10.1016/j.ijpe.2006.11.029

The Adaptable Mobile Solution with the Remote-Controlled Omnidirectional Platform

Olisaemeka Barineka Sunny-Ossai, Gloire Wa Lwamba Sibadia, and Henry Maduka Onyebuchi

1 Introduction

This chapter establishes the groundwork for the research by presenting the project's setting, aims, and importance. The text examines the rationale for creating an omnidirectional mobile platform, highlighting the practical and theoretical deficiencies the research intends to address. Furthermore, this chapter outlines the study issues and scope, establishing the foundation for ensuing debates.

This chapter underscores the prospective contributions of the study to the domain of robotics and its wider ramifications for sectors dependent on nimble and efficient robotic systems. The domain of robotics has undergone significant progress in recent decades, evolving from basic automated devices to advanced systems proficient in executing intricate tasks. Initial robots were engineered for repetitive industrial functions, frequently constrained to set motions or predetermined actions. Robots have grown into dynamic beings capable of independently adapting to their environments and executing complex functions, thanks to advancements in microcontrollers, sophisticated sensors, and artificial intelligence. Omnidirectional mobile robots have attracted considerable interest due to their distinctive capability to navigate effortlessly in any direction without changing their orientation. This capacity is especially beneficial in settings that demand high agility and precision, such as production floors, warehouses, and medical facilities. The history of robotics illustrates a significant evolution from ancient automata to contemporary advanced systems. The name "robot" originated from Karel Čapek's 1920 play *Rossum's Universal Robots* (R.U.R.) and became broadly linked to robots engineered to execute human activities. The contemporary notion of robotics accelerated in the mid-twentieth century due to the emergence of industrial automation. George Devol's creation of the Unimate in 1954, frequently regarded as the inaugural industrial robot, represented a

O. B. Sunny-Ossai (✉) · G. W. L. Sibadia · H. M. Onyebuchi
Department of Mechatronic Engineering, Near East University, Nicosia, Türkiye
e-mail: phanphraze@gmail.com

A. Azizi (ed.), *Intelligent Mechatronic Control Systems*, Emerging Trends in Mechatronics, https://doi.org/10.1007/978-981-95-3925-3_7

crucial advancement. The Unimate, engineered for executing repetitive and perilous activities in manufacturing, transformed industries such as automotive production, where it was utilized for welding and material handling. The inaugural generation of industrial robots (1950–1967) comprised programmable machines characterized by restricted autonomy and communication capability. These early robots utilized pneumatic actuators and mechanical logic systems to execute basic tasks such as loading and unloading. The advancement of technology during the second generation of robots (1968–1977) incorporated servo-controllers and microprocessors, facilitating more intricate functions such as continuous-path motion and fundamental environmental interaction. This era witnessed the inception of emblematic designs such as the Stanford Arm and the Programmable Universal Machine for Assembly (PUMA) robot, which established the foundation for contemporary robotics. The third generation of robots (1978–1999) introduced a paradigm shift through the incorporation of sensors, sophisticated control systems, and rudimentary artificial intelligence. These robots can engage with their surroundings, adjust to varying conditions, and execute high-precision activities such as assembly and arc welding. Innovations such as Selective Compliance Assembly Robot Arm (SCARA) robots and Delta robots have broadened the scope of robotics applications beyond conventional production, especially in the electronics and consumer goods industries. The swift breakthrough in robotics was propelled by concurrent developments in processing power, materials science, and control theory. The 1970s oil crisis, which demanded more industrial efficiency, also expedited the implementation of robotic systems. By the 1980s, Japan had established itself as a global leader in robotics, spearheading advancements in design and manufacturing. Robots such as the FlexPicker and coordinated multi-robot systems illustrate the increasing complexity and adaptability of industrial automation. As robots advanced in intelligence and capability, they commenced integration into new sectors, including health care, space exploration, and service industries. The late twentieth century established the groundwork for the fourth generation of robots, distinguished by collaborative abilities, deep learning, and sophisticated computational intelligence. Currently, robotics advances technological limits, transforming industries and redefining human–machine interaction. The advancement of robotics exemplifies humanity's unwavering quest for efficiency, safety, and creativity. From the mechanical automata of antiquity to contemporary intelligent machines, each advancement signifies a deep comprehension of engineering, creativity, and the transformative capacity of technology. This historical arc emphasizes both the landmarks attained and the boundless opportunities that await.

A major issue in robotics has been attaining efficient mobility. Conventional wheeled or tracked robots frequently encounter significant constraints regarding agility and the capacity to maneuver in confined or turbulent environments. Researchers and engineers have investigated omnidirectional mobility to surmount the limitations of traditional movement methods. Omnidirectional platforms utilize specialized wheel configurations, such as mecanum wheels or omni wheels, enabling independent control of movement across several axes. This remarkable discovery allows robots to execute translational and rotational movements concurrently, offering exceptional flexibility and control. The advancement of omnidirectional platforms is a pivotal achievement in robotics, transforming mobility in intricate and dynamic settings. Originating in the late twentieth century, these platforms developed concurrently with progress in control theory, actuator technology, and mechanical design. Their viability was evidenced by initial prototypes that showcased the benefits of omnidirectional mobility in high-precision applications, including automated guided vehicles (AGVs) in manufacturing. In contrast to conventional wheeled systems that necessitated turning arcs or limited movement patterns, omnidirectional platforms provided exceptional flexibility of motion, facilitating effortless navigation in confined or crowded environments. The advancement of these systems was further accelerated by the incorporation of sophisticated sensing technologies like light detection and ranging (LiDAR) and vision-based navigation. Through the integration of real-time mapping, object detection, and obstacle avoidance, omnidirectional platforms have become essential in sectors such as warehousing, health care, and logistics. In contemporary distribution centers, these devices can adeptly navigate among personnel and machinery, streamlining operations and improving safety. The integration of mechanical agility and cognitive sensing has established omnidirectional platforms as vital instruments in settings requiring superior maneuverability and adaptability, facilitating their extensive utilization across several domains. Omnidirectional robots have proved essential in numerous applications. In manufacturing, they facilitate accurate material handling and assembly in restricted areas, minimizing the necessity for extensive turning radii. In logistics and warehousing, activities are optimized by enabling the efficient transfer of commodities. Health care environments have adopted omnidirectional robots for functions such as patient transport and medical supply distribution, where seamless and accurate navigation is crucial. The emergence of omnidirectional platforms underscores the overarching trend in robotics toward enhanced adaptability and specialization. By transcending conventional mobility constraints, these systems create novel opportunities for robotic applications in both structured and unstructured settings. With ongoing advancements in research and development, omnidirectional robots are set to significantly influence the future of automation, enhancing efficiency and fostering creativity across all sectors.

The progression of robotics from basic mechanical systems to advanced autonomous platforms has facilitated advancements such as omnidirectional movement.

These platforms signify a substantial advancement in robotic design, meeting the essential demand for adaptability and accuracy across diverse applications. Omnidirectional robots enhance navigation and operation in intricate situations, hence broadening robotic capabilities and transforming machine interactions with their surroundings.

1.1 The Primary Objective of This Initiative

The primary impetus for the creation of a remote-controlled omnidirectional mobile platform (RC-ODP) robot arises from the increasing demand for adaptable and efficient robotic systems across multiple industries. Conventional wheeled robots frequently encounter considerable constraints and challenges in maneuverability, which might impede their efficacy in dynamic and confined settings. These challenges become particularly apparent in situations such as congested warehouses, manufacturing floors with limited space, or production facilities with complex configurations. Resolving these problems necessitates a novel approach that offers improved mobility and accuracy. This research aims to enhance the operational efficiency of robotic systems by utilizing the benefits of omnidirectional mobility to address existing constraints.

The necessity for robots that can travel effortlessly in any direction without altering their orientation has become increasingly essential. Omnidirectional robots are ideally adapted for contexts where agility and spatial awareness are essential. In logistics and warehousing, the capacity for lateral or diagonal movement can substantially decrease the time required for navigation and repositioning, resulting in more efficient and seamless operations. In health care environments, omnidirectional platforms facilitate smoother and more precise movement for operations like patient transport and surgical equipment handling, hence improving safety and reliability. This project seeks to provide a platform that fulfills practical requirements while establishing a new benchmark for versatility and adaptability in robotic mobility. The exploration and application of omnidirectional mobility create prospects for enhancing research in robotics, control systems, and artificial intelligence. Creating a platform with omnidirectional capabilities necessitates addressing intricate engineering issues, including the construction of advanced wheel mechanics (e.g., mecanum or omni wheels), the optimization of control algorithms, and the integration of sophisticated navigation sensors, if feasible. This initiative addresses these issues, thereby enhancing the field of robotics and facilitating advancements in autonomous navigation, collaborative robotics, and human–robot interaction.

This initiative aligns with the global transition toward automation and Industry 4.0. As enterprises progressively implement intelligent systems to improve production and decrease labor expenses, robots with sophisticated mobility features will be essential in fulfilling these requirements. The incorporation of omnidirectional platforms into automated systems can afford enterprises a competitive advantage, allowing them to respond to dynamic market demands with agility and accuracy. This project aims to establish itself as a significant contributor to this disruptive movement, providing a strong and scalable answer for contemporary automation difficulties.

The primary aim of this project is to overcome the practical constraints of conventional robots while furthering theoretical studies in robotic mobility. This project, by integrating innovative design with practical applications, has the potential to profoundly impact industries dependent on efficient and precise robotic systems, advancing the evolution of robotics as a transformational force in technology and society.

- **Main goal of the remote-controlled platform**: The primary goal of this project is the design and building of a remote-controlled platform cell, which can be deployed to various regions of the country to facilitate efficient operations. The project aims to tackle global resource management difficulties in the industrial and manufacturing sectors by offering realistic and innovative solutions that enhance safety, efficiency, and productivity.
- **Industry meeting and innovation requirements**: In sectors where precision and weight management are critical, prevailing transportation techniques encompass physical labor and equipment delivery. These solutions, notwithstanding their efficacy, impose constraints such as mobility limitations, insecurity in confined areas, and significant security issues. The objective is to address such deficiencies by implementing a platform that employs sophisticated mobility and innovative management strategies to redefine asset management. The platform's comprehensive mobility enables it to maneuver in confined locations without rotation or displacement. This capability addresses a critical requirement in industries like construction, where work sites are congested with outdated trucks that are challenging to manage properly. The program prioritizes flexibility, enabling the platform to adjust to various workloads and work environments, rendering it a versatile instrument for many applications.
- **Enhance safety and diminish manual labor**: Numerous studies indicate that lifting and inappropriate handling are primary contributors to physical injuries and accidents in the industrial industry. The platform enhances safety by automating the transportation process, thereby minimizing direct human involvement in high-impact operations. The platform's remote capabilities enable employees to operate from dangerous locations, including building sites with moving machinery. This emphasis on safety corresponds with international standards prioritizing worker protection and accident prevention, establishing the platform as a novel service for society.
- **Integrating academic information with practical application**: Another objective extends beyond merely developing a functional product; there is the aim to reconcile theoretical knowledge with real-world implementation. This project offers a chance to implement engineering ideas, including mechanical design, control systems, and logic programming, to address real-world challenges. By surmounting problems like weight distribution, stability, and terrain unpredictability, the team enhanced not only the technical skills but also problem-solving capabilities. With thorough testing and ongoing enhancement, this project seeks to illustrate how interdisciplinary proficiency in mechatronics, robotics, and automation can provide efficient engineering solutions. This practical experience is essential for preparation for future professions in manufacturing, automation, and industrial design, where these talents are highly sought after.
- **Promote sustainable development**: The platform is engineered with sustainability as a priority, embodying the objective of conforming to eco-friendly

methods. The mobile directional remote-control system, in contrast to traditional fuel-operated devices, incorporates energy-efficient components and emphasizes durability to minimize waste. This sustainability strategy guarantees that the platform addresses the industry's urgent requirements while also aiding the worldwide initiative to mitigate environmental degradation. This initiative facilitates the shift to sustainable practices in the building and industrial sectors by implementing energy-efficient and reusable solutions. It underscores the potential of academic innovation to advance long-term societal objectives, including the reduction of carbon footprints and the promotion of sustainable development.

- **Establishing a framework for future innovations**: A significant objective of this project is to establish a foundation for forthcoming advancements in robotics and automation. The insights and techniques established in this effort may stimulate future advancements in analogous domains. Spherical mobility systems can be utilized for autonomous vehicles, medical transport robots, or disaster response apparatuses. By emphasizing flexibility and adaptability, the system will serve as a paradigm for further investigation and enhancement. This concept broadens the project's scope beyond immediate industrial application and establishes it as a catalyst for ongoing innovation.

1.2 Main Objectives of the Project

The main objectives of this project are:

- The design and cultivation of a remote control (RC) platform conducive to omnidirectional development, considering factors such as dimensions, mass, materials, and energy source to attain maximal performance and mobility.
- The examination of the incorporation of contemporary technology, including sensors and communication systems, to enhance the platform's performance and adaptability to various scenarios.
- Development and construction of an advanced software and control system for the platform, emphasizing accurate mobility and movement. This involves creating algorithms to control the platform's movement in all directions and ensure immediate responses to user requests.
- The identification of methods to enhance the platform's stability and performance. Construction of the RC platform's framework and functional elements, prioritizing durability, utility, and ease of assembly.
- The utilization of computer-aided design (CAD) tools to create accurate designs that consider structural integrity, weight distribution, and maintenance accessibility. Integration of modular components to facilitate modifications and updates as required.
- Development of a multifunctional platform applicable to many domains, including indoor navigation, surveillance, and logistics. Examination of the incorporation of novel elements, such as payload capacity, camera mounts, and sensor arrays, to enhance the platform's functionality for diverse tasks.

- The carrying out of comprehensive testing and validation to confirm that the platform satisfies performance criteria and user expectations across many situations and contexts.
- The evaluation of opportunities for future expansion and integration with emerging technologies to maintain the platform's relevance and competitiveness in the ever-evolving field of robotics.

1.3 Challenges to Be Encountered in the Project

- **Restricted access to materials and components**: A primary issue in the development of an omnidirectional remote sensing platform in Northern Cyprus is the constrained availability of distinct and individual materials. The requisite high-quality sensors, microcontrollers, and motors for constructing the platforms are not locally accessible. The shortfall necessitates the delivery of these commodities, potentially resulting in delays, increased costs, and operational challenges. Additionally, several elements necessary for constructing the platform, such as lightweight yet durable metal, omnidirectional wheels, and appropriate electrical components, are either prohibitively expensive or challenging to source locally. The absence of reliable local vendors heightens reliance on international providers, interrupts sales, and postpones developmental phases.
- **Design and simulation challenges**: Assessing platform performance is a crucial aspect of the design process, particularly for novel projects. Nevertheless, access to simulation tools and resources may be restricted in Northern Cyprus. Advanced software like SolidWorks, Matrix Laboratory (MATLAB), or Analysis System (ANSYS) is frequently used for simulating mechanical and control systems; however, access may be limited due to licensing restrictions or financial limitations. Access to simulation software notwithstanding, developing a realistic model of your platform necessitates considerable computational power. Restricted access to high-performance computing equipment diminishes the efficiency and precision of simulations, prolonging the cross-section design and testing procedures. Furthermore, the absence of local proficiency in advanced simulation technologies demands that teams invest time in acquiring knowledge and utilizing these tools.
- **Constraints of testing and sample**: The sample component of the project presents a significant challenge. Producing physical prototypes requires access to manufacturing facilities equipped with sophisticated machinery, like computer numerical control (CNC) machines, laser cutters, and 3D printers. Nevertheless, these regions frequently exhibit deficiencies in Northern Cyprus, compelling teams to either compromise on model quality or depend on foreign manufacturers, thereby extending project timelines and expenses. Evaluating the prototype presents an additional hurdle. For instance, assessing soil performance across various terrains must encompass test sites with diverse surface conditions. Nonetheless, the absence of these resources locally may restrict the team's capacity to verify the platform's performance.

- **Financial limitations and budgetary concerns**: Financial limitations frequently pose significant issues for such initiatives. The expenses associated with acquiring premium materials, importing machinery, licensing simulation software, and procuring prototype equipment might rapidly surpass the designated budget. The volatility of import duties and delivery expenses complicates financial forecasting. These budgetary limitations may compel all parties to reach concessions, such as utilizing substandard equipment or depending on free or low-capacity software solutions. Although these methods may decrease expenses, they will impact the performance, longevity, and overall quality of the platform.
- **Limitations in knowledge and expertise**: Constructing a remote-controlled omnidirectional mobile platform necessitates proficiency across various fields, including mechanical design, electronics, robotics, and control systems. The project team has a basic understanding, but addressing complex technical challenges—like building an efficient omnidirectional wheel system or programming accurate control algorithms—might necessitate more assistance from industry experts or academia. Access to specialists with expertise in mechatronics and robotics may be restricted in North Cyprus. The absence of local mentorship may impede problem-solving and necessitate the team's dependence on self-directed learning, which can be inefficient and protracted.
- **Logistical and administrative challenges**: The complexities of importing materials and components into North Cyprus are exacerbated by administrative and logistical difficulties. Customs laws, shipping delays, and elevated import taxes can impede business schedules and escalate expenses. Moreover, relationships with international suppliers necessitate meticulous planning to guarantee that equipment arrives in optimal shape and within the specified timeframe. Restricted local alternatives intensify these logistical difficulties, providing teams with no flexibility to adjust to unforeseen delays or shortages.
- **Adjust to environmental circumstances and operations**: Developing a platform capable of functioning in varied terrains necessitates comprehension of the environmental conditions it will face. Nevertheless, the restricted experimental facilities in North Cyprus hinder the simulation of these conditions. Teams must formulate assumptions on factors such as surface tension, material strength, and load distribution, which may not entirely correspond to actual conditions. The disparity between simulated environments and reality may result in variations in platform performance, necessitating further refactoring and optimization. The absence of access to tools and diverse testing environments exacerbates the situation, as teams depend on cognitive processes instead of data.
- **Chronology/agenda**: This project timeline describes the methodical strategy used to finalize the RC omnidirectional mobile platform. Commencing last spring, the project advanced through critical phases, encompassing research, design, and control system development. Thorough testing and validation were performed over the fall semester to guarantee dependability and performance. The final documentation and reports are being finalized before year-end to properly conclude the project. Figure 1 gives a chart of the project timeline with the various tasks involved.

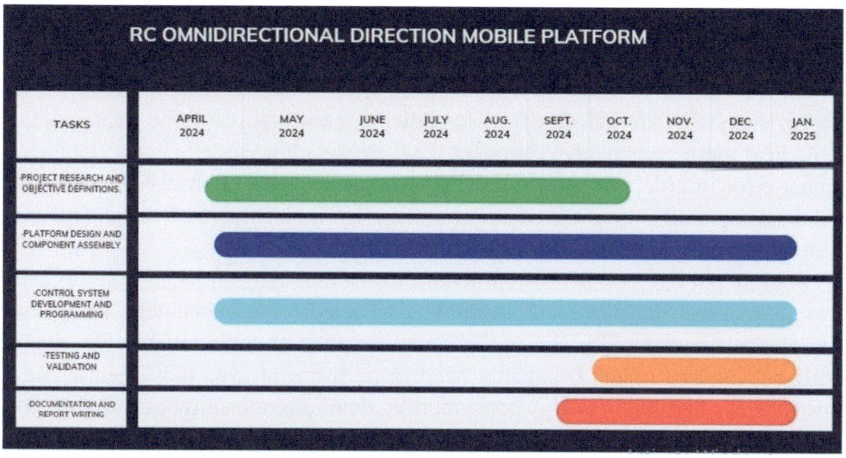

Fig. 1 The estimated timeline for completion of the task on the project

2 Review of the Literature

A thorough analysis of current robotics research is given in this chapter, with an emphasis on omnidirectional platforms. To discover cutting-edge designs and technology, this analysis examines significant developments in control theory, sensor integration, and mobility systems. To illustrate the necessity of the suggested study, this chapter also identifies shortcomings with the existing methods, such as difficulties with precision, power efficiency, and navigation. The chapter creates a solid theoretical foundation and places the study in the larger context of robotics innovation by synthesizing this data.

2.1 The Remote-Controlled Omnidirectional Mobile Platform (RC-ODP) Literature Review

Over the past few decades, the field of robotics has experienced extraordinary growth, driven by a variety of technological advancements, the growing need for automation, and the intense search for novel solutions to challenging and complicated jobs. The advent of omnidirectional platform robots and remote-controlled robots stands out among the numerous advancements in robotics since they represent a critical turning point in the discipline. Without having to adjust their orientation, these robots have the amazing and unusual ability to move fluidly in any direction. These robots' exceptional mobility and flexibility in a variety of settings and terrains are made possible by their advanced wheel layout and control systems. Omnidirectional robots are especially beneficial and widely sought after in a variety of industries and domains, such as industrial automation, healthcare, logistics, defense, manufacturing, research, and so forth, due to their high functionality and versatility. For example, these robots can move supplies precisely, maneuver through busy areas, and adjust to changes in the terrain's layout on building sites. Similar to this, omnidirectional platforms are used in the healthcare industry for telemedicine, patient support, the transportation of medical supplies, and other purposes. This greatly improves efficiency and lowers human error in crucial circumstances. These robots are being investigated for use in defense applications where accuracy and agility are critical, such as surveillance, reconnaissance, and the disposal of explosive ordnance.

Omnidirectional mobility's significance stems mostly from its capacity to overcome issues that the primary conventional wheeled robot encounters, such as its restricted maneuverability and inefficiency in confined environments. To change direction, modern robots frequently need to perform intricate movements, which causes delays and raises energy consumption. Omnidirectional robots, on the other hand, overcome these constraints by providing instantaneous direction changes, allowing for more efficient operations and improved resource optimization. This feature is particularly important in settings like production lines, self-driving delivery systems, etc., that require a high degree of precision. Despite its enormous potential, remote-controlled omnidirectional robot platforms have several drawbacks, such as expensive development costs, energy efficiency issues, and the requirement for sophisticated control algorithms to guarantee dependable and seamless operations. Additionally, integrating these robots into the current systems frequently necessitates a great deal of customization, which may prevent mainstream adoption. These difficulties highlight how crucial it is to conduct continuous research and innovation to improve the technology and broaden its range of applications.

2.2 The Review's Objective and Scope

This literature review's primary goal is to present a thorough analysis of remote-controlled omnidirectional mobile platform robots, including an investigation of their historical evolution, technological developments, applications, difficulties, and potential future trends. The review will provide insight into the key inventions that have influenced the evolution of these robots by exploring their historical development and expansion. The state-of-the-art technologies that make omnidirectional mobility possible, like the well-known mecanum wheels, omnidirectional drive systems, and sophisticated sensors, will also be covered. The wide range of uses for these robots will also be covered in this overview, emphasizing how they are revolutionizing several different areas and businesses. Examining case studies and actual applications will highlight the useful advantages and constraints of omnidirectional movement. Additionally, this review will cover ways to overcome the various major obstacles that these robots provide, such as the technological, financial, and social elements of the robot. Lastly, the review will look at future developments and trends, providing information on possible innovations and breakthroughs.

This review aims to increase knowledge of remote-controlled omnidirectional mobile platform robots and their contribution to the advancement of robotics by offering a thorough examination of these diverse subjects. Researchers and engineers will find this knowledge to be a great and valuable resource, which will help guide future advances and encourage the adoption of this technology in a variety of applications.

2.3 *Initial Design Concepts and Developments for Omnidirectional Robots*

This project cannot begin without talking about the mecanum wheel, which is its primary component. Bengt Ilon, a Swedish engineer, created the mecanum wheel, one of the first omnidirectional wheels, according to a study by Rojas R. This revolutionary invention of traditional wheels allows for omnidirectional movement without the need for a traditional steering system and does not require the normal translational movement that regular wheels do. Four-mecanum-wheeled vehicles can go in either direction, providing increased efficiency and versatility. "Uranus," the first robot to use these wheels, was created at Carnegie Mellon University. Uranus was ingenious, but its performance on uneven surfaces was limited since it lacked a suspension mechanism. Future designs were based on early research that presented a kinematic model and feedback control algorithm for robots using mecanum wheels, such as that done by Muir and Neuman (1987a, 1987b).

With important contributions from Diegel et al. (2002), Koestler and Braunl (2004), and Efendi et al. (2006), other projects have since exploited and enhanced the mecanum wheels to achieve omnidirectional mobility. Mecanum wheels have significant drawbacks despite their benefits. One frequent problem is slippage, which is a reduction in motion that causes irregularities in longitudinal and lateral travel distances that change according to the terrain or ground conditions. This slippage creates serious issues for the wheels' employment in practical applications, as do the horizontal vibrations they experience during lateral travel and directional dependency when navigating obstacles. The decreased precision of conventional dead-reckoning techniques, such as their heavy reliance on rotary shaft encoders, when used on mecanum wheel vehicles, is another significant drawback. Visual dead-reckoning systems, which use onboard cameras to record and examine ground movement, were used to address this. This method, which was influenced by the optical mouse, uses image processing to determine relative speed and direction. These systems are not completely stand-alone solutions, despite their effectiveness. These ideas led to the development of a hybrid strategy that combined conventional odometry devices with visual input to increase positioning accuracy. This approach provides a more reliable framework for attaining accurate navigation by integrating sensor data, improving reliability, and reducing the drawbacks of mecanum wheels [1].

2.4 An Overview of Omnidirectional Wheels' Development

- **The omnidirectional wheel by Grabowiecki (1919)**: The first omnidirectional wheel ever produced was this one. J. Grabowiecki patented it in 1919. The transversal rollers on this wheel enabled multidirectional movement. Future advancements in omnidirectional motion were made possible by this early invention [1].
- **Ilon wheel (1970s)**: This was created in the 1970s by Swedish inventor Bengt Ilon. This design's use of a circular shape and many rollers greatly decreased friction and enhanced mobility. For early omnidirectional robots, the Ilon wheel was a common option [1].
- **Killough rollers**: Bradbury patented the Killough rollers in 1980. They improved omnidirectional movement efficiency and reduced friction by using two truncated spheres. Killough rollers have a variety of industrial uses and were frequently utilized in robotic contests.
- **The mecanum wheels**: In 1973, engineer Bengt Ilon of the Swedish business Mecanum AB created the mecanum wheel. In honor of him, the wheel is occasionally referred to as the Ilon wheel. Mecanum's invention was purchased by the US Navy, who at first only utilized the technique to load cargo onto their ships. The technology wasn't made accessible to the general public or commercial enterprises until 1997. It had rollers positioned at a 45-degree angle to the axis of the wheel. This architecture made it possible to travel precisely and smoothly in any direction without having to readjust. Because of their current simplicity of use and adaptability, mecanum wheels have become common in many omnidirectional mobile platforms and robots [2].

2.5 Developments in Technology

Omnidirectional mobile robots have become adaptable instruments for a variety of uses thanks to technological developments in sensors, actuators, and control systems. The capacity of robots to move smoothly in any direction has been transformed by numerous inventions, including the mecanum omnidirectional wheels and complex motor control algorithms. These diverse technologies make robots an exceptional asset in industries like manufacturing, logistics, and health care since they not only improve maneuverability but also allow them to function well in constrained or changing situations.

2.5.1 Advanced Systems for Motor Control

Omnidirectional robots are now far more sensitive and efficient due to the development of sophisticated motor control technologies. To guarantee a seamless transition between directions and reduce energy consumption, modern algorithms even allow for precise coordination of wheel motions. According to Ahmed Neaz et al.'s study, integrating intelligent robots into industrial production processes has the potential to greatly increase productivity and lessen human hardship. Such robots must, however, have a sufficient awareness of their surroundings and be able to maneuver through small hallways without running into either stationary or moving impediments to function well in human situations. An omnidirectional automobile mobile robot has been developed in this study to carry out industrial logistics duties in busy and changing situations. Both high-level and low-level algorithms have been incorporated into a control system, and each control system now includes a graphical user interface. To operate the motors with the proper degree of precision and resilience, a very effective microcontroller—myRIO—has been used as a low-level computer. With the help of many LiDAR sensors, an inertial measurement unit (IMU), and odometry data produced by wheel encoders, a Raspberry Pi 4 has also been used in combination with a distant PC for high-level decision-making tasks, including path planning, localization, and mapping the experimental environment. For software programming, the robot operating system (ROS) was used to develop the higher-level software architecture, while LabVIEW was used for the low-level computer. The creation of medium- and large-category omnidirectional mobile robots with autonomous mapping and navigation skills is made possible by the suggested methods covered in this work [3].

A new active visual simultaneous localization and mapping (SLAM) technique for omnidirectional robots is presented in the study by Elia Bonetto et al. The main objective of the system is to minimize energy consumption and optimize for information gain while allowing the robot to concurrently localize itself and map a new environment. The technique uses a multilayered strategy. Paths to informative goal locations are generated by the top layer. Using the most recent map and local feature data, the second and third layers actively replan and carry out the route. Utility functions that take into consideration impediments and the robot's location are incorporated into the system. The robot's omnidirectional design enables continuous camera heading control for optimal coverage of the surrounding area. By drastically cutting down on trip time, this proactive strategy improves autonomy and uses less energy. Using Festo Didactic's Robotino, the system was thoroughly evaluated in simulations and real-robot studies, demonstrating that it provides comparable coverage with up to 39% shorter distances than state-of-the-art techniques. The data and code are publicly accessible [4].

A new calculated voltage-based control rule for an omnidirectional mobile robot with four mecanum wheels is presented by Armando Saenz and colleagues. Using the Euler–Lagrange formulation, the research started by creating a dynamic model of the robot that included actuator dynamics and motion constraints using Lagrange multipliers. Motor armature voltage is the input of the final model. This method does not require wheel speed output from motor encoders, in contrast to many other controllers now in use. Instead, it only requires position and velocity feedback from a multicam vision system. A high-order sliding modes differentiator is used to estimate the linear and angular velocities. The direct Lyapunov approach was used to demonstrate the control law's stability, and the effect of mechanical parameter uncertainties was examined.

Last but not least, experimental findings show how successful this strategy is [5].

2.5.2 Sensor Integration

Omnidirectional robots' ability to sense and engage with their immediate environment depends heavily on sensors. LiDAR, ultrasonic sensors, and cameras are commonly utilized for mapping out the environment, detecting obstacles, and navigating. Robots can carry out activities more effectively and independently thanks to vision-based technologies, such as stereo and depth cameras, which give detailed spatial information. These robots can adjust to changing circumstances because sensors and sophisticated processing units are integrated to allow for real-time decision-making.

The difficulties in developing open-source aerial robotics systems with omnidirectional visual perception are discussed in a study by Peize Liu and colleagues. In contrast to off-the-shelf options, the system's utilization of a multi-fisheye camera set for omnidirectional FoV requires extensive hardware development, calibration, and structure design. Tartankalibr and kalibr-toolbox are used by OmniNxt to simplify the calibration procedure and increase success rates. To provide 500 Hz low-noise IMU data for visual-inertial odometry (VIO), which is essential for precise localization and mapping, the system also incorporates a coin-sized, high-performance flight controller (Nxt-FC). The software framework, which uses only 27% of CPU and 40% of GPU resources, is resource-efficient and includes Omni-VINS for VIO and Omni-Depth for real-time dense mapping. The modular architecture of the system makes it simple to adjust to multiple CNN models in Omni-Depth, guaranteeing flexibility and adaptability to a range of platforms and accuracy needs. Lastly, OmniNxt's design facilitates simple sensor swapping, as evidenced by the integration of Intel L515 RGB-D and D435 stereo cameras [6].

The design and deployment of an omnidirectional mobile robot for educational purposes are described in detail in the document by Ioan Doroftei. Its sensor integration for obstacle avoidance and navigation is a crucial feature. Without requiring a traditional steering system, the robot can move in any direction thanks to its mecanum wheels. However, the precision of conventional odometry utilizing rotary shaft encoders is hampered by these wheels' propensity for slippage. The robot uses visual dead reckoning to get around this. Using an onboard video camera, this method continually records ground images, which are then processed to ascertain the robot's direction and speed in relation to earlier frames. By adding to the odometry system, this visual data increases position accuracy. Furthermore, bumper switches and a pair of ultrasonic sensors mounted on a servo to enable 360° scanning offer obstacle detection for self-navigating vehicles. Data from these sensors is integrated into the robot's control system to allow for remote control operation, line-following mode, and autonomous obstacle avoidance [7].

2.5.3 Feedback Control and Kinematic Modeling

Significant progress has been made in the mathematical modeling of omnidirectional movement, which offers a framework for accurate control. Feedback control systems ensure precision and stability by modifying the robot's behavior based on sensor inputs. To solve problems like wheel slippage, methods like dead reckoning and visual odometry were developed, increasing dependability in a variety of terrains.

The kinematic modeling approach for Uranus, an omnidirectional wheeled mobile robot (WMR) being developed at Carnegie Mellon University's Robotics Institute, is described in the work by Patrick F. Muir and Charles P. Neuman. Kinematic equations of motion are derived, and transformation matrices are specified by assigning coordinate systems. Actuator velocities can be calculated from robot velocities and vice versa, thanks to the proposed sensing forward and actuated inverse solutions. Whereas the detected forward solution determines robot velocity based on sensed wheel locations and velocities, the actuated inverse solution determines wheel velocities based on robot velocities. The creation of algorithms for wheel slip detection based on discrepancies in the kinematic equations, kinematic feedback control, and real-time robot position calculation (dead reckoning) using shaft encoder signals is an important component. To correct for errors, the feedback control system uses the actuated inverse solution to set actuator velocities after calculating the robot position using the sensed forward solution and comparing it to the desired position. The system uses absolute positioning techniques (computer vision, ultrasonic ranging, and laser range finders) to update the robot's position and is intended to stop if wheel slip is detected [8].

To increase productivity and lessen human hardship, Sandro A. Magalhães' study investigates the incorporation of intelligent robots in industrial production. Introduced is a specially made omnidirectional automobile mobile robot that can maneuver through tight spaces and dodge obstructions in changing conditions. With the use of a graphical user interface, the control system integrates low-level and high-level algorithms. A Raspberry Pi 4 and a remote PC are used to carry out high-level decision-making, such as mapping, path planning, and localization, while a myRIO microcontroller manages the low-level control, guaranteeing motor precision and resilience. These systems make use of information from wheel encoders, IMUs, and LiDAR sensors. Robot Operating System (ROS) is used for higher-level software architecture, while LabVIEW is used for low-level programming. For the development of medium-scale and large-scale autonomous omnidirectional robots for industrial logistics, this framework provides a scalable solution [9].

- **A quick comparison of several strategies**

Trade-offs in omnidirectional robot designs are highlighted by a comparative study of the wheel design, control schemes, and sensor systems from above. Generally speaking, mecanum wheels are preferred for their ease of use and adaptability, although they can have significant levels of friction and slippage on rough terrain. Killough rollers, on the other hand, reduce these problems but require more intricate construction. Although omni balls and spherical wheels provide incredibly smooth motion, their large and complex mechanism limits their use. Additionally, control techniques present several unique issues. For example, PID controllers are easy to install and perform well in steady-state conditions, while model predictive control (MPC) provides greater flexibility at the expense of high computational demands. A promising approach that strikes a compromise between complexity and performance is hybrid control techniques. While LiDAR is excellent at mapping and obstacle identification, it may not perform as well in reflective situations. Vision-based approaches in sensor systems provide high-resolution data, but they are resource-intensive and susceptible to bad lighting. The integration of data from several sources, or sensor fusion, is a strong way to improve the dependability and flexibility of robotics systems in a variety of settings.

2.6 Use Cases and Applications

Omnidirectional mobile platforms' efficiency, agility, and diversity have made them important in a wide range of industries. They are useful in many different industries, including manufacturing, logistics, health care, and so on, because of their capacity to precisely navigate complex settings. These robots offer creative solutions that revolutionize and enhance contemporary workflows by tackling issues like labor shortages, space limitations, and the requirement for accuracy. Here are a few specific instances of its application in the mentioned industries.

2.7 Applications in Industry

The industrial sector makes extensive use of these robots. An advanced omnidirectional mobile robot designed and implemented for material handling in various production situations is presented in the work by Hongfu Li, Jiang Liu, Changhuai Lyu, Daoxin Liu, and Yinsen Liu. The robot's omnidirectional locomotion, made possible by a four-wheel-drive mecanum wheel chassis, allows for smooth navigation across constrained and dynamic factory spaces. To adapt to uneven surfaces, its design incorporates a modular suspension mechanism, which lowers vibration and improves load stability. In addition to the platform, a robot arm with an electric claw and a depth camera allowed for the recognition and grabbing of objects using machine vision algorithms such as the Hough circular transform. The control architecture of the system is a key component of its innovation. A fuzzy PID control system is used by the robot to dynamically modify parameters to rectify trajectory deviations brought on by uneven surfaces. Compared to conventional PID control, this method significantly increases navigation precision. With real-time feedback that improves localization and maneuverability, the mobile platform is directed by radio-frequency identification (RFID) tags and magnetic strips. Communication between the robot arm and mobile base, obstacle avoidance, and motor control may all be seamlessly integrated with a modular embedded system built on the Real-Time Thread architecture. The robot is capable of more than just transportation. With the aid of reliable image processing techniques and hand-eye calibration, it automatically recognizes, selects, and positions materials at predetermined locations. By bringing the camera and the robot arm's coordination system into alignment, calibration guarantees very accurate object spatial localization. Reliability and scalability are key design elements of the control system, which incorporates communication protocols including Modular Bus–Remote Terminal Unit (Modbus-RTU) for arm coordination and Controller Area Network open (CANopen) for motor drivers. The system's effectiveness was confirmed by prototype testing, which demonstrates its capacity to manage intricate jobs in industrial workflows and achieve a positioning accuracy of within 5.5 mm. This robot exemplifies the potential to increase automation, decrease manual labor, and boost productivity in manufacturing, logistics, and a host of other industrial applications. The study offers a scalable solution for contemporary industrial problems by fusing powerful sensing technology, adaptable control algorithms, and innovative mechanical design [10].

2.8 Medical Applications and Health Care

The SESTO robotic bed mover, created by research team Zhau Guo, is a perfect example of how the healthcare industry has welcomed the integration of robotic technologies to meet its different jobs and difficulties. The physical strain that healthcare personnel endure during patient transfers—a procedure that is known to contribute to work-related musculoskeletal disorders—is lessened because of this novel technology. The omnidirectional mobility platform of the SESTO robotic bed mover makes use of the active split offset caster (ASOC) mechanism to provide smooth movement in tight areas. The robot converts traditional stretchers into power-assisted devices that may be operated by a single person by attaching to hospital beds that are already in place. Utilizing a force/torque sensor-based control interface, this configuration offers a user-friendly haptic experience that simulates manual pushing while amplifying an input force for effortless mobility. In comparison to manual and powered stretchers, clinical experiments showed that the bed mover may significantly reduce muscular activity, especially in the back muscles, while halving the amount of labor required for bed transfers. According to an electromyography (EMG) study, users reported significantly reduced activation levels for muscles such as the latissimus dorsi and erector spinae, as well as less strain on their muscles during important maneuvers like turning and parking. This ergonomic enhancement has enormous potential to alleviate long-term ailments, including lower back discomfort, which is a common problem among medical workers.

The omnidirectional capabilities of the SESTO robotic bed mover surpass those of conventional motorized stretchers, which primarily offer forward propulsion. Because of its exceptional agility, it can be precisely controlled in confined settings like intensive care units, where the weight and complexity of beds loaded with equipment pose significant obstacles. The system is a useful addition to contemporary healthcare settings because of its scalability and versatility, which address both labor efficiency and the physical health of healthcare personnel. The study notes limitations, such as the need for additional real-world testing under various hospital circumstances and patient weights, even though the results show promise. Notwithstanding these limitations, the SESTO robotic bed mover is an important and noteworthy advancement in the use of robotics in healthcare, supporting larger initiatives to boost worker safety, operational effectiveness, and patient care in a field that is becoming more and more demanding [11].

2.9 Application of Disaster Response and Public Safety

Additionally, employ omnidirectional robots to protect human life. The design, development, and testing of the OmniMaxbot, an autonomous omnidirectional robot for Cameco Corporation's Port Hope Conversion Facility, are described in Nicholas Charabaruk's 2012 thesis. Transporting uranium ash cans is the main duty of the OmniMaxbot, which reduces worker exposure to dangerous substances. Two prior capstone projects—one on the design of the mobile base and the other on the forklift assembly and first power system—are built upon by this one. Due to the unreliability of encoder data with Mecanum wheels, Charabaruk's work included implementing laser scan matching for odometry data, completely redesigning the power distribution system using lithium-ion polymer (LiPo) batteries, and creating software to control the robot autonomously using the Robot Operating System (ROS). He chose and combined ROS programs for local planning (base_local_planner), global planning (global planner), and SLAM (hector_mapping), setting them up for best results. The OmniMaxbot's ability to autonomously navigate to a pickup location, pick up a mock ash can, transport it to a drop-off location, deposit the can, and return to a standby location while avoiding collisions with static and dynamic obstacles was proven through extensive testing, which included individual component, subsystem, and full system tests. Several ROS packages and nodes, such as those for managing the motor drivers and processing information from the different sensors (such as laser range finders, Kinects, and stereo vision cameras), as well as the overall control strategy for autonomous navigation, are also developed by the thesis. For can and drop-off position detection, augmented reality (AR) codes were found to be a more straightforward approach than object recognition. Key findings showed that laser scan matching was successful in producing more accurate odometry data, but encoder-based odometry using Mecanum wheels was inaccurate. The thesis ends by suggesting enhancements such as a simultaneous battery charging system and upgraded sensor systems for increased safety and efficiency, as well as the creation of a high-fidelity prototype for testing in the real-world workplace [12].

2.10 Applications for Security

Another application for omnidirectional robots is security. Significant improvements in security-related tasks are among the many applications for which the Cybersecure Omnidirectional Autonomous Mobile Robot (OAMR) has been developed and integrated, according to research by Brijesh Patel and colleagues. The OAMR's Mecanum wheels provide it with unmatched mobility, which makes it perfect for navigating in tight spaces like safe facilities, public spaces, and smart industries. These robots' omnidirectional movement skills allow them to monitor regions, maneuver through confined spaces, and effectively react to possible security threats. When paired with a wide range of sensors, including LiDAR, cameras, and inertial measurement units (IMUs), the OAMR is highly effective at identifying obstacles, keeping an eye on the environment, and delivering the real-time situational awareness that security operations require. The robot makes use of a strong cybersecurity framework to increase its efficacy in security applications. The OAMR guarantees secure communication and identity identification through the use of digital certificate trust chains, public key infrastructure (PKI), and a zero-trust architecture. This guards against illegal access and hacking attempts, which is crucial in settings where high standards of data integrity and operational dependability are necessary. Its function in protecting sensitive areas is reinforced by the layered security design, which incorporates blockchain-based identity verification, frequent authentication checks, and limited device access. Furthermore, the incorporation of cutting-edge SLAM (simultaneous localization and mapping) technology permits accurate mapping and navigation, enabling the OAMR to efficiently improve security patrols and adjust to changing conditions. The OAMR provides a wide range of possibilities for public safety. It is perfect for surveillance and patrolling wide areas like malls, airports, and vital infrastructure because of its mobility and real-time monitoring capabilities. The autonomous nature of the robot eliminates the need for human intervention in high-risk or dangerous areas by enabling it to function both independently and as a component of a coordinated security system. Additionally, its smooth integration with current networks and systems offers flexibility in responding to different security circumstances, such as emergency response and crowd control. The study emphasizes the OAMR's versatility and scalability, highlighting its potential to completely transform security operations in smart settings. Future advancements like improved sensor fusion and AI integration should increase its capabilities and establish it as a fundamental component of sophisticated security systems in delicate and dynamic environments [13]. According to the aforementioned uses, omnidirectional does play a significant and astounding function in a variety of disciplines and industries.

2.11 Omnidirectional Platform Robots: Difficulties and Restrictions

Omnidirectional robots have several benefits in terms of mobility and adaptability, but their performance and general acceptance are hampered by several technical issues. Among the fundamental concerns are.

2.12 Problems with Power Consumption

According to Linzhu Yue's research, a major obstacle in the creation and functioning of the omnidirectional robot was its high-power consumption. For these robots to move precisely and make decisions in real time, energy-intensive parts like motors, sensors, and processors are needed. To sustain omnidirectional mobility, mecanum wheels and similar systems require constant energy input, which frequently results in rapid battery depletion. The requirement for continuous electricity can restrict operating efficiency in situations where robots work in demanding conditions, like industrial automation or surveillance. High-capacity batteries are frequently used to address this problem, but they add weight, which makes the design more difficult and limits mobility. To solve this problem and guarantee extended operational capability without sacrificing performance, developments in energy-efficient systems, lightweight battery technologies, and enhanced power management solutions are needed [14].

2.13 Issues with Stability

According to the previously described study, stability was still a major problem for the omnidirectional robot, especially when it was moving in an unexpected direction or on uneven terrain. Despite having a wide range of motion, mecanum wheels frequently have trouble keeping steady ground contact, which can cause slippage or a reduction in movement accuracy. This problem is made worse by the intricate control algorithms needed to precisely synchronize wheel motions, which raises the need for computing power and design expenses. There are also extra challenges when navigating in real-world settings. To guarantee dependability, complex sensor integration and real-time processing are required due to obstacles, different terrains, and dynamic changes in the environment. Although sensor fusion methods that integrate information from LiDAR, cameras, and inertial measurement devices can improve navigation accuracy, their proper operation necessitates complex algorithms. To overcome these obstacles, wheel designs must be improved, adaptive control techniques must be used, and reliable kinematic models that can handle these operational difficulties must be created [14].

2.14 Getting Around in Complicated Situations

Despite advancements in sensor integration and algorithms, omnidirectional robots still face substantial hurdles while navigating complex settings, according to a report by Liyana Wijayathung, Alexander Rassau, and Douglas Chai. Effective autonomous operation depends on precise mapping and real-time obstacle avoidance, but these tasks are sometimes made more difficult by environmental conditions. The dynamic nature of some settings, reflecting surfaces, and inadequate lighting can all affect how reliable sensors are. Inertial measuring units (IMUs), LiDAR, and cameras are frequently used for perception, although each has drawbacks. For example, cameras have trouble in low light or strong glare, and LiDAR might make mistakes in mapping and object detection when it encounters reflecting or transparent objects. These difficulties are made worse by unstructured settings like catastrophe sites, congested cities, or thick vegetation. Robots traversing such terrains have to deal with uneven surfaces, obstacles that change quickly, and the requirement for accurate decision-making in the face of uncertainty. To create precise environmental maps, simultaneous localization and mapping (SLAM) algorithms, which are frequently used for navigation, need reliable input from several sensors. However, changes in occlusions, shadows, or illumination can make SLAM systems less effective. Furthermore, dynamic agents like people or cars add difficulty because robots have to swiftly adjust to their movements while still being safe and effective [15]. To solve these problems, sophisticated sensor fusion techniques are becoming more and more popular. Robots can obtain a more complete picture of their environment by merging input from several modalities, such as LiDAR for depth perception and cameras for contextual comprehension. Convolutional neural network (CNN)-based deep learning models are used to improve sensor fusion, which allows for improved object segmentation and categorization in the environment. Nevertheless, these models are resource-intensive and computationally demanding, which may restrict their use in real-time applications.

Using multimodal scene awareness to enhance navigation is an additional strategy. Robots can more accurately distinguish between traversable and non-traversable locations by combining data from RGB cameras, depth sensors, and LiDAR. Semantic segmentation techniques, for instance, label areas of the environment and differentiate between dynamic items, open routes, and impediments. These methods are crucial for producing useful maps that facilitate trustworthy route planning. A significant technical challenge still exists, nevertheless, in processing these high-dimensional data streams in real-time. Additionally encouraging are recent advances in dense depth estimates and completion. These methods improve the robot's depth perception in situations with low visibility, for example, where conventional sensors have trouble. Researchers have created algorithms that can produce dense and precise depth maps by combining sparse LiDAR data with camera images. This makes it possible for robots to move more confidently through challenging terrain, especially when they encounter obstacles that are hard to identify using single-sensor methods. In the end, overcoming the difficulties of navigating in intricate situations calls for a blend of cutting-edge hardware, intricate algorithms, and ongoing innovation. Important advancements include creating adaptive control systems, enhancing the effectiveness of deep learning models, and using reliable sensor fusion approaches. These developments guarantee the safe and dependable operation of omnidirectional robots in demanding and dynamic environments, in addition to improving navigation capabilities. To fully realize the potential of autonomous navigation technology, future studies must pursue these directions with an emphasis on scalability and practical application [15].

2.15 Ethics in Robotics: A Concern

Using a two-tiered search method across several databases, the scoping review et al. Battistuzzi L. investigated the body of research on ethical concerns surrounding the deployment of rescue robots in crises. Only six of the 429 papers that were initially found matched the inclusion criteria when strict selection procedures were applied. These studies, which were mostly published in scientific publications between 2008 and 2020, examined a variety of ethical topics, including responsibility, privacy, safety, trust, labor replacement, unfairness and discrimination, and unrealistic or exaggerated expectations. While some studies used qualitative methods engaging stakeholders to offer deeper insights, the majority of the studies used simulations or hypothetical scenarios to predict ethical implications. With few thorough studies addressing the ethical aspects of rescue robot deployment, the findings revealed a sizable gap in the literature. This deficiency highlights the pressing need for more thorough investigations and preventative actions to guarantee that these issues are sufficiently resolved before the widespread use of rescue robots. Building trust and safety in disaster response contexts will require tackling ethical issues, including fair deployment, reasonable public expectations, and strong privacy protections as these technologies develop further [16].

2.16 Prospects for Omnidirectional Robots in the Future

3D LiDAR: Tsung-Ying Lin and Jih-Gau Juang's research work investigates the use of 3D point cloud maps and image recognition for an omnidirectional wheeled mobile robot's (WMR) indoor navigation. The robot maps its surroundings in three dimensions using a LiDAR camera. The You Only Look Once, version 4 (YOLOv4) object detection network with this map enables the robot to recognize characteristics like doorknobs and door numbers and display them on the map. The study emphasizes the benefits of omnidirectional wheeled mobile robots (WMRs) for interior navigation, especially in negotiating difficult and confined environments. The robot then uses this combined map and image data to determine its location and plot a course to a predetermined point. The authors compare this to conventional robots that use a variety of sensors to navigate and avoid obstacles. The ROS (Robot Operating System) framework integrates an NVIDIA Jetson AGX Xavier for computation, a DYNAMIXEL 2XC430-W250-T servo motor for head control, and an Intel RealSense LiDAR camera L515 for depth and visual data.

The procedure of creating a 3D point cloud map is described in full in the paper. This includes obtaining a depth map, calculating the point cloud using the pinhole camera model, and applying filters (voxel grid and average) to minimize noise and data size. YOLOv4's real-time capabilities and advantages over other object detection algorithms are highlighted in this explanation of its application for object detection. YOLOv4's training dataset comprised pictures of room numbers, doorknobs, and doorplates. The robot can find itself and move to predetermined locations thanks to the combination of object recognition and the 3D point cloud map. The experimental findings show that the suggested approach is successful in accomplishing indoor navigation. Accurate object recognition and path planning are made possible by the technology, which also efficiently shrinks the point cloud map and speeds up processing. The goal of the upcoming work is to incorporate increasingly difficult tasks, such as using elevators and unlocking doors. According to the study's findings, combining 3D LiDAR with image recognition offers a straightforward yet efficient indoor robot navigation solution, opening the door for increasingly complex autonomous mobile robots [17]. Artificial intelligence: Eirale and colleagues' essay describes the creation of Marvin, a new assistive robotic platform intended for home settings, with a focus on how it uses state-of-the-art AI for vision and voice control. Marvin discusses the growing demand for trustworthy helpers for the elderly and people with limited mobility, especially in light of population aging and isolation brought on by pandemics. The robot uses PoseNet for pose estimation, classifying a person's posture (standing, sitting, or lying) to detect potential emergencies [18]. Marvin's omnidirectional base, which uses four mecanum wheels, offers agile mobility and effective obstacle avoidance, which are essential for navigating cluttered domestic environments.

Super capacitors: The study by Wang Y investigates the potential application of crumpled in-plane micro-super capacitors (MSCs) in omnidirectional robotics. For next-generation robotics, miniaturized energy storage is essential, and MSCs have the advantages of being compact, safe, long-lasting, and rapid to charge. However, the capacity of current 2D MSCs is limited, and 3D versions have slow kinetics and fabrication issues. In this work, an inflatable balloon is used as an omnidirectional elastic substrate to present a revolutionary prestretching construction technique. This makes it possible to create densely packed, crumpled CNT electrode architectures without the need for complicated fittings. The inflation and deflation of the balloon offer a straightforward but efficient method for constructing structures at the micro- and maybe nanoscale. The crumpled form improves capacitance by greatly increasing the surface area exposed to the electrolyte. A single crumpled MSC outperformed previously published CNT-based microdevices by achieving a high areal capacitance of 13.5 mF cm^{-2} and a volumetric capacitance of 9.3 F cm^{-3}. Experiments and theoretical simulations using molecular dynamics and mesoscopic transport theory have verified that the compact crumpled structure's higher conductance and decreased ion adsorption energy are the causes of the better performance. With configurable electrode designs that enable series and parallel integration to obtain higher voltages or capacities, the method enables the development of extremely small devices (0.1 mm^2). Bending, kneading, poking, and knotting are all tolerated by the resultant devices, which exhibit exceptional mechanical stability and flexibility. Applications in wearable, flyable, and underwater power systems are made possible by this adaptability. Using a 3D printed mask, the researchers also show how to integrate MSCs on a spherical surface, producing a large-scale spherical integrated device with a high output voltage. The devices are appropriate for application in tough environments because they function effectively in extreme circumstances, such as low temperatures and high stress. This study presents a new and adaptable technique for creating high-performance MSCs with a variety of possible uses, especially in the creation of sophisticated omnidirectional robots that need reliable and small power supplies [19]. Soft robotics: To help with medical X-ray scans, this research study describes the creation of an omnidirectional soft robot platform. The robot's flexible pneumatic rubber legs provide benefits, including transparency and compliance, enabling safe patient engagement during treatments. This design attempts to get over the drawbacks of conventional soft robots, which include slow speeds and erratic movements.

Although silicone-based soft robots can be transparent and human-compliant medical assisting technologies, they have several drawbacks, such as being slow, having long and imprecise gait trajectories, and requiring control parameters to be changed for motion diversion. With eight symmetrically positioned legs, the platform's octopedal design allows for omnidirectional movement, including rotation, diagonal movement, and straight-line movement. The radiologist does not need to be present during the examination because the robot is controlled remotely through a computer interface. {Source: "The soft robot platform design can be seen in Fig. 2. It consists of eight legs in a symmetrical, octopedal arrangement, with the eight legs surrounding the bottom of the robot and no leg at the center." A maximum forward speed of 6.90 mm/s and a maximum payload of 70 g were demonstrated during testing. {Source: "The platform is controlled remotely, eliminating radiologist intervention during the medical examination." One of the most significant developments in medical assistive technology is the robot's capacity to carry a pneumatic pillow, which is used to compress the patient's abdomen during fluoroscopy. {Source: "Tests have confirmed its omnidirectional locomotion ability; it has a maximum speed of 6.90 mm/s in forward locomotion and a maximum payload of 70 g." This omnidirectional, soft robotic platform, according to the researchers, could increase the effectiveness and comfort of X-ray exams. The use of soft robotics for omnidirectional locomotion in medical applications highlights a significant future trend in robotics, offering improved safety, precision, and efficiency in patient care [20].

3 Methodology

3.1 Research Procedure

A thorough explanation of the system design and implementation process is provided in this chapter, which dives into the project's technical and methodological components. It covers both the software architecture, with an emphasis on control algorithms, communication protocols, and navigation strategies, and the hardware elements, such as wheel mechanics, power systems, and sensor arrays. To make sure that the methodology is exact and repeatable, this chapter also describes the testing methods used to assess the robot's performance. This part acts as a technical manual for comprehending the project's execution and the reasoning behind the techniques selected.

Any technical project must have a methodology section, which provides a methodical and thorough explanation of how the aims and objectives were met. This approach acts as a road map that leads the reader through the methodical procedures and choices used for a project as complex as the design and development of a remote-controlled omnidirectional mobile platform robot. It makes the work clear and transparent so that others can comprehend, duplicate, or expand upon it. This part explores the theoretical and practical factors that shaped the project's design concepts, selection of hardware and software elements, control schemes, and incorporation of contemporary technology.

The approach examines the project's interdisciplinary nature, where engineering, computer science, and robotics all come together, in addition to its technical features. This section shows the fundamental building elements of the robot by detailing the choice of materials, power sources, and parts such as the Arduino Uno, mecanum wheels, and the HC-05 Bluetooth module. It also emphasizes how crucial software systems are, such as the communication protocols and control algorithms that guarantee the platform runs smoothly. To provide exact mobility and adaptability in a variety of settings.

This section's focus on testing and validation processes is another important feature. The approach gives a thorough overview of how difficulties were recognized and resolved by detailing the procedures followed to model, assess, and improve the robot's performance. After a thorough analysis, solutions are created to lessen the effects of problems, including stability, power efficiency, and communication reliability. The methodology also takes into account the iterative design process, where ongoing feedback influenced changes to enhance functionality and guarantee that the platform fulfilled performance requirements.

The technique part basically acts as a link between the theoretical and the applied. In addition to explaining the robot's design and development process, it offers insightful information about the difficulties and breakthroughs that influenced its development. This part provides a strong basis for comprehending the project's results and ramifications by outlining the methodological approach, which also supports the project's legitimacy and dependability.

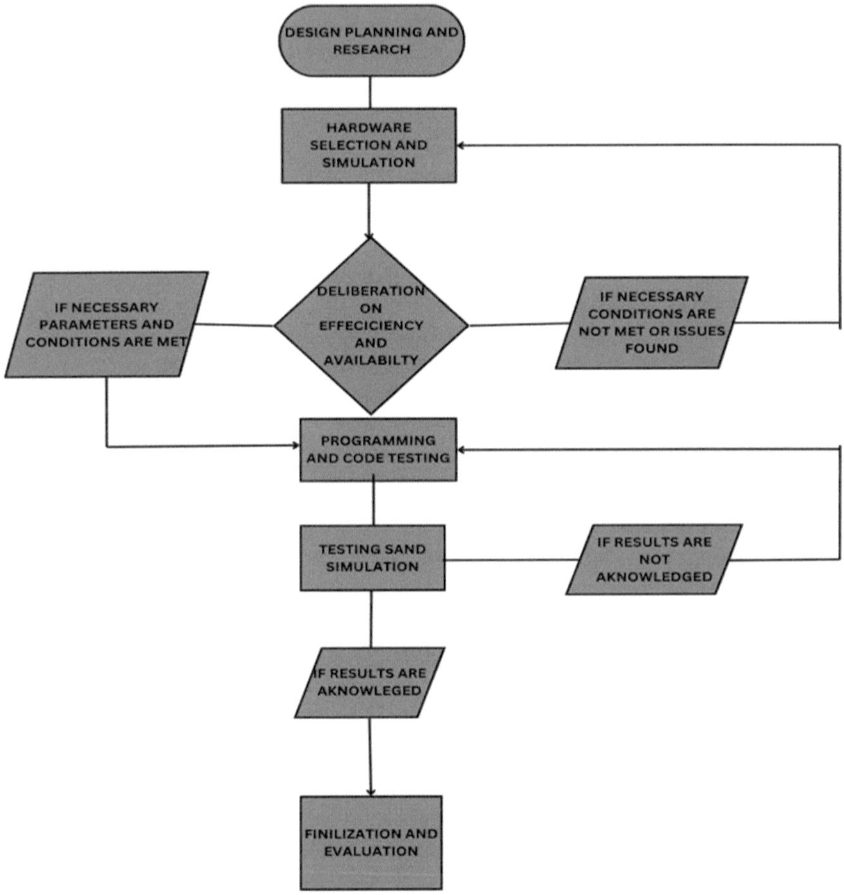

Fig. 2 Workflow for the project

3.2 *Project Workflow*

See Fig. 2.

3.3 *Research and Design Planning*

The design planning and research phase of the project workflow is where the fundamental ideas for the robot's requirements and functionality are established. Understanding the goals of the remote-controlled omnidirectional robot and the function of feedback components is crucial during this stage. To investigate accessible hardware components, communication protocols, and feedback mechanisms, research is necessary. The objective of the project is to build a robot that can move in any direction utilizing Bluetooth control through a remote control console. Feedback mechanisms will also be included to enable dynamic modifications depending on data that is available in real time. Information on motor driver integrated circuits (like the L298N), sensors that could be utilized for feedback (like distance sensors or encoders), and wireless communication techniques (like Bluetooth and Wi-Fi) are all essential during the research phase. Learning the control algorithms that can handle commands (like basic forward/backward or more complex movement with sensor input) and investigating programming libraries for Bluetooth connection (like SoftwareSerial or BluetoothSerial) are also crucial. This phase serves as the project's foundation since it offers an organized method for choosing hardware, creating software, and integrating all the components.

3.4 Simulation and Hardware Selection

The goal of the hardware selection and simulation phase is to choose and virtually test the appropriate parts to make the robot a reality. Choosing the parts that will be represented in Proteus, such as the Arduino Uno or Arduino Mega, the L298N motor driver, the HC-05 Bluetooth module, DC motors, and any sensors, is part of this phase because the project is simulated. The Arduino board was chosen in accordance with the project's I/O and computational requirements. For instance, an Arduino Mega might be better than an Arduino Uno if more sensors are added or if more motor control pins are needed. Before these parts were physically constructed, Proteus was used to model their wiring and connections. The L298N motor driver was selected for the motor control simulation because of its simplicity and dependability in controlling DC motors. The HC-05 module replicated Bluetooth connectivity, enabling wireless control signal delivery. In this stage, meticulous wiring and virtual component configuration guarantee that the simulation faithfully replicates the desired physical setup.

3.5 Code Testing and Programming

The emphasis switches to creating and improving the software that will operate on the Arduino during the programming and code-testing stages. This entails programming the motor control logic to enable the robot to move in all directions and building the code to manage the wireless communication between the Arduino and the HC-05 Bluetooth module. Additionally, if sensor data is used, the code includes feedback loops. To operate the motor driver, the control logic reads Bluetooth signals (such as forward, backward, left, and right) and transforms them into digital outputs. The code must contain methods for error correction and feedback processing if the feedback components—such as encoders or distance sensors—are integrated. To test whether the application correctly understands Bluetooth commands and drives the motors as intended, the code is uploaded to the Arduino in the simulation environment. Debugging tools such as Proteus's virtual serial monitors and serial monitor provide real-time code and communication logic monitoring and troubleshooting.

3.6 Simulation and Testing

The testing and simulation stage starts when the programming is finished. This stage is essential for assessing the robot's reaction to feedback and control inputs. You can mimic the robot's behavior with Proteus, making sure that the robot moves as predicted when Bluetooth orders are transmitted from the remote control (or a Proteus mimicked Bluetooth interface). Sending the proper commands via the Bluetooth interface allows you to test all movement directions, including forward, backward, left, right, and diagonals. If feedback sensors are used, this step verifies that the Arduino interprets sensor data accurately and modifies the robot's motions appropriately. For example, the robot's speed may be changed in response to real-time feedback if motor revolutions are measured by an encoder. Before completing the program for a real-world application, this step allows simulations to be run several times to test various scenarios, such as moving through limited locations or avoiding obstacles.

3.7 Completion and Assessment

In the last stage, finalization and evaluation, the project is concluded by assessing and improving the robot's functionality and appearance. All code is rigorously verified during finalization to guarantee stability and that the robot reacts to all orders in a variety of scenarios. Debugging and improving the code will fix any problems that were discovered during the testing process, such as unexpected behavior or poor communication. The Proteus simulation was completed by making sure that every part was wired correctly and that the robot's behavior was realistic. Documenting the project—including the design choices, system flow, programming logic, and test results—is another aspect of evaluation. Reflection on the project's objectives and accomplishments is part of the final assessment. If further functions, such as path following or obstacle avoidance, are thought of but not yet implemented, they are listed as possible future enhancements.

3.8 Design of the System

The omnidirectional mobile platform robot's system design places a strong emphasis on a scalable and modular architecture to guarantee scalability for a range of uses and the smooth integration of future enhancements. The main goal of the design is to integrate mechanical, electrical, and computing subsystems to create a sturdy platform with accurate and adaptable mobility. The Arduino Uno microcontroller, which serves as the main control unit, the chassis, mecanum wheels, motor drivers, power system, and communication modules are the main components of the design.

3.9 Design of the Chassis

The robot's chassis serves as its structural base and is made to support and house all of its vital parts while preserving stability and longevity. The chassis, which is made of lightweight materials like polycarbonate or aluminum, guarantees robustness without sacrificing mobility. Because of its balanced weight distribution layout, it won't tilt when transporting payloads or making sudden direction changes. The chassis's modular design makes future upgrades, maintenance, and assembly easier. The robot may be adjusted to meet various operating needs thanks to mounting points included for extra parts like sensors or cameras. To ensure dependable functioning over extended periods, the chassis is also made to shield delicate parts, like the Arduino Uno and motor drivers, from external elements like dust or knocks.

3.10 Mobility and Mecanum Wheels

The mecanum wheels are essential to the robot's ability to move in all directions. The robot can produce force vectors in several directions thanks to these wheels' slanted rollers, which are positioned at a 45-degree angle around their circle. The robot may move forward, backward, sideways, diagonally, or rotate in still without changing its orientation by precisely coordinating the pace and direction of each wheel's rotation.

This mobility is especially helpful in situations that call for a high degree of dexterity, such avoiding obstacles or navigating tight places. With their angles facilitating seamless transitions between various movement directions, the wheels' positioning on the chassis guarantees peak performance.

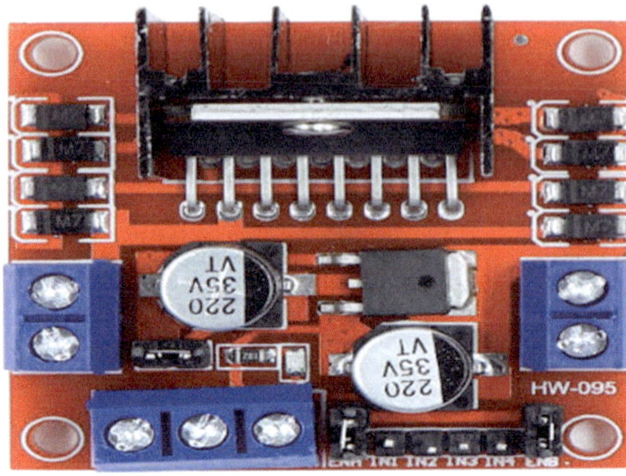

Fig. 3 The L298N motor driver [21]

3.11 Actuation and Motor Drivers

Because they control the voltage and current delivered to the motors, the L298 motor drivers are essential to the system architecture. Each motor may be independently controlled by these dual H-bridge drivers, enabling exact changes in speed and direction. The Arduino Uno sends pulse-width modulation (PWM) signals to the motor drivers, which use these signals to control the rotational direction and speed of the wheels in response to human input. This subsystem dynamically modifies the output to each motor to guarantee the seamless execution of intricate actions like lateral sliding or in-place rotation. The motor drivers' integrated thermal protection prevents overheating during prolonged operation, and they are made to manage the torque requirements of DC motors even under varied load situations. Figure 3 shows a figure of the L298N driver, which was the chosen driver used in the project.

3.12 System of Power

All subsystems are intended to receive constant and dependable energy from the power system. The Arduino Uno and motor drivers are powered by a LiPO battery, and steady voltage levels are maintained using regulated circuitry. The purpose of capacitors is to reduce voltage variations that can impair communication dependability or motor performance. The robot's versatility is increased by the system's modular power design, which enables interoperability with alternate energy sources such as rechargeable lithium-ion batteries. This guarantees continuous functioning in a range of situations, from low-demand applications to heavy-duty jobs.

Fig. 4 Hc-05 Bluetooth module [22]

3.13 System of Communication

The HC-05 Bluetooth module, which allows wireless data transmission between the robot and the remote control, was the focal point of the communication system. To receive control inputs and send feedback, the module interfaces directly with the Arduino Uno via universal asynchronous receiver-transmitter (UART) connection.

With response rates of less than 100 ms, the HC-05 module guarantees low-latency communication and gives real-time control over the robot's motions. For the majority of usage situations, its operating range is adequate, reaching up to 10 m indoors and up to 15 m outdoors. Even when there is possible interference from other wireless devices, the module's error correction methods keep the connection steady. The robot receives wireless commands for rotation and movement from the remote control. After processing these commands, the Arduino Uno uses kinematic equations to convert them into wheel velocities. Additionally, the HC-05 module facilitates bidirectional communication, which enables the robot to communicate with the operator in both directions and provide status updates, including sensor readings and battery levels. By keeping the operator updated on the robot's operational state, this feedback loop improves usability. Figure 4 shows an image of the HC-05 Bluetooth module.

3.14 Integration and Control Unit

The Arduino Uno microcontroller, the primary control unit, is at the center of the system. It uses kinematic equations to determine the necessary wheel velocities, receives input commands from the communication module, and produces PWM signals to power the motors. The Arduino Uno is perfect for adding new parts or

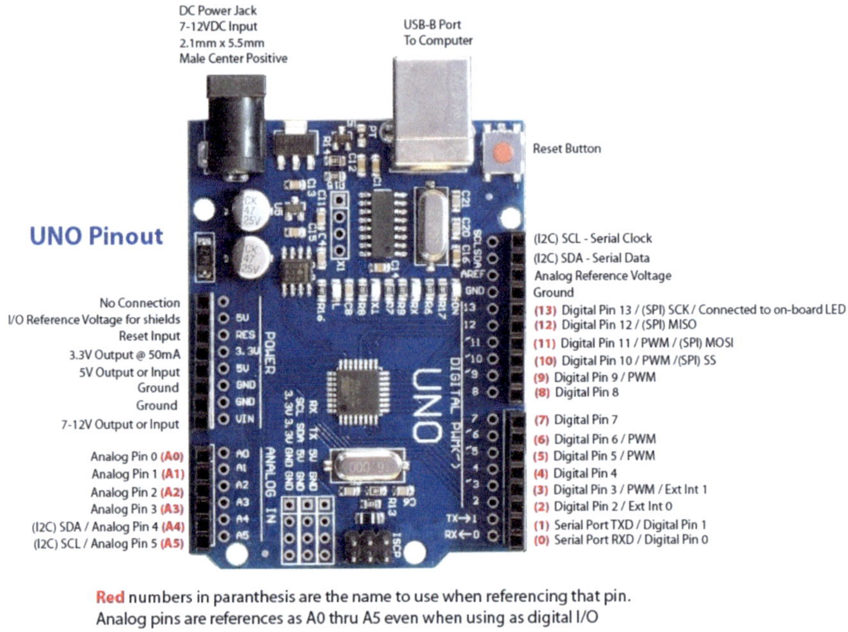

Fig. 5 Arduino Uno R3 [23]

improving the robot's control algorithms because of its simple design and wide range of library compatibility. With the Arduino Uno serving as the central hub for motor drivers, sensors, and communication modules, the integration of all subsystems was accomplished through a neat wiring architecture. Each subsystem works in unison thanks to this integrated architecture, which enhances the robot's total performance. Because individual components may be changed or improved without affecting the overall design, the system's modularity also makes troubleshooting and updates easier. Figure 5 contains the image of the Arduino Uno R3 with each of its pins and ports labeled.

3.15 Design Scalability

The system design's modularity enables the robot to develop in tandem with new technologies. It is possible to add more functions, like sophisticated sensors or cameras, without having to make major structural adjustments. This adaptability guarantees that the platform will continue to be useful and relevant for a wide range of applications, including logistics, surveillance, and indoor navigation.

In conclusion, modularity, accuracy, and flexibility are given top priority in the omnidirectional mobile platform robot's system design. With an emphasis on preserving scalability for upcoming improvements, each subsystem is meticulously integrated to provide dependable performance. The robot is guaranteed to be well-equipped to handle the demands of diverse operational scenarios thanks to this design approach.

3.16 Core Software Tools and Frameworks

The platform's software ecosystem combines specialized tools and libraries to enable seamless development, simulation, and control, such as.

3.17 The Arduino IDE

The main programming environment for creating, compiling, and uploading code to the ESP32 microcontroller is the Arduino IDE. The Arduino IDE is well-known for its ease of use and straightforward design, making it a platform that both novice and experienced developers may use.

Important features and selection criteria: The IDE's user-friendly interface makes creating and debugging code easier.

Broad library support: It provides an extensive library ecosystem that includes libraries for sensor data processing, motor control, mecanum wheel kinematics, and communication protocols, including Bluetooth and Wi-Fi.

Cross-platform compatibility: Developers may use the Arduino IDE on a variety of operating systems thanks to its compatibility with Windows, macOS, and Linux.

Serial monitor: By showing data transmitted from the ESP32, such as wheel speeds or sensor readings, the integrated serial monitor enables real-time troubleshooting.

The Arduino IDE was utilized in this project to create the control algorithms that process joystick inputs, regulate the robot's movement, and manage wireless connections. To improve functionality, libraries like Adafruit_Sensor, which handles sensor data, and AccelStepper, which controls motors, are smoothly incorporated into the IDE.

3.18 Suite for Proteus 8.9 Design

An industry-standard simulation tool for virtually designing and testing electrical circuits is the Proteus 8.9 Design Suite. Without requiring actual parts, it enables developers to model the robot's hardware configuration, including other components like the ESP32 microprocessor, TB6612FNG motor drivers, and various sensors.

Important features and selection criteria: Circuit modeling: Proteus's real-time circuit modeling capability helps developers spot and fix hardware problems early on in the development cycle.

Extensive component library: It is simple to essentially duplicate the project's hardware configuration thanks to the software's extensive library of components, which includes microcontrollers, motor drivers, and sensors.

Hex file integration: To test how the microcontroller communicates with other parts, developers can import the compiled hex files produced by the Arduino IDE into the Proteus simulation.

Visualization of outputs: Proteus can replicate wireless communication, sensor data, and motor motions to give a visual depiction of the robot's operation.

3.19 KiCAD

For the robot's electronic components, printed circuit board layouts and schematics are made using KiCAD, an open-source PCB design program. Although the focus of this research is on software and simulation, KiCAD plays a crucial role in creating a scalable design for practical application.

Key features and selection criteria: Schematic capture: KiCAD offers an easy-to-use interface for creating intricate circuit diagrams, guaranteeing accurate representation of all connections.

PCB layout: It enables designers to create multi-layered PCBs with accurate component placement, trace routing, and power distribution considerations.

Cross-platform support: KiCAD is compatible with various operating systems, just like the Arduino IDE, giving developers freedom.

Free and open-source: KiCAD is an open-source program that provides expert-level functionality without the price tag of proprietary software.

3.20 Frameworks and Libraries

Apart from the primary software tools, several libraries and frameworks make development easier and improve the robot's capabilities. For example, the Mecanum Wheel Control Library is used to calculate and carry out omnidirectional movement using the mecanum wheels. Sensor libraries: Adafruit_Sensor, NewPing, and other libraries are used to process data from encoders, ultrasonic sensors, and IMUs.

Wi-Fi and Bluetooth libraries, such as WiFi.h and BluetoothSerial.h, are examples of ESP32-specific libraries that allow wireless communication between the joystick and the robot.

PID controller library: For applying proportional-integral-derivative control to guarantee accuracy and stabilize the robot's movements.

3.21 Algorithm of Control

The omnidirectional mobile platform robot's control system is a complex and vital part that makes it possible for it to move precisely and steadily in all directions. Fundamentally, the system uses sophisticated inverse kinematics-based algorithms to calculate the necessary wheel velocities for a particular motion. The inverse kinematics model takes into account the rotating velocity around the center of the robot as well as the linear velocity in the X and Y axes, which stand for left–right and forward–backward movements, respectively. The wheel velocity equations for a four-wheel mecanum configuration were developed to guarantee smooth omnidirectional travel. Using formulas that take into account these velocity components and are modified according to the wheel's actual location and orientation, the front-left, front-right, rear-left, and rear-right wheel velocities are determined. The platform is extremely adaptable in dynamic surroundings because of these computations, which allow the robot to execute intricate movements like sliding laterally, spinning while advancing, or moving diagonally.

The ESP32 microcontroller, if used for the project, would process joystick inputs, which represent user commands, and map them to velocity values to carry out these computations. Directional movement is controlled by the analog X and Y signals on the joystick, while rotation is initiated by a digital signal. Wheel velocities were calculated using these normalized inputs and sent into the inverse kinematics equations, which subsequently converted them into pulse-width modulation (PWM) signals. Each wheel's speed and direction are adjusted by the motor drivers, which are controlled by the PWM signals. Even when changing directions quickly or rotating, the robot will move precisely and smoothly thanks to this synchronized control. In addition to wireless connection and sensor integration, the ESP32 microcontroller's dual-core processing capabilities enable the effective completion of key control tasks.

Feedback methods are also incorporated into the control system to improve stability and precision. The motors' rotational speeds are measured by wheel encoders, which give real-time input on the wheels' actual velocities. A proportional-integral-derivative (PID) controller would be used to adjust any differences between this data and the intended velocities determined by the control algorithms. To minimize error and guarantee exact conformity to the planned motion, the PID controller would dynamically modify the PWM signals. Furthermore, an inertial measurement unit (IMU) would be incorporated to track the robot's angular velocity and direction. This information is essential for preserving stability, especially when the robot is performing intricate movements or traversing uneven ground, where its equilibrium may otherwise be jeopardized. Additionally, to give spatial awareness, obstacle detection sensors would be integrated, such as LiDAR modules or ultrasonic sensors. By detecting and avoiding obstructions in its path, these sensors enable the robot to navigate safely and avoid collisions.

The control system enables sophisticated features like trajectory tracking and path planning for autonomous operations in addition to simple mobility. Trajectory tracking makes sure the robot follows the course precisely, while path planning algorithms determine the best way for the robot to go to its objective while avoiding obstacles. To compensate for variations brought on by outside influences, the system continuously modifies wheel velocities to match the intended trajectory. Additionally, adaptive control methods were put in place, which allowed the robot to modify its behavior in response to changes in the payload or the surroundings. For example, the control system uses feedback from sensors to adjust wheel velocities and maintain stability if the robot faces uneven terrain or an increased load. The incorporation of sensor fusion techniques further improves the precision of the control system. The robot can accurately determine its position, orientation, and velocity by integrating information from encoders, IMUs, and obstacle-detecting sensors. These data streams are processed and integrated using methods like Kalman filtering, which lowers noise and increases the accuracy of the robot's navigational judgments. These sophisticated control features were put into practice and thoroughly tested using Proteus simulations, which verified the robot's behavior in a simulated setting before actual deployment. The Arduino IDE's serial display was used for real-time debugging during testing, enabling developers to view sensor data, computed velocities, and input values. modifications made to improve the overall performance of the system and optimize the PID controller parameters.

The omnidirectional mobile platform robot's control system is essentially a well-balanced combination of sophisticated features, real-time feedback, and mathematical models. Its resilience and versatility are highlighted by its capacity to perform accurate omnidirectional movements while retaining stability and adjusting to changes in the surroundings. The control system guarantees that the platform can satisfy the requirements of many applications, ranging from manual navigation to completely autonomous operations, through meticulous design, implementation, and testing.

3.22 Systems of Communication

A critical component enabling real-time interaction between the omnidirectional mobile platform robot and its operator is the communication system. This system facilitates seamless command transmission and feedback exchange, ensuring precise and efficient remote control. At the heart of this setup is the HC-05 Bluetooth module, a versatile wireless communication solution known for its reliability and low-latency performance. Its straightforward integration and robust architecture make it ideal for applications requiring dependable short-range wireless control, even in environments with limited bandwidth. Unlike Wi-Fi-based systems, the HC-05 establishes a direct Bluetooth link between the robot and the operator's device (e.g., a smartphone, laptop, or dedicated remote). This peer-to-peer connection eliminates the need for an intermediary network, simplifying deployment while maintaining responsiveness. The module supports Serial Port Protocol (SPP), allowing for straightforward data exchange via a virtual serial interface. Commands from the operator—such as movement instructions—are transmitted as serial data packets, which the robot's onboard microcontroller interprets and executes. Similarly, sensor feedback (e.g., battery status or obstacle detection) is relayed back to the operator in real time.

To enhance communication efficiency, a lightweight messaging protocol (such as a custom ASCII-based or binary packet structure) can be implemented over Bluetooth. Unlike MQTT, which relies on a broker-based architecture, this approach minimizes overhead by transmitting only essential data. For instance, movement commands can be encoded as compact strings (e.g., F100 for "move forward at speed 100"), while sensor updates are formatted for quick parsing. This method ensures low latency without taxing the robot's processing resources. Security is maintained through pairing authentication, where only trusted devices with the correct passkey can establish a connection. While Bluetooth Classic (used by the HC-05) lacks the advanced encryption of modern BLE or Wi-Fi, its inherent 128-bit AES security (in some firmware versions) and controlled physical range reduce the risk of interference or unauthorized access. For additional protection, data payloads can include checksums or simple encryption for critical commands.

The HC-05's integration with the robot's control system is streamlined by its UART interface, allowing the main microcontroller (e.g., an Arduino or STM32) to handle both communication and motor/sensor tasks without conflict. Since Bluetooth operates as a secondary peripheral, the primary processor dedicates its resources to real-time control while delegating wireless communication to the HC-05's dedicated firmware. This separation of responsibilities ensures stable performance even during continuous operation. Beyond basic remote control, the system can support diagnostic features, such as transmitting runtime logs or sensor calibrations to a paired device. While not as scalable as IoT cloud platforms, this direct link suffices for field debugging and performance tuning. Future expansions could incorporate dual-module setups (e.g., HC-05 + Wi-Fi) or hybrid protocols for broader functionality.

In summary, the robot's communication system—powered by the HC-05 Bluetooth module—delivers a robust, low-latency wireless link for real-time control and feedback. Its simplicity, reliability, and adaptability make it well-suited for applications where direct operator interaction takes precedence over complex network infrastructure.

4 Results and Evaluation

4.1 Findings and Assessment

Results from the development and testing of the remote-controlled omnidirectional mobile platform robot are presented in the results section, which also includes a thorough analysis of the data collected and a detailed review of the robot's performance, assessment techniques, and experimental setting. Key performance indicators, including mobility, accuracy, stability, communication dependability, and power efficiency, were the focus of the testing phase to assess the platform's capabilities and its capacity to fulfill the project's goals. To highlight the improvements and innovations made possible by this project, these outcomes are further contrasted with pre-existing solutions, highlighting the design's distinctive contributions.

4.2 Chassis Development

Using robust 3D computer-aided design (CAD) tools, the team was able to produce an intricate and useful model of the robot's configuration. The mechanical specifications of the robot, such as the requirement for omnidirectional movement made possible by the mecanum wheels, were clearly understood from the beginning of the design process. The chassis had to be small, light, and sturdy enough to hold the necessary parts, including the Arduino board used as well as the motor drivers, Bluetooth module, and power supply, while still making room for the four-mecanum wheels.

The chassis was made to maintain a low and balanced center of gravity to guarantee smooth and controlled motions for the robot. To avoid tipping or instability during movement, component placement and weight distribution had to be carefully considered. Mounting locations for the motors and motor driver electronics were also included in the chassis design. These parts were thoughtfully designed to guarantee that the motors are mounted firmly while enabling simple upkeep and replacement as needed. To prevent tangled wiring from impairing the robot's performance and mobility, the mounting design also featured slots and holes for cable management.

In SolidWorks, the chassis design was built utilizing a mix of assemblies, cuts, and extrusions to guarantee a practical and effective arrangement. Overall, the chassis's dimensions were kept small, but there was enough room for all required parts to avoid crowding, which could cause overheating or obstruct the motors' ability to operate as intended. The chassis's sidewalls are made with cutouts to lighten the weight without sacrificing structural soundness. These cutouts were set carefully to prevent weakening the chassis, especially where the wheels and motors were installed.

Furthermore, the team examined the stresses and strains on the chassis under load circumstances using SolidWorks' Simulation tool. This made it easier for us to spot any possible weak spots in the chassis that might give way or distort when the robot's parts are heavily loaded or when it moves quickly. By strengthening the sections that

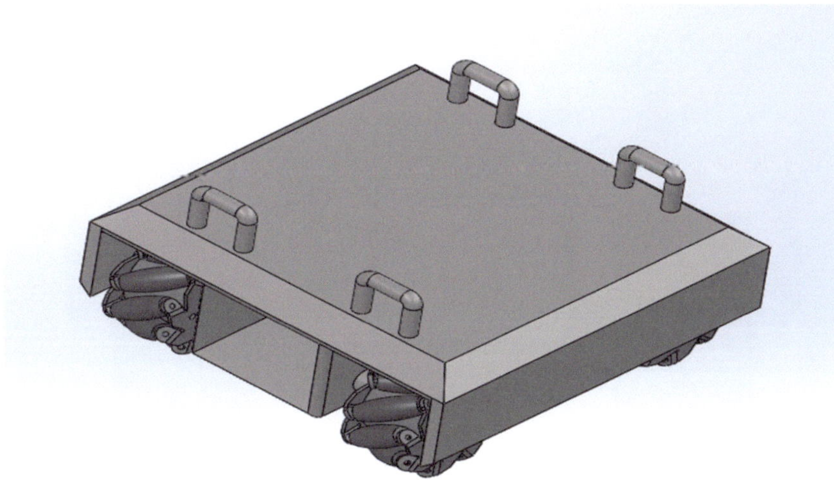

Fig. 6 Global view of final project

needed more reinforcement, the team was able to alter the design and guarantee the finished product's dependability and durability. The team also produced accurate blueprints and manufacturing instructions after finishing the design and simulations, which were then utilized to build the genuine chassis in a real-world setting. Even though the project was simulated, this SolidWorks step made sure that the team completely understood the mechanical design and enabled the team to improve their strategy before thinking about actual execution.

In addition to helping the team optimize the chassis's mechanical design, this virtual development method made it easier to align the electrical components and guarantee that all of the motors, wiring, and Bluetooth modules would fit inside the allotted dimensions. It was simpler to cooperate and make changes in response to criticism or fresh information, thanks to SolidWorks' clear picture of the finished output. Even though the team did not physically build the chassis for this project, the thorough SolidWorks model was crucial in confirming the design choices and guaranteeing that every part would fit together properly, laying the groundwork for further development or real-world prototyping. Below are the global, right, and top views of the chassis from SolidWorks are shown in Figs. 6, 7, and 8, respectively.

4.3 *Omnidirectional System Design and Analysis*

To accomplish precise and flexible mobility, a complex interaction of mechanical, electrical, and computing components is incorporated into the design of the

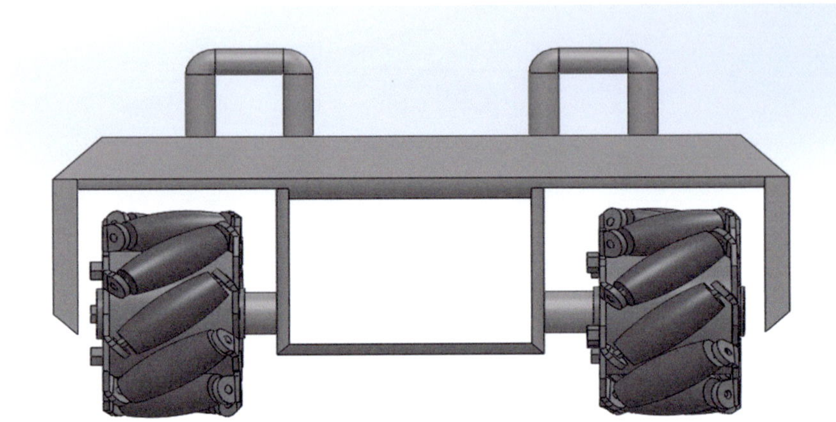

Fig. 7 The illustration of the design for the right view

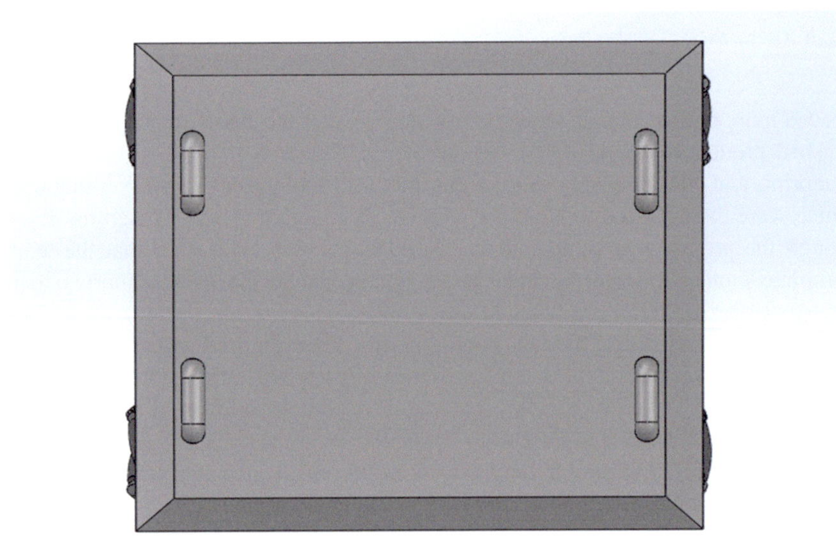

Fig. 8 The top view of the design

remote-controlled omnidirectional mobile platform robot. Two Arduino Uno boards, Mecanum wheels, four DC motors, two L298 motor drivers, two HC-05 Bluetooth modules, and ten control buttons on the remote control are among the meticulously chosen parts that make up the robot. These elements came together to develop a unified system design that guarantees reliable operation and smooth operator-robot communication. True 360° mobility is made possible by the creative use of Mecanum wheels, which enable the platform to move forward, backward, sideways, diagonally, and rotate in place. This is very useful in situations when high maneuverability is required.

4.4 *Roles of Components and System Architecture*

The robot's modular system architecture makes assembly, debugging, and future upgrades easier. Two Arduino Uno boards are at the heart of it; one is located on the robot to control the motor and sensors, while the other is on the remote to transmit commands. The microcontrollers use HC-05 Bluetooth modules to communicate, control inputs and outputs, and perform kinematic calculations. The four DC motors, each attached to a mecanum wheel, can be independently controlled thanks to the employment of two L298 motor drivers.

4.5 *Wheels of Mecanum*

The platform's omnidirectional movement is made possible by the mecanum wheels' distinctive design, which features rollers positioned at 45-degree angles around their circle. The robot may generate resultant forces in any desired direction by precisely regulating the speed and direction of rotation of each wheel. The technology is very adaptable for navigating in limited spaces or surroundings with plenty of obstacles, since this is accomplished by vector addition of the forces produced by the wheels.

DC Motors and Motor Drivers: One of the four H-bridge circuits in the two L298 motor drivers controls each of the four DC motors that power the mecanum wheels. These drivers enable speed modulation and forward and reverse rotation by modifying the voltage and current provided to the motors in response to control signals received from the Arduino Uno. PWM signals produced by the Arduino regulate the motor speeds, enabling accurate motion and seamless direction changes.

4.6 *Bluetooth Modules for HC-05*

The robot and remote control may communicate wirelessly thanks to the HC-05 Bluetooth modules. The robot interprets and converts the commands from the remote into motor motions that correspond to directional movements. In addition to enabling feedback methods like sending the robot's status back to the remote, this bidirectional communication guarantees a real-time response.

4.7 Control Buttons

There are ten buttons on the remote control, each of which is assigned a distinct movement command. These comprise clockwise and counterclockwise rotation, forward and backward motion, sideways left and right, upper-right and upper-left diagonals, and lower-right and lower-left diagonals. This user-friendly design makes operation easier and gives you exact control over the robot's motions.

4.8 System Architecture

1. **Omnidirectional Mobile Platform**

The mobile robotic system, which is in charge of movement and control functions, is represented in this section. The following elements make up the platform:

- **Motors**: The system has four integrated motors that enable omnidirectional mobility. They move the robot by acting as actuators.
- **Electronic drivers**: These parts regulate the motors by providing the required voltage and current in response to instructions from the microcontroller.

  As the central processing unit, the microcontroller decodes Bluetooth commands and adjusts the motor drivers accordingly.
- **Bluetooth Module 1**: This module makes it easier for the remote control device to send wireless signals.

  All components of the mobile platform, such as the motors, motor drivers, and microprocessor, are powered by the battery.

2. **Unit for Remote Control**

The remote control device serves as the robot's operating interface. Among its constituents are:

- **Control buttons**: The user can submit input commands using these buttons.

  The Arduino Uno is a microcontroller that uses Bluetooth to send control signals after processing input from the buttons.
- **Bluetooth Module 2**: This module makes sure that the robot's platform's Bluetooth module can communicate wirelessly.

  The remote control system, which includes the Arduino Uno and the Bluetooth module, is powered by the battery.

3. **The Flow of Communication**

The following steps are involved in the communication between the mobile platform and the remote control:

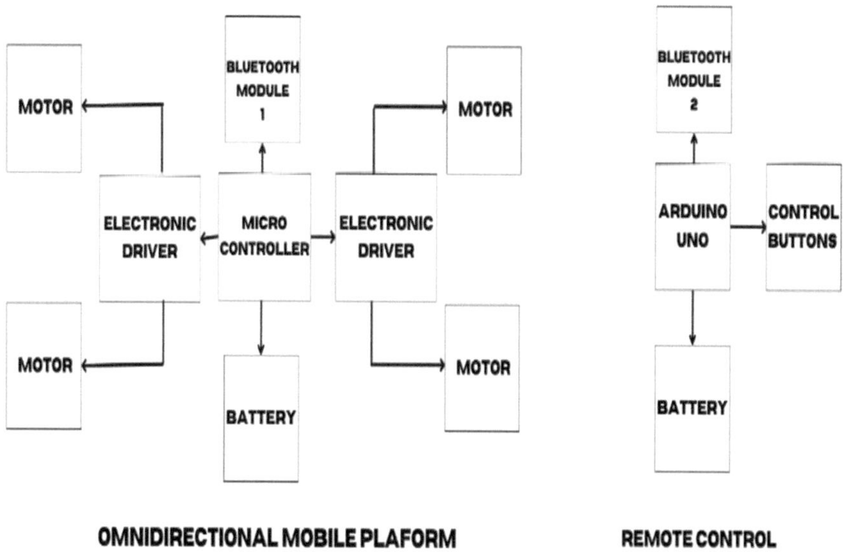

Fig. 9 System architecture

1. The operator uses the remote control unit to press control buttons.
2. The Arduino Uno uses Bluetooth Module 2 to wirelessly transmit commands after processing button input.
3. These signals are received by the mobile platform's Bluetooth Module 1.
4. After processing the commands, the microcontroller instructs the electronic drivers to turn on the motors.
5. In response to the commands, the motors carry out the intended movement.
6. Figure 9 shows the system architecture through which the flow of communications works.

4.9 Background Theory and Kinematic Model

Through the use of a kinematic model that controls the link between wheel velocities and the robot's overall motion, the platform can move in all directions. Each wheel's input velocity and the position and orientation of the mecanum wheels determine the system's kinematics.

Three key elements of motion are taken into consideration to regulate the robot's movement:

- **Linear velocity in the x-direction (V_x)**: Determines left–right movement.
- **Linear velocity in the y-direction (V_y)**: Determines forward–backward movement.
- **Angular velocity (ω)**: Determines rotational motion about the robot's center.

The relationship between the wheel velocities (V_{FL}, V_{FR}, V_{RL}, and V_{RR}) and these motion components is expressed through the following equations:

$$V_x = \frac{1}{4}(V_{FL} + V_{FR} + V_{RL} + V_{RR})$$

$$V_y = \frac{1}{4}(-V_{FL} + V_{FR} + V_{RL} - V_{RR})$$

$$\omega = \frac{1}{4}\left(\frac{VFL}{d} - \frac{VFR}{d} - \frac{VRL}{d} - \frac{VRR}{d}\right)$$

Here:

- $\mathbf{V_x}$ and $\mathbf{V_y}$ are the linear velocities in the x and y directions, respectively.
- $\mathbf{\Omega}$ is the angular velocity (rotational speed).
- **D** is the distance from the center of the robot to each wheel.
- $\mathbf{V_{FL}}$, $\mathbf{V_{FR}}$, $\mathbf{V_{RL}}$, and $\mathbf{V_{RR}}$ represent the velocities of the front-left, front-right, rear-left, and rear-right wheels, respectively.

These equations allow the system to calculate the required speed and direction for each wheel to produce a desired overall motion. For example:

- To move forward (Vy > , Vx = zero, $\omega = 0$), the front and rear wheels on both sides rotate in the same direction.
- To move sideways left (Vx < zero, Vy = zero, $\omega = 0$), the front-left and rear-right wheels rotate backward, while the front-right and rear-left wheels rotate forward [24].

4.10 Calculations and Implementation

- In practice, the Arduino Uno on the robot calculates the required wheel velocities using joystick or button inputs from the remote. The inputs are first mapped to Vx, Vy, and ω, which are then fed into the kinematic equations. The resulting wheel velocities are converted into PWM signals sent to the motor drivers, which adjust the motor speeds accordingly.
- For instance, when the "Upper-Right Diagonal" button is pressed, the system calculates Vx > zero and Vy > zero, with ω = zero. Using the equations, the velocities are distributed among the wheels to create diagonal motion. Similarly, for clockwise rotation ($\omega > 0$, Vx = 0, Vy = 0), the front-left and rear-right wheels rotate forward, while the front-right and rear-left wheels rotate backward.
- The integration of the theoretical model into the system's programming ensures precise execution of movements. The system also incorporates feedback loops using encoders to monitor wheel speeds and IMUs to track orientation, allowing real-time adjustments for enhanced stability and accuracy.

4.11 Performance Measurements and Attributes of Each Component

- **Arduino Uno**

The Arduino Uno microcontroller functions as the primary control unit for the robot, overseeing motor control, communication, and the execution of kinematic computations. It was chosen for its simplicity, versatility, and interoperability with various sensors, actuators, and communication modules.

- **Microcontroller**

The Uno, with the ATmega328P microcontroller, operates at a clock speed of 16 MHz, delivering adequate computational capacity for managing the robot's control logic, real-time input processing, and communication functions.

- **Digital and Analog Terminals**

The Uno comprises 14 digital I/O pins (6 of which accommodate PWM) and six analog input pins, facilitating the incorporation of various components like as motor drivers, Bluetooth modules, and sensors. In this system, the digital pins are utilized to operate the motor drivers, whereas the analog pins are reserved for prospective enhancements, including sensor integration.

The Arduino Uno accommodates a supply voltage range of 7–12 V and functions internally at 5 V, guaranteeing compatibility with L298 motor drivers and additional components. The integrated voltage regulator enables it to manage minor variations in input voltage without compromising performance.

4.12 Direct Current Motors

The four DC motors serve as the principal actuators for the mecanum wheels, supplying the mechanical force required for omnidirectional motion. Their performance is characterized by essential specifications including voltage, current, torque, and speed.

- **Voltage specification:** The motors, functioning at 12 V, are compatible with L298 motor drivers and the power supply, guaranteeing efficient energy conversion and dependable performance.

The continuous current rating is 1.5A, with peak current requirements reaching 2A during high-torque applications. This capability guarantees that the motors can manage substantial loads or surmount friction during directional shifts.

The stall torque of 3.2 Nm delivers the requisite force to manage payloads of up to 2 kg while ensuring stable and precise movement.

The no-load speed of 150 RPM guarantees that the robot attains an appropriate velocity for practical applications. Under standard load conditions, the speed averages approximately 120 RPM, ensuring control and precision.

The motors demonstrate durability and have little temperature accumulation during moderate workloads. During extended high-load conditions, heat dissipation is essential to avert performance decline.

4.13 L298 Motor Drivers

The L298 motor drivers play a critical role in regulating the voltage and current supplied to the DC motors, enabling precise control over speed and direction.

- **Voltage range**: The drivers operate within a range of **5 V to 35 V**, making them compatible with the Arduino Uno and the 12 V DC motors used in the system.
- **Current handling**: Each channel can deliver up to **2A of continuous current**, sufficient for driving the motors even during peak load conditions.
- **Thermal performance**: Equipped with heat sinks, the motor drivers dissipate heat effectively, ensuring consistent performance during extended use.
- **Control signals**: The drivers are controlled using **PWM signals** generated by the Arduino Uno, enabling smooth acceleration, deceleration, and precise speed modulation. The ability to independently control the forward and reverse motion of each motor is critical for implementing the omnidirectional kinematic model.

4.14 HC-05 Bluetooth Modules

The HC-05 Bluetooth modules are responsible for wireless communication between the remote control and the robot, enabling intuitive and responsive control.

- **Operating voltage**: The modules operate at **3.3 V** and include built-in voltage regulation, allowing them to interface seamlessly with the Arduino Uno's 5 V logic levels.
- **Data rate**: Supporting a maximum data rate of **2.1 mbps**, the modules ensure rapid transmission of commands and feedback, with a practical baud rate of **115,200** for stable operation.
- **Range and stability**: Indoors, the modules maintain a stable connection over a range of **10 m**, which extends to **15 m** in open environments. Error correction mechanisms ensure reliable communication even in interference-prone settings.
- **Latency**: With a latency of less than **100 ms**, the modules enable real-time command execution, which is essential for precise robot control.

4.15 Mecanum Wheels

Mecanum wheels are fundamental to the robot's omnidirectional mobility, enabling movement in any direction, rotation in position, and execution of intricate maneuvers.

- **Design**: Each wheel features angled rollers set at 45 degrees, creating force vectors that facilitate multidirectional movement in conjunction with the other wheels.
- **Material**: The wheels are fabricated with a robust plastic or aluminum hub and high-traction rubber rollers, guaranteeing durability and adequate grip.
- **Load capacity**: Each wheel can sustain a load of up to 5 kg, offering sufficient capacity for the robot's chassis, components, and payload.
- **Performance**: The wheels demonstrate superior capability on smooth surfaces, providing accurate and responsive movement. Although performance on irregular or slick surfaces is somewhat diminished, modifications to velocity and control algorithms alleviate these constraints.

4.16 Simulation of Electrical Components Using Proteus System Dynamics

4.16.1 System Dynamics Post-Simulations

The dynamic performance of the remote-controlled omnidirectional mobile platform robot was meticulously examined to assess its response times and stability across diverse settings. The capacity for rapid response and stability is essential for guaranteeing efficient operation and accuracy, especially in settings where the robot may face abrupt directional shifts or external disruptions. The analysis encompassed evaluations of responsiveness to control inputs and the system's capacity to adapt to and recover from motion interruptions. The robot demonstrated remarkably swift response times to control inputs, with a recorded latency of under 100 ms from the issuance of a command on the remote control to its execution by the robot. The low latency is mostly due to the effective connectivity enabled by the HC-05 Bluetooth module and the refined control algorithms executed on the Arduino Uno. The swift response guaranteed that maneuvers like diagonal transitions, in-place rotations, or rapid directional shifts were performed seamlessly and without noticeable delays. This ability is especially advantageous for jobs necessitating high accuracy, such as maneuvering in confined areas or evading dynamic impediments. The robot's stability was assessed by adding external disturbances, such as quick directional changes, abrupt halts, and uneven terrain. The control algorithms, utilizing real-time feedback from encoders and the IMU, adeptly modified wheel velocities to mitigate these disturbances. During swift directional alterations, the robot preserved its equilibrium and closely followed its intended path, illustrating the effectiveness of its kinematic model and feedback systems. The system's capacity to swiftly recover from transient imbalances, such as when facing minor barriers, further underscores its dynamic stability. The results validate that the robot's design and control system are sufficiently robust to manage real-world operational problems while ensuring consistent performance.

The team conducted a transient examination of the electrical and electronic components of the system in Proteus and gathered all necessary components, and carefully attached them to their respective terminals and pins. Subsequently, the corresponding codes to each Arduino module were uploaded. Subsequently, a virtual serial port to enable the connection of the Bluetooth modules was established. Afterwards, a simulation of the complete circuitry was done to analyze its temporal behavior in reaction to diverse input signals supplied to the motors. Ultimately, an examination of the outcomes, concentrating on the voltage waveforms across various segments of the platform, was conducted.

4.17 *Demonstrations of Movement*

The simulations below demonstrate three movement patterns of the mobile platform, with an emphasis on the two front wheels. Figures 10, 11, and 12 show the results of the simulation when the platform is at rest, when it is moving forward, and when it is moving in a clockwise direction. Figure 10 shows no voltage fluctuations for all four wheels, while Figs. 11 and 12 show the voltage fluctuating simultaneously for forward movement and inversely for clockwise movement for the two wheels on the side, respectively.

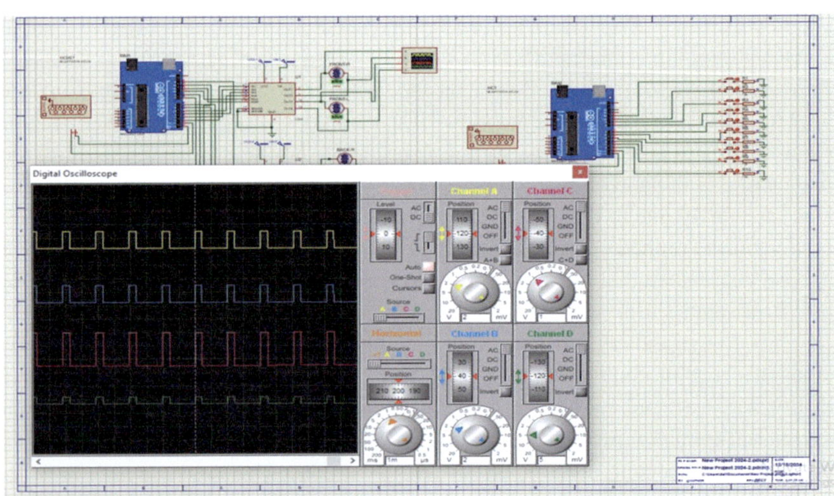

Fig. 10 When the platform is at rest

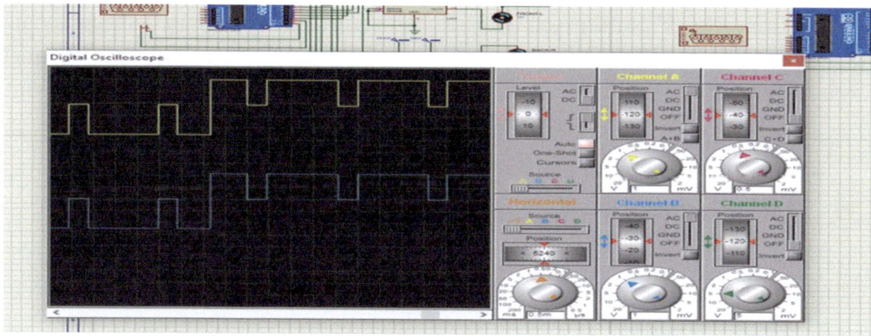

Fig. 11 When the platform is moving forward

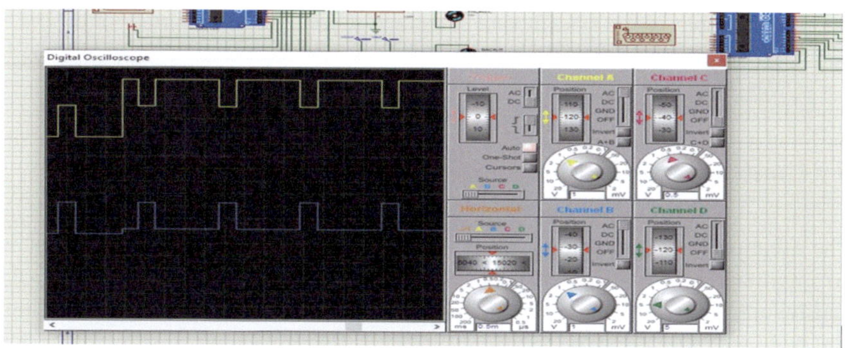

Fig. 12 When the platform is in clockwise rotation

4.18 Communication Validation

The Bluetooth connection between the remote and the platform was thoroughly confirmed. The platform successfully sent and accurately received commands from the remote (which included Arduino Uno, HC-05, and buttons). The simulation showed low latency in command execution, ensuring a quick response to user input.

4.19 Motor and Wheel Performance

The performance of the L298 motor drivers and DC motors in the simulation was thoroughly assessed.

Speed and direction control: The motor drivers received the correct PWM signals and produced the desired outputs for motor speed and direction. This was confirmed by monitoring consistent and accurate motor behavior changes in response to input commands.

Mecanum wheel simulation: The simulation precisely captured the behavior of the Mecanum wheels, displaying precise omnidirectional motion. This includes seamless transitions in all directions, proving the efficacy of the wheel arrangement.

4.20 Power System Performance

For the purpose of ensuring that the system operates reliably and securely, the simulated power requirements and stability of the system were thoroughly analyzed. The distribution of power to essential components, including the Arduino, HC-05 modules, motor drivers, and motors, remained consistent and within safe working ranges during the entirety of the simulation. By doing so, it is demonstrated that the power supply of the system is resilient and reliable under a variety of different operating situations.

5 Conclusion

The creation of a remote-controlled omnidirectional movable platform robot is an important accomplishment in the field of robotics, tackling critical issues in agility, efficiency, and versatility. This capstone project aims to design, develop, and build a robot capable of navigating complicated surroundings with great precision and responsiveness. By utilizing advanced components, control algorithms, and modeling tools, the project met its objectives and demonstrated the potential for substantial applications in several industries. Several major conclusions arose over the course of this investigation. The robot's usage of mecanum wheels permitted complete omnidirectional movement, allowing for smooth transitions between linear and circular motion. The combination of two Arduino Uno microcontrollers enabled robust management and communication, with one operating the platform and the other handling the remote control interface. The HC-05 Bluetooth modules ensured reliable wireless communication, allowing for real-time control and feedback.

The research also showed the value of rigorous design and simulation. The team was able to test and confirm the electrical and electronic components before physical implementation using Proteus' transient analysis. This method not only saved time and resources but also provided useful insights into the system's behavior under various scenarios. Performance parameters such as speed, accuracy, power efficiency, and responsiveness proved the system's efficacy and dependability. The successful construction of this robot has far-reaching ramifications for the field of robotics and its applications. The capacity of the robot to maneuver in tight places and dynamically adapt to changing circumstances can improve operational efficiency and save labor costs in logistics and manufacturing. In health care, robots can help with patient care and telemedicine, offering remote support and lowering the workload on healthcare staff.

The project's findings also highlight the possibility of additional advances in robotic mobility. Future research can build on this work by investigating new control algorithms, sensor integration, and power management strategies, resulting in even more powerful and adaptable robots. Simulation tools like as Proteus will continue to play an important role in the design and testing of complex robotic systems, assuring their performance and reliability in real-world applications. Despite the favorable results, this initiative had a few drawbacks. The experimental setting is limited to controlled surroundings, which may not accurately replicate the obstacles encountered in real-world scenarios. Furthermore, long-term durability and dependability tests were outside the scope of this experiment, leaving room for future research.

Future studies should focus on overcoming these constraints by testing the robot in a variety of uncertain scenarios. Improving the control algorithms to enable better autonomous decision-making and obstacle avoidance can expand the robot's capabilities. Furthermore, researching advanced materials and power sources can improve the robot's durability and operational longevity, making it more appropriate for actual applications. Reflecting on the journey of this capstone project, I can say that it was an important learning experience that greatly improved my understanding of robotics, control systems, and simulation tools. The difficulties faced and solutions devised have provided valuable insights into the intricacies of building and executing advanced robotic systems. This project not only improved my technical capabilities but also honed my critical thinking, problem-solving ability and provided a solid platform for future robotics research and development.

Finally, the remote-controlled omnidirectional mobile platform robot created for this capstone project represents substantial advances in robotic mobility and control. The project's findings add vital knowledge to the field of robotics and set the way for future developments and applications. The insights and abilities that were obtained during this project will surely have an impact on my future robotics ventures and beyond. As the field of robotics evolves, the insights gathered from this project will serve as a platform for future research and development, resulting in innovations that can alter numerous industries and improve people's lives.

References

1. Rojas R, A short history of omnidirectional wheels
2. Omniwheels Gaining Popularity in Robotics—IEEE Spectrum. Accessed 21 Dec 2024. https://spectrum.ieee.org/omniwheels
3. Neaz A, Lee S, Nam K (2023) Design and implementation of an integrated control system for omnidirectional mobile robots in industrial logistics. Sensors 23(6):3184. https://doi.org/10.3390/S23063184
4. Bonetto E, Goldschmid P, Pabst M, Black MJ, Ahmad A (2021) iRotate: active visual SLAM for omnidirectional robots. Rob Auton Syst 154. https://doi.org/10.1016/j.robot.2022.104102
5. Saenz A, Santibañez V, Bugarin E, Dzul A, Ríos H, Villalobos-Chin J (2021) Velocity control of an omnidirectional wheeled mobile robot using computed voltage control with visual feedback: experimental results. Int J Control Autom Syst 19(2):1089–1102. https://doi.org/10.1007/S12555-019-1057-6/METRICS

6. Liu P, Feng C, Xu Y, Ning Y, Xu H, Shen S (2024) OmniNxt: a fully open-source and compact aerial robot with omnidirectional visual perception. Accessed 01 Dec 2024. http://arxiv.org/abs/2403.20085
7. Choi W-K, Jeon H-T, Kim S-J (2007) Multiple sensor fusion and motion control of snake robot based on soft-computing. INTECH Open Access Publisher
8. Muir PF, Neuman CP
9. Magalhães SC, Moreira AP, Costa P (2022) Omnidirectional robot modeling and simulation. https://doi.org/10.1109/ICARSC49921.2020.9096069
10. Li H, Liu J, Lyu C, Liu D, Liu Y (2023) Design and implementation of omnidirectional mobile robot for materials handling among multiple workstations in manufacturing factories. Electronics 12(22):4693. https://doi.org/10.3390/ELECTRONICS12224693
11. Guo Z, Yee RB, Mun KR, Yu H (2017) Experimental evaluation of a novel robotic hospital bed mover with omni-directional mobility. Appl Ergon 65:389–397. https://doi.org/10.1016/J.APERGO.2017.04.010
12. Charabaruk N (2015) Development of an autonomous omnidirectional hazardous material handling robot
13. View of Cybersecure Omnidirectional Autonomous Mobile Robot (OAMR) for Indoor Manipulation of Roll-to-Roll Flexible Printed Circuits (FPC). Accessed 02 Dec 2024. https://iroboticsjournal.org/index.php/irobotics/article/view/169/113
14. Yue L et al (2024) A fast online omnidirectional quadrupedal jumping framework via virtual-model control and minimum jerk trajectory generation. http://arxiv.org/abs/2407.00658
15. Wijayathunga L, Rassau A, Chai D (2023) Challenges and solutions for autonomous ground robot scene understanding and navigation in unstructured outdoor environments: a review. Multidisciplinary Digital Publishing Institute (MDPI). https://doi.org/10.3390/app13179877
16. Battistuzzi L, Recchiuto CT, Sgorbissa A (2021) Ethical concerns in rescue robotics: a scoping review. Springer Science and Business Media B.V. https://doi.org/10.1007/s10676-021-09603-0
17. Lin TY, Juang JG (2023) Application of 3D point cloud map and image identification to mobile robot navigation. Meas Control (United Kingdom) 56(5–6):911–927. https://doi.org/10.1177/00202940221136242/ASSET/IMAGES/LARGE/10.1177_00202940221136242-FIG19.JPEG
18. Eirale A, Martini M, Tagliavini L, Gandini D, Chiaberge M, Quaglia G (2022) Marvin: an innovative omni-directional robotic assistant for domestic environments. Sensors 22(14). https://doi.org/10.3390/s22145261
19. Wang Y et al (2022) Fixture-free omnidirectional prestretching fabrication and integration of crumpled in-plane micro-supercapacitors. Sci Adv 8(21). https://doi.org/10.1126/SCIADV.ABN8338
20. Ribuan MN, Wakimoto S, Suzumori K, Kanda T (2016) Omnidirectional soft robot platform with flexible actuators for medical assistive device. Int J Autom Technol 10(4):494–502. https://doi.org/10.20965/ijat.2016.p0494
21. L298N Dual H Bridge DC & Stepper Motor Driver Module. Accessed 17 July 2025. https://www.thingbits.net/products/l298n-dual-h-bridge-dc-stepper-motor-driver-module
22. Your Robotics Store in Cyprus. Accessed 22 Dec 2024. https://robo.com.cy/products/bluetooth-hc-05-ble-module
23. https://www.twinschip.com/Arduino-UNO-R3-SMD
24. Rijalusalam D, Iswanto I (2021) Implementation kinematics modeling and odometry of four omni wheel mobile robot on the trajectory planning and motion control based microcontroller. J Robot Control (JRC) 2. https://doi.org/10.18196/jrc.25121

Correction to: Developing a Data Envelopment Analysis Model for Performance Evaluation of Green Supply Chain Management

Farzad Zaare Tajabadi and Sahand Daneshvar

Correction to:
Chapter 6 in: A. Azizi (ed.), *Intelligent Mechatronic Control Systems*, Emerging Trends in Mechatronics, https://doi.org/10.1007/978-981-95-3925-3_6

The original version of this book was inadvertently published with an incorrect second author affiliation in Chapter 6 which has now been corrected. The book and the chapter have been updated with the changes.

The updated version of this chapter can be found at
https://doi.org/10.1007/978-981-95-3925-3_6

A. Azizi (ed.), *Intelligent Mechatronic Control Systems*, Emerging Trends in Mechatronics, https://doi.org/10.1007/978-981-95-3925-3_8

MIX
Papier aus verantwortungsvollen Quellen
Paper from responsible sources
FSC® C105338

If you have any concerns about our products,
you can contact us on
ProductSafety@springernature.com

In case Publisher is established outside the EU,
the EU authorized representative is:
Springer Nature Customer Service Center GmbH
Europaplatz 3, 69115 Heidelberg, Germany

Printed by Libri Plureos GmbH
in Hamburg, Germany